I0489366

Collected Papers on Christofredo Jakob

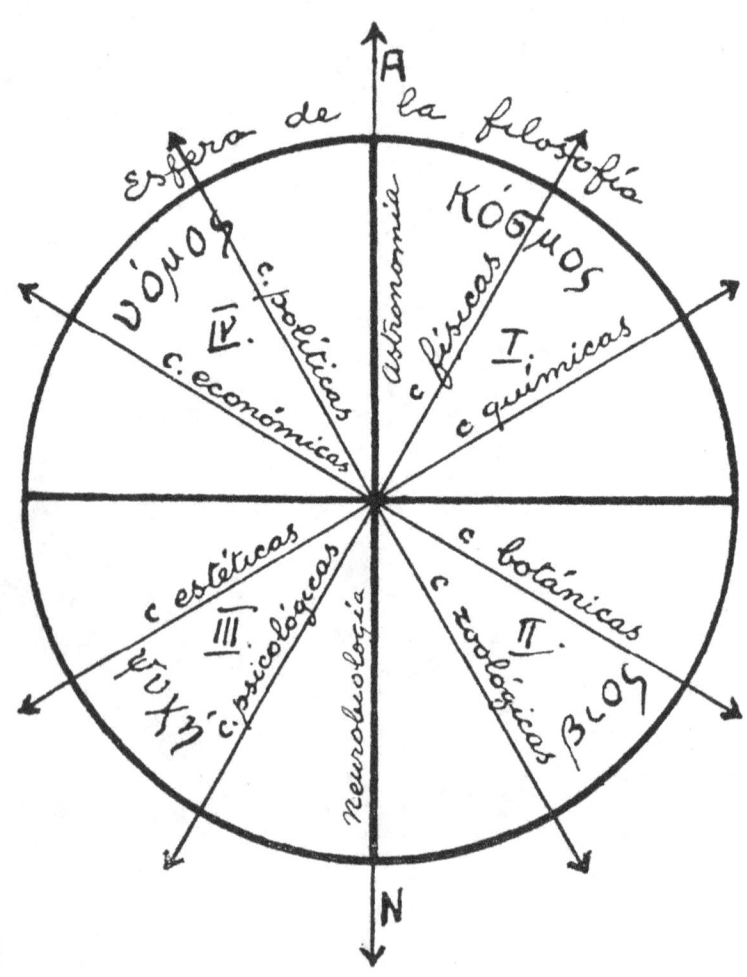

Collected Papers on Christofredo Jakob

Lazaros C. Triarhou, M.D., Ph.D.
Professor of Neuroscience, University of Macedonia, Greece

CORPUS CALLOSUM
Thessalonica · Indianapolis

Collected papers on Christofredo Jakob.

A Corpus Callosum book published by

Lazaros C. Triarhou, MD PhD
Professor of Neuroscience
University of Macedonia
Egnatia 156
Thessalonica 54636 (Greece)

Copyright © 2017 by L. C. Triarhou
All rights reserved. No part of this book may be used or reproduced, in whole or in part, including illustrations, in any form by any electronic or mechanical means (beyond that copying permitted by Sections 107 and 108 of the U.S. Copyright Law and excerpt by reviewers for the public press), without permission in writing from the author.

LIBRARY OF CONGRESS CONTROL NUMBER: 2017909937

Triarhou, Lazaros Constantinos
    Collected papers on Christofredo Jakob / Lazaros C. Triarhou. — 1st ed.
    p. cm.
    Includes bibliographical references.
    LCCN: 2017909937
    ISBN-13: 978-1-5482-5024-9
    ISBN-10: 1-5482-5024-4

    1. Jakob, Christofredo (Christfried), 1866-1956. 2. Neuropathologists—Argentina. 3. Neuroanatomists—Germany. 4. Neurobiology—History. I. Title.

Printed in the United States of America

The Corpus Callosum logo is based on a lithograph by Jakob in *Icones Neurologicae* (1897).

Contents

Preface

In the present volume I have compiled sixteen articles that my colleagues and I published over the last twelve years on selected works of the neurobiologist Christofredo Jakob (1866–1956). Thus, the present volume complements the other two published volumes, entitled, *The Geographical Exploration Papers of Doctor Christofredo Jakob* (2017), and *The Complete Published Works of Doctor Christofredo Jakob: A Comprehensive Illustrated Catalog* (2017). The latter lists all his books and articles with full bibliographic information.

Christfried Jakob was born in Wörnitzostheim, Bavaria; in 1899, he moved to Argentina, his adopted country of vocation, where his first name was 'castillianized' to Christofredo. Physician by training, he specialized in neuroanatomy and neuropathology, and was appointed professor of neurobiology at the Faculty of Humanities and Educational Sciences of the National University of La Plata and the Faculties of Medicine and of Philosophy and Letters of the University of Buenos Aires. He retired in 1945. He also directed the Laboratory of Pathology of the Neuropsychiatric Clinic of Hospicio de Las Mercedes in Buenos Aires.

In all, Jakob published 52 books and 260 papers on the development, evolution, and pathology of the nervous system. He made landmark discoveries on the 'visceral brain' and formulated his concept three decades before the 'circuit of Papez'. In 1923 he suggested a tripartite system of neuronal hierarchical levels that anticipated the 'triune brain' of MacLean of 1969. His other contributions include the dual evo-devo origin and ubiquitous sensory-motor function of the cerebral cortex, language pathophysiology, a structural-functional context of cerebellar neurobiology, consciousness, and neurophilosophy.

I remain grateful to the colleagues who participated in this project, namely, Drs. Manuel del Cerro, Kyrana Tsapkini, Ana B. Vivas, Zoe D. Théodoridou, Athanasios Koutsoklenis, Anny Tzouma, and Daniel S. Margulies. I also gratefully acknowledge the journals and publishers where the papers appeared originally.

Lazaros C. Triarhou
June 2017

COLLECTED PAPERS ON CHRISTOFREDO JAKOB

J Neurol (2007) 254:124–125
DOI 10.1007/s00415-006-0307-8

Lazaros C. Triarhou
Manuel del Cerro

Christfried Jakob (1866–1956)

Received: 20 March 2006
Received in revised form: 12 May 2006
Accepted: 28 May 2006

Prof. L.C. Triarhou, MD, PhD (✉)
Economo-Koskinas Wing for Integrative
and Evolutionary Neuroscience
Dept. of Educational and Social Policy
University of Macedonia
Egnatia 156, Bldg. Z-312
54006 Thessaloniki, Greece
Tel.: +30-2310-891-387
Fax: +30-2310-891-388
E-Mail: triarhou@uom.gr

Prof. M. del Cerro, MD
Dept. of Neurobiology & Anatomy and
Ophthalmology
University of Rochester
Rochester NY 14642, USA

JON 2307

Neurologists are familiar with Bavarian-born Alfons Maria Jakob (1884–1931), Professor of Neurology and Psychiatry at the University of Hamburg, after whom Jakob-Creutzfeldt disease is named. Another Jakob, altogether unheeded in the English biomedical literature, is the neuropathologist Christfried (Christofredo) Jakob, 18 years senior to Alfons, also Bavarian-born, and considered to be the father of Neurology in Argentina, his adopted homeland.

Commanding an impressive background knowledge in philosophy, literary studies and classical music, Jakob dedicated his life to the study of unknown areas of comparative neurobiology with special emphasis on the natural history of the cerebral cortex, descriptive and pathological anatomy and neuropathology [6], psychiatry, neuropsychology [7], neurophilosophy, embryology, zoology, botany, hydrobiology, palaeontology and geography.

Christfried Jakob was born on 25 December 1866 in Wörnitzostheim. He studied medicine at the University of Erlangen from 1886–1890. For his thesis he studied aortitis syphilitica, under Friedrich Albert von Zenker (1825–1898). In the early 1890s he was assistant to Adolf von Strümpell (1853–1925) at Erlangen Medical Clinic and privately practised medicine in Bamberg [9, 10].

Jakob published his first book in 1895, an atlas of the normal and pathological anatomy of the nervous system [1], with lithographs and woodcuts made from original drawings and photographs, and cross-sections of the brain showing different layers with the aid of superimposed flaps. The book was duly translated into English, Russian, French and Italian and went through a second edition in 1899. In 1897 Jakob published an atlas of methods of clinical investigation [2], an epitome of internal medicine, which was also translated into French and English.

In 1899, through the mediation of Domingo Cabred (1859–1929), the Argentinian Professor of Psychiatry, Jakob went to Argentina to direct the Laboratory of the Psychiatric and Neurological Clinic of Hospicio de las Mercedes at the University of Buenos Aires [9, 10]. He was attracted by the prospect of collecting 300 brains per year for pathoanatomical study, as opposed to 2–3 brains he had access to in Germany. His three-year contract was later extended until 1910. For a dozen years, Jakob produced works in anatomy [4, 5, 8], neurology, psychopathology and anthropology [3].

He returned to Germany to continue his neurohistological work and to oversee publication of his Atlas of the human brain [4] and the German version of his Atlas of comparative neuroanatomy [8]. In 1911 he presented his idea on the ubiquity, across species, of the sensory-motor dual function of the cerebral cortex [5]. In his monograph [4], Jakob proposed the existence of a visceral brain, antedating by more than a quarter of a century the North American neuroanatomist James W. Papez (1883–1958).

In 1912 Jakob returned to Argentina, where he would spend the remaining 44 years of his life, as director of the Laboratory of the Hospital Nacional de Alienadas (mental asylum for women), affiliated with the University of Buenos Aires and the Universidad Nacional de La Plata. Jakob's numerous textbooks in Spanish include 'Elements of Neurobiology' (1923), 'Pathological Anatomy and Physiology' (1924), several lecture courses on the organization of the CNS of higher vertebrates, and the three masterful volumes of the Atlas of *Folia Neurobiológica Argentina* [6] covering the phylogeny, ontogeny, topographical anatomy and pathological anatomy of the human brain.

Jakob's papers cover a wide range of interests. In evolutionary neuroscience he made important contributions by studying the comparative neurohistology of some of Patagonia's rare mammals, including *Grypotherium domesticum*, the pichiciego (*Chlamyphorus truncatus*) and opossum (*Didelphys azarai, Metachirus crassicaudatus*), and reptilians, such as the yacaré (*Caiman latirostris*) [8]. In 1905 he provided the first anatomical description of the brains of South American Indians [3]. In addition he published articles on 'The frontal lobes and higher mental functions' (1906), 'Cerebral histology and psychology' (1911), 'Psychology and its relation to cortical biology' (1913), 'The harmonious development of intelligence and the brain in the child' (1913) and 'The frontal lobe' (1943).

His neuropathological and neuropsychiatric studies include reports on acute alcoholic neuritis, hemiplegia and hemianaesthesia with crossed oculomotor paralysis, neurosyphilis, motor aphasia and its localization, familial progressive spastic paraplegia (von Strümpell disease), primary fibrochondro-osteoma of the brain, familiar olivo-ponto-cerebellar atrophy, pituitary tumours, cerebral arteriosclerosis, schizophrenia, forms of Schilder, Hallervorden-Spatz and Pick diseases, and a syndrome consisting of dementia and a distinct type of combined bilateral cerebellar atrophy, involving the depth of the marginal sulcus.

Jakob retired from his academic duties in 1945, leaving a legacy of 20,000 examined brain specimens, 30 monographs and 200 papers, and having established the Argentinian Neurobiological School (*Escuela Neurobiológica Argentina*) through one of the most important neurobiological laboratories worldwide. He passed away in Buenos Aires on 6 May 1956 at the age of 90.

References

1. Jakob C (1895) Atlas des gesunden und kranken Nervensystems nebst Grundriss der Anatomie, Pathologie und Therapie desselben. Verlag von J. F. Lehmann, München
2. Jakob C (1897) Atlas der klinischen Untersuchungsmethoden nebst Grundriss der klinischen Diagnostik und der speziellen Pathologie und Therapie der inneren Krankheiten. Verlag von J. F. Lehmann, München
3. Jakob C (1905) Contribution à l'étude de la morphologie des cerveaux des Indiens. Rev Museo de La Plata 12:59-72
4. Jakob C (1911) Das Menschenhirn: Eine Studie über den Aufbau und die Bedeutung seiner grauen Kerne und Rinde. I. Teil. Tafelwerk nebst Einführung in den Organisationsplan des Menschlichen Zentralnervensystems. J. F. Lehmann's Verlag, München
5. Jakob C (1912) Ueber die Ubiquität der senso-motorischen Doppelfunktion der Hirnrinde als Grundlage einer neuen biologischen Auffassung des kortikalen Seelenorgans. Münch Med Wochenschr 59:466-468
6. Jakob C (1939-1941) El cerebro humano (Folia Neurobiológica Argentina, Atlas I-III). Aniceto López Editor, Buenos Aires
7. Jakob C (1941) La función psicogenética de la corteza cerebral y su posible localización. An Inst Psicol (Buenos Aires) 3:63-80
8. Jakob C, Onelli C (1913) Atlas del cerebro de los mamíferos de la República Argentina: Estudios anatómicos, histológicos y biológicos comparados, sobre la evolución de los hemisferios y de la corteza cerebral. Imprenta de Guillermo Kraft, Buenos Aires
9. Meyer L (1981) Cristofredo Jakob. A veinticinco años de su muerte. Acta Psiquiát Psicol Amér Lat 27:13-14
10. Orlando JC (1966) Christofredo Jakob: su vida y obra. Editorial Mundi, Buenos Aires

Historical Note

Eur Neurol 2006;56:176–188
DOI: 10.1159/000096424

Received: May 6, 2006
Accepted: July 12, 2006
Published online: October 19, 2006

European Neurology

Semicentennial Tribute to the Ingenious Neurobiologist Christfried Jakob (1866–1956)

1. Works from Germany and the First Argentina Period, 1891–1913

Lazaros C. Triarhou[a]   Manuel del Cerro[b]

[a]Economo-Koskinas Wing for Integrative and Evolutionary Neuroscience, Department of Educational and Social Policy, University of Macedonia, Thessaloniki, Greece; [b]Departments of Neurobiology and Anatomy, and Ophthalmology, University of Rochester, Rochester, N.Y., USA

Key Words
Christfried Jakob · Comparative neuro-anatomy · Evolutionary neuroscience · Cognitive neuroscience · Circuit of Papez · Neural correlates of consciousness · History of neuroscience

Abstract
This study, and the companion paper that follows, pays homage to the life and work of Christfried (also Christian or Christofredo) Jakob, a German-born neuropathologist who adopted Argentina as his country of vocation. Rated by von Economo and Koskinas among the three most important pre-1925 cortical neuro-anatomists, alongside Ramón y Cajal, Jakob is little known in the English literature. He has left an impressive record of publications, 30 richly illustrated monographs and 200 articles that span over a vast array of neurological themes, including cortical development and evolution, and the visceral brain. The present paper reviews works from his German years and the first visit to Argentina in 1899–1910. The companion paper covers his works (all in Spanish) during his 'second Argentina period', after 1913.

Copyright © 2006 S. Karger AG, Basel

Introduction

This year marks half a century since the death of Christfried Jakob (fig. 1), a German-born neuropathologist who spent the best part of his prolific scientific life in Argentina. There is a remarkable paradox between the magnitude and relevance of his contributions to neuroscience on the one hand, and the complete absence of any reference whatsoever in the contemporary English medical literature on the other. Aimed at rectifying this lingering omission and the regrettable obscurity to which works not written in English seem to be condemned, the present study and the companion paper that follows survey the diverse contributions by a multifaceted brain scientist of the twentieth century.

Jakob is considered the father of neurobiology [1], neurology [2], and forensic histopathology [3] in Argentina, his chosen country of residence and vocation (his forename was first 'gallicized' to Christian and later 'castillianized' to Christofredo). Physician, philosopher, artist and educator, he is likely the only neuro-anatomist in the world to have a lake named after him: Lago Jakob, which he explored in 1934, is located 1,600 m above sea level, near Bariloche in the Argentinian Nahuel Huapí region of Western Patagonia (approx. 41°S, 71°W).

KARGER

Fax +41 61 306 12 34
E-Mail karger@karger.ch
www.karger.com

© 2006 S. Karger AG, Basel
0014–3022/06/0563–0176$23.50/0

Accessible online at:
www.karger.com/ene

Lazaros C. Triarhou, MD, PhD
University of Macedonia
Egnatia 156, Bldg. Z-312
GR-54006 Thessaloniki (Greece)
Tel. +30 2310 891 387, Fax +30 2310 891 388, E-Mail triarhou@uom.gr

Fig. 1. a Adolf von Strümpell (1853–1925). **b** Christfried Jakob (1866–1956). Strümpell, the son of the philosopher Ludwig Strümpell, was professor and director of the medical clinic in Erlangen between 1886–1903, in Breslau between 1903–1909 and in Leipzig between 1910–1925; his photo and signature are from the textbook of pathology [11]. Jakob's photo is from Orlando [7], courtesy of the Library of Congress; Jakob's signature facsimile is from the 1924 textbook of pathological anatomy and physiology (complete reference given in the companion paper [58]), courtesy of the Staatsbibliothek Berlin.

The only mentions of Jakob we found in the general English literature are two books [2, 4]. On the contrary, references to his life and works abound in Argentinian print [1, 3, 5] and electronic (http://electroneurobio.se-cyt.gov.ar) scientific journals. Two frequently consulted and documented biographies are those written by his disciples Luis López Pasquali [6] and Jacinto Carlos Orlando [7].

Jakob's published works exceed the 8,000 page mark [8] and comprise 30 monographs and about 200 articles, the result of apparently studying over 20,000 brains (as Jakob himself relates on November 15, 1939, in the preface to the second volume of his massive *Folia Neurobio-lógica Argentina* atlas). Jakob has been called a giant of science of the calibre of Cajal. In their monumental opus on cerebral cyto-architectonics, von Economo and Koskinas [9] express the view that future research on the cortex would have to be based on the fundamental works of three investigators: Theodor Kaes (1852–1913), Santiago Ramón y Cajal (1852–1934) and Christfried Jakob; they go on pronouncing Jakob's ideas on cortical phylo-ontogeny 'ingenious'.

Biographical Note

Christfried Jakob was born to Gottfried Jakob and Babette Körber on Christmas Day (hence his Christian name), Tuesday, December 25, 1866, in Wörnitzostheim (48.83°N, 10.65°E), east of Nördlingen-im-Ries, Bavaria, Germany. Jakob entered medical school at the University of Erlangen in 1886 and graduated on July 18, 1890, with a prize of 1,000 DEM, offered by the faculty to the most distinguished student [5, 7]. He completed his doctoral thesis on aortitis syphilitica [10] under Friedrich Albert von Zenker (1825–1898). In 1892 Jakob was second and in 1893 first assistant at the Erlangen Medical Clinic headed by Professor Adolf von Strümpell (1853–1925) (fig. 1), author of many well-known works, including a two-volume textbook of pathology [11] and founding editor of the *Deutsche Zeitschrift für Nervenheilkunde*. By 1895, Jakob had started a private medical practice in Bamberg. Both of his mentors had served as Prorectors at the University of Erlangen, von Zenker during the academic year 1869–1870, and von Strümpell in 1892–1893 (www.uni-erlangen.de/inforcenter/uniarchiv/materialien/rektoren).

In 1898, through the initiative of Domingo Cabred (1859–1929), Professor of Psychiatry in Buenos Aires, who had embarked on a European trip on a lookout for ways to promote neuro-anatomy in psychiatric research, Jakob was summoned to a 3-year contract with the national government of Argentina to direct the Laboratory of the Psychiatric and Neurological Clinic at Hospicio de las Mercedes, after von Strümpell refused the offer [7]. A key motive was the prospect of obtaining 300 brains annually for pathological study, when in Germany the corresponding number was 2–3 brains.

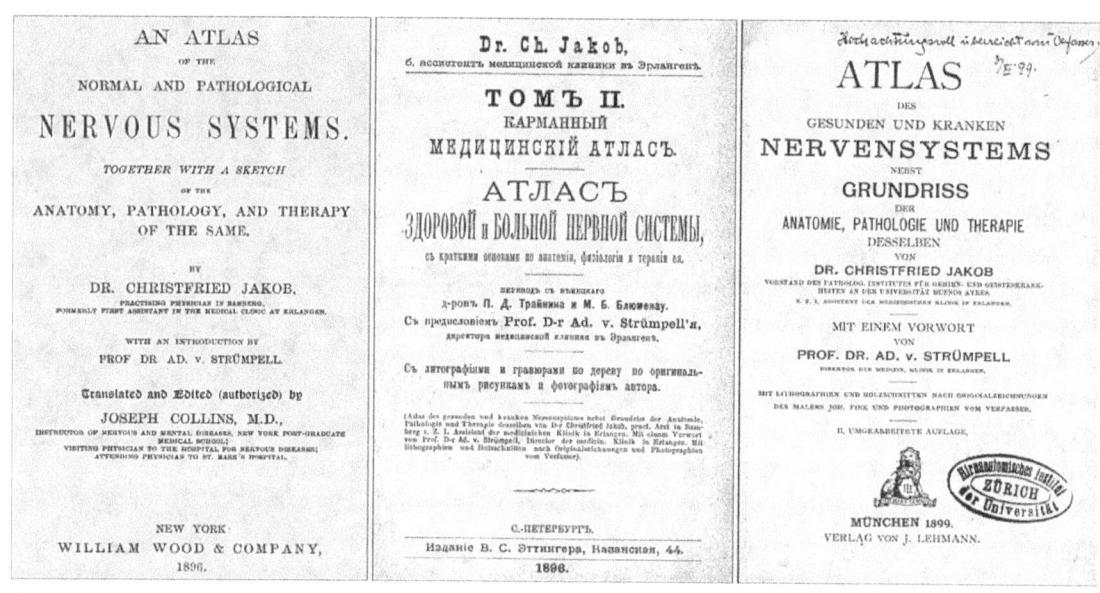

Fig. 2. Title pages of various editions of the *Atlas of the Nervous System* [12, 13]. English (left) and Russian (centre) translations of the first edition, and second German edition (right), the latter with the author's handwritten inscription.

Jakob left Germany via Hamburg and arrived in Buenos Aires on July 17, 1899. For the dozen years that followed, he produced works in neurology, psychopathology, biology, anthropology and paleontology [3]. His collaborators included psychiatrist-philosopher José Ingenieros (1877–1925), paleontologist Florentino Ameghino (1854–1911), naturalist Clemente Onelli (1864–1924) and anthropologist Roberto Lehmann-Nitsche (1872–1938). Jakob's initial contract had been renewed through 1910. Sometime afterwards, he went back to Germany to stay there for about 2 years before his second – and permanent – move to Argentina.

Monographs

Jakob published his first book at the age of 29, an atlas of the normal and pathological anatomy of the nervous system [12] with 78 plates and a preface by von Strümpell; it was speedily translated into English, Russian and French (fig. 2). In 1899, the work went through a second German edition [13] with 84 plates, appearing in French in 1900 and English in 1901. The handbook was illustrated with

black-and-white woodcuts and colour lithographs made from Jakob's original drawings (fig. 3, 4). Amidst the many figures, one (fig. 4c) appears to depict the beading of myelinated nerve fibres, a phenomenon revived in the 1960s thanks to the efforts of Sidney Ochs [14].

The atlas was a clear exposition of the neurological knowledge available at the time, and a testimony to the profound versatility of its author in histological techniques employed to study the brain [8]. Jakob concluded the book by mentioning five approaches, which he thought could help achieve a better knowledge of the structure of the nervous system: (1) serial section reconstruction of adult human brains; (2) the study of pathological cases; (3) comparative anatomy and embryology; (4) developmental studies, and (5) experimentally induced lesions [12, 13].

In his preface, von Strümpell wrote:

'Dr. Jakob has been occupied in a most industrious manner with the normal and pathological anatomy of the nervous system. Having an extensive collection of histological preparations, which he prepared according to the most reliable research methods, along with his uncommon drawing talent, the author was able to compile this

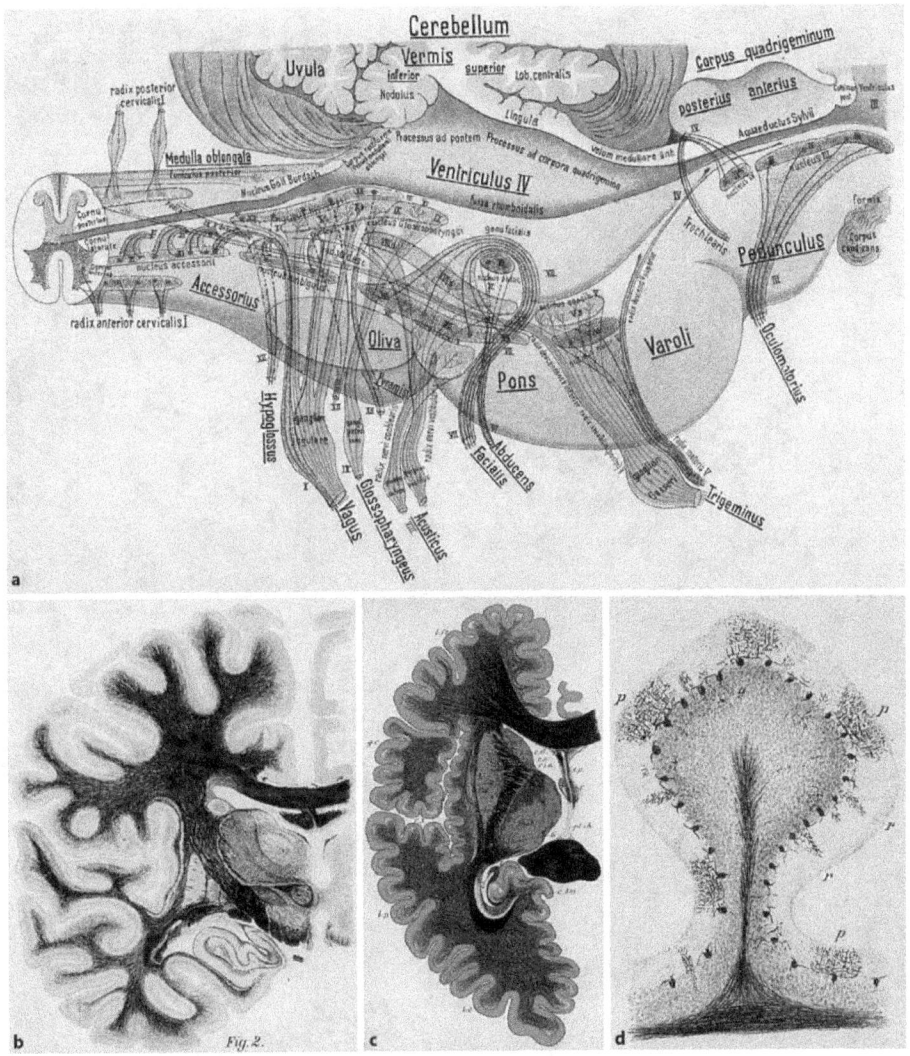

Fig. 3. a Lateral view of the brainstem showing in schematic form the position of the cranial nerve nuclei and the course of cranial nerves [13]. **b** Section of the cerebral hemisphere through the central convolutions [13]. **c** Horizontal section through the entire left cerebral hemisphere at the middle of the basal ganglia [12]. **d** Cerebellar convolution, silver stain [13].

atlas with great care. Any impartial observer will be convinced, as I am, that illustrations accomplish all that one would expect of them to convey. They present the actual conditions lucidly, and depict virtually all the numerous and important discoveries brought forth in studying the nervous system. The student and practising physician wishing to keep in touch with advances made in this field of medical science has the opportunity, with this atlas, of a clear conception of the present state of neurology with little trouble. There is perhaps no other branch of medi-

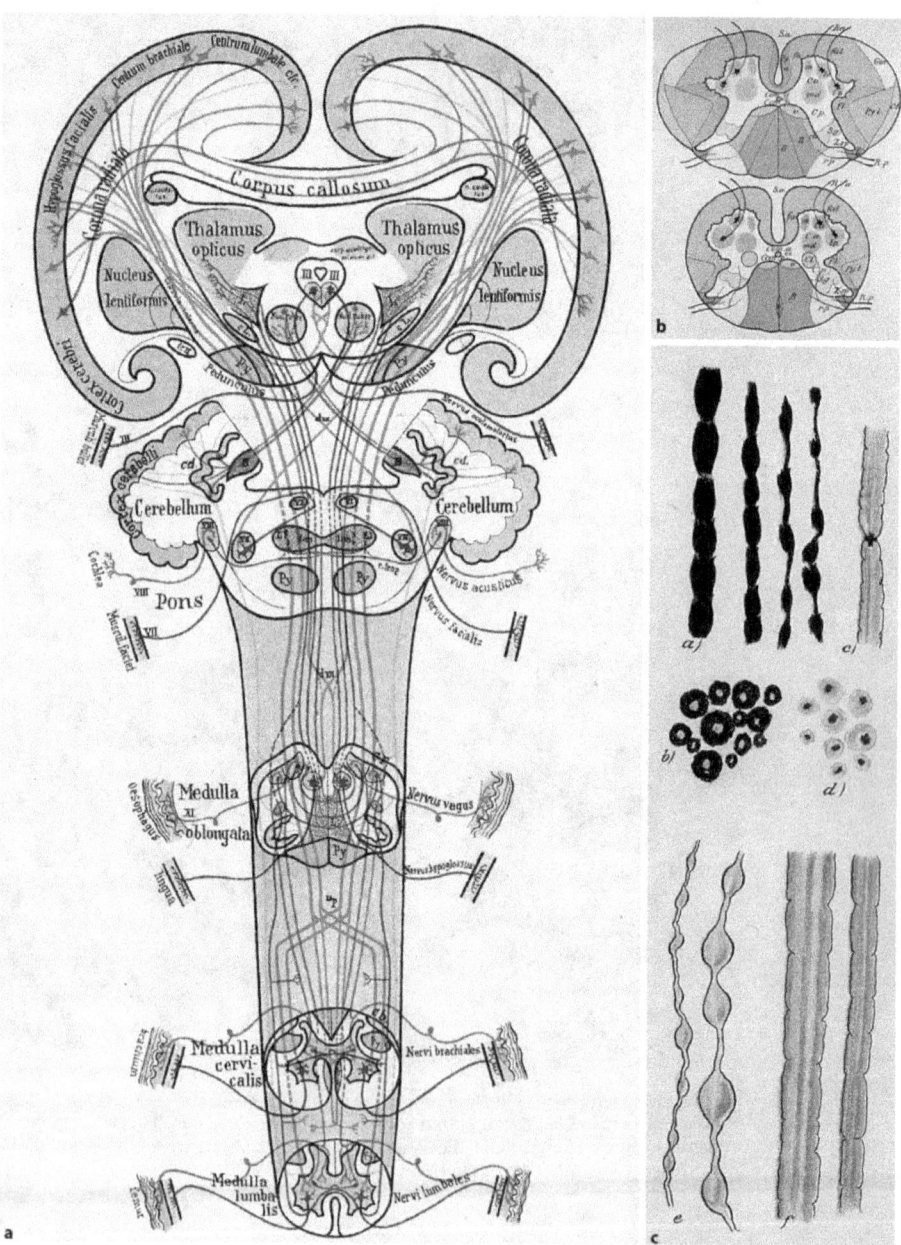

Fig. 4. a Schematic representation of the most important nerve tracts from a clinical point of view, in transverse sections of the cerebral hemispheres, crura, pons, cerebellum, medulla oblongata, cervical and lumbar enlargement, showing motor, sensory, rubral and cerebellar pathways [13]. **b** Diagrammatic representation of the position of the cervical and lumbar enlargements of the spinal cord [13]. **c** Various isolated nerve fibres in longitudinal and transverse section [13]; 'isolated fibres in the recent state, swollen from immersion in salt solution', images that resemble beading phenomena of myelinated nerve fibres observed in stretched mammalian nerve [14].

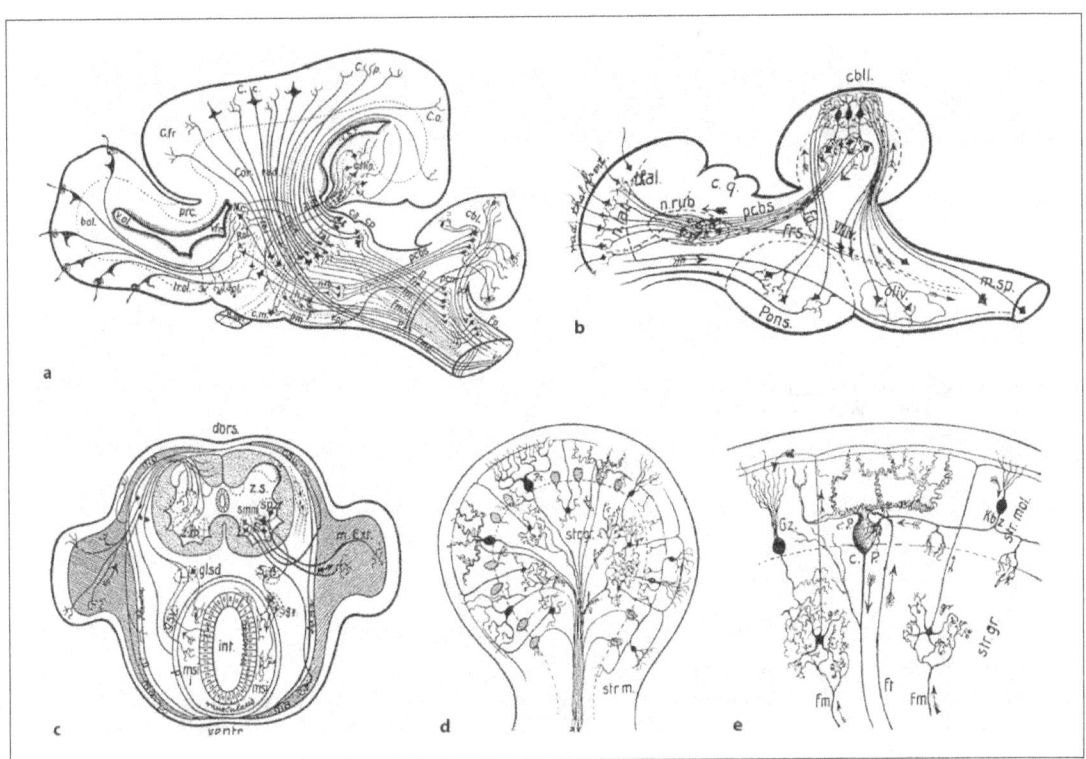

Fig. 5. a Projection pathways of frontal, central, parietal, occipital and temporal sectors in a sagittal section of the hemisphere of a mammalian brain [18, 22]. **b** Anatomical connections between the cerebellum, brainstem and spinal cord [17]. **c** Topographic plan of the lower nervous system and its reflex pathways, with the spinal nuclei, sympathetic ganglia and nerves, and the visceral innervation [17, 22]. **d, e** Histological structure of the cerebellar cortex [17].

cine than neuropathology, where the intimate relationship between clinical pathology on the one hand and normal and pathological anatomy on the other is so apparent and consistent. The treatment of normal and pathological anatomical facts, in conjunction with detailed, not schematic, illustrations render the work highly didactical. The author spared no pain devoting himself with untiring industry to achieving a really worthy and lasting goal' [12, 13].

In 1897, Jakob published an atlas of methods of clinical investigation, with an epitome of clinical diagnosis and special pathology and treatment of internal diseases [15], which was translated into French in 1898 and 1899 and English in 1899. In 1897, von Strümpell and Jakob produced an epitome of *Icones Neurologicae,* with 13 folded plates, 80 × 100 cm in size [16]; the plates were later re-

edited and expanded by Friedrich Müller and Hugo Spatz in a 1926 edition.

The two classic German monographs of Jakob which are extensively alluded to by von Economo and Koskinas [9] are *The Human Brain* [17] and *From Animal Brain to Human Brain* [18] (large volumes, 30 × 40 cm in size). The former contains a 50-page introduction on the organizational plan of the central nervous system, profusely illustrated with 51 figures and 90 photomicrographic plates of cell and fibre staining. The latter work begins with the quote 'sun and brain are the creators of our world'; it contains 40 pages of text with 54 figures, followed by 48 plates. The two works delineate anatomical, histological, developmental and evolutionary aspects of the nervous system (fig. 5).

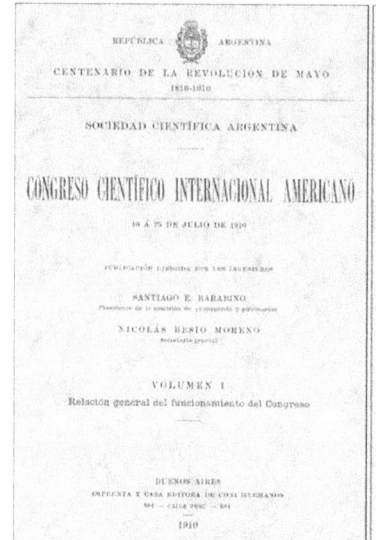

Fig. 6. The book of abstracts of the 1910 Scientific Congress where Jakob (note the spelling of the forename, 'Christian') presented an overview of his work in a plenary lecture on 12 July [20], and in a joint presentation with Clemente Onelli on 21 July [21], shortly before leaving for Germany.

Jakob had already made a name for himself through his earlier atlases [12, 13, 16]. Both of the new 1911 works presented elegant documentation. *The Human Brain* [17] begins with a systematic, objective and faithful exposition of the histotopography of the grey matter through depictions of the spinal cord, medulla and brainstem, diencephalon and cerebral cortex. There is a section on cortical development. Particular emphasis was placed on detailed descriptions of the relations of the diencephalon with cortical areas, based on retrograde cell degeneration in the thalamus, covering its morphology and physiology. The author communicated numerous new vistas that deviated from previous opinions, e.g. on the projection from the mammillary bodies to the anterior nucleus of the thalamus, and the supracallosal gyrus or 'first sagittal pre-segment' as a 'visceral centre', based on clinical-pathological evidence and experimental data from operations in dogs and apes. Information is given with the purpose of studying the biological foundations of mental activities, such as memory, will, expression and imagination, in association with the underlying cortical structure and function.

The Animal Brain [18], co-authored with Onelli, constitutes a far-sighted comparative neuro-anatomy and neurohistology with a constant consideration of functional differentiation, in diverse species of the South American fauna little studied until then. Biological details were given on the species, with special consideration for sensory organs and cerebral cortical functions, which form the basis for the morphological and biological understanding of the human cerebrum – 'our noblest organ, to which we humans owe more than any other creature on earth'. Jakob concludes with a description of the *Primatentyp* ('primate type') and compares the orangutan cortex with the human.

In later studies on the opossum brain, Gray [19] credits Jakob and Onelli [18] with providing descriptions of opossum species from South America. The external form of the brain of *comadreja overa* or *Didelphis azarae* and the four-eyed opossum *Metachirus crassicaudatus* [18] bears striking similarities to the Virginian opossum, and the transitional zone to the occipital cortex presents certain features of Gray's area peristriata. Further, brain regions of the Virginian opossum resemble in histological structure piriform, temporal, and prefrontal regions of South American edentates, such as the long-nosed armadillo of the Dasypus genus.

In 1910, Jakob made two keynote presentations at the *Congreso Científico Internacional Americano* (fig. 6): a plenary lecture on dysgeneses and ageneses of the human central nervous system [20] and a joint presentation with Onelli, director of the Buenos Aires Zoological Garden, on the comparative anatomy and phylogeny of the mammalian brain, based on the cortical biology of species of

[18]

Argentina's fauna [21]. That presentation was the prelude to *The Animal Brain* [18] and the *Atlas del cerebro de los mamíferos de la República Argentina* [22], subsidized by funds from the congress.

The 1913 Argentinian edition [22] constitutes a gem of the comparative neuro-anatomy literature, combining ontogenetic and phylogenetic concepts. The text of the German edition [18] was tripled with chapters on the biology of each species, classification, and morphological details on cortical sulci and convolutions. It is an invaluable work, covering the morphology of 40 characteristic mammals of South America, including exotic species such as the puma *(Felis puma)*, *aguará-guazú (Chrysocyon jubatus)*, *tucu-tucu (Ctenomys magellanicus)*, *carpincho (Hydrochoerus hydrochaeris)*, *guanaco (Lama huanacus)*, *oso* and *osito hormiguero (Myrmecophaga jubata* and *Tamandua tetradactyla)*, *perezoso (Bradypus tridactylus)* and *mulita (Dasypus hybridus)*.

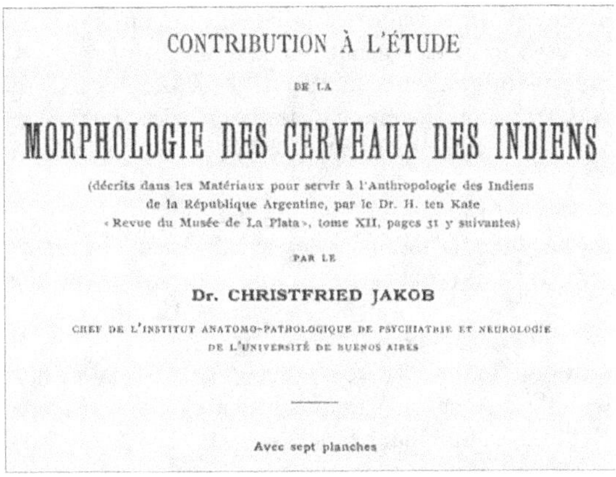
CONTRIBUTION À L'ÉTUDE

DE LA

MORPHOLOGIE DES CERVEAUX DES INDIENS

(décrits dans les Matériaux pour servir à l'Anthropologie des Indiens de la République Argentine, par le Dr. H. ten Kate «Revue du Musée de La Plata», tome XII, pages 31 y suivantes)

PAR LE

Dr. CHRISTFRIED JAKOB

CHEF DE L'INSTITUT ANATOMO-PATHOLOGIQUE DE PSYCHIATRIE ET NEUROLOGIE DE L'UNIVERSITÉ DE BUENOS AIRES

Avec sept planches

Fig. 7. Frontispiece of the article on the external morphology of the brains of South American Indians [25].

Spectrum of Published Articles

In addition to the 30 books that Jakob published (8 of those by 1913), he has left a record of close to 200 articles, dating from 1893 to 1949. We have compiled a comprehensive list based on four main sources: the Index Medicus volumes from 1891 to 1957, the index of the *Archives of Psychiatry and Criminology* edited by José Ingenieros, Jakob's own listings of his papers in his various books, and the list appended in Orlando's biography [7].

About one third of the articles, written in German, French and Spanish, were published before 1913, and about two thirds thereafter, all in Spanish. Some 40 papers deal with anatomy, histology, embryology, phylogeny and evolution, 70 with general pathology and neuropathology, and 50 with philosophy, neurophilosophy, neuropsychology, language functions and the frontal lobe. A 1911 German article is on 'The problem of the impending rise of meat price and its solution' [23]. There are 11 published reviews [24] of books authored by D. D'Arman, A. Cramer, F. Courmont, E. Frohse, E. Kraepelin, A. Mahaim, W. Osler, E. Perregaux, G. Pianese, S. Ramón y Cajal and H. Unverricht.

Brain Morphology in South American Indians

In 1905, Jakob [25] provided the first anatomical description of the brains of South American Indians [4, 26]. Four brains formed the basis of that communication, which had been written in French (fig. 7): two male (Yahgan and Gennaken-Huilliche) and two female (Inacayal and Alakalouf), from the Patagonian pampa and from Tierra del Fuego. As controls, he had by that time a collection of over a thousand brains, as well as the 'precious atlas of Retzius, which one may consider as a treasure for morphological studies' [25]. Before describing his results, Jakob states: 'If, to avoid unnecessary length, I designate such and such descriptions, I do not want to denote deviations of such a kind as non-typical or as inferior. There is not yet, neither will be probably for a long time, an absolute understanding of the opinions on a greater or a lesser inferiority of the different variants.' The study, accompanied by 24 photographs of the external form of the brain in 7 plates, gives data on overall appearance (size, development, convolutions), morphological variations and morphometric details for each lobe of each cerebral hemisphere, including diameters, patterns of gyri and the branching of sulci.

Features that stand out are: brain 1, an extraordinary development of the cap of the third frontal convolution on both sides, and the first temporal on the right: no stigma of inferiority; brain 2, massive with large convolutions, typically eurygyrencephalic, a pronounced development of the inferior parietal convolutions, and an unusual formation in the rostral portion of the calcarine sulcus, found, according to Cunningham, only in 2.3% of cerebral statistics; brain 3, dimensions somewhat smaller with light hypoplasia of the left frontal lobe (without any pathological lesions), and a development perhaps on the lower end of the average size of its convolutions, but without offering evident detail anomalies in their configuration; brain 4, a pronounced type of eurygyrencephaly, with large convolutions and fairly rich in secondary convolutions, and without any finding of atypical dispositions.

In conclusion, Jakob finds the four brains perfectly at the height of the mean development of European brains; some features are above, other under, the mean line, that is to say, they vary in an ideal way, like all brains, without any notable alteration that had not been encountered in European brains. 'These observations are in agreement with the fact that all nations, which are today considered civilized, would find themselves, for the last 2,000 years, in the same state, plus or minus, as these Indians; further, the so-called mass culture is nothing but a methodical suppression of individual physiological functions; it is nothing but a development of inhibitory centres dictated by the laws of family, society and state.'

Neuropathological and Neuro-Anatomical Articles

During his tenure as assistant at Erlangen, Jakob published a case of acute alcoholic neuritis in a 5-year-old child [27], a neuro-anatomical contribution to the theory of combined systemic diseases of the spinal cord [28], a case of hemiplegia and hemi-anaesthesia with crossed oculomotor paralysis [29] and a contribution to the understanding of cortical and thalamic pathology [30].

His neuro-anatomical and pathological studies from his first period in Argentina include an early article on the state of leucocytosis in infectious diseases [31] and an exchange of commentaries [32] on the histopathology of *coup de chaleur* with Abel Ayerza and Horacio G. Piñero – who, in their 1899 classic paper [33], call Jakob a 'respectable authority'.

Some other works from that period include an article on the development of the cerebral cortex [34] and topical papers on aphasia and the 1906 controversy of Pierre Marie and Jules Déjèrine on the localizationist-holistic debate regarding language centres in the brain [35, 36].

In 1909, Jakob published a case of familial progressive spastic paraplegia or von Strümpell disease [37]. Apparently, the number of neuropathological studies on hereditary spastic paraplegia in the literature is sparse, 15 articles between 1886 and 2001 [38]; Jakob's paper ranks fifth in chronological sequence after the 1886 and 1904 papers of A. Strümpell and the 1904 and 1906 papers of L. Newmark [38].

In poring over a diverse range of topics, Jakob pioneered some important neurobiological concepts. A brief overview follows.

Dual Origin and Ubiquitous Sensorimotor Function of the Cerebral Cortex

At the second annual meeting of the International Society for Medical Psychology and Psychotherapy, organized in 1911 by Oskar Vogt in Munich, Jakob proposed the principle of the dual evolutionary origin and ubiquitous sensorimotor function of the cerebral cortex [39, 40], based on comparative studies of primates and species such as *Caecilia lumbricoides*, an unusual legless amphibian of the Gymnophiona (Apoda) order that resembles a giant earthworm, and *Amphisbaena darwini*, the blind viper. In the article 'Psychology and its relation to cortical biology' [41], he defended the view that all cortical regions contain receptive elements. Most sensory pathways end up in what he calls the 'outer fundamental cortical layer' (small and medium-sized pyramidal cells), which ontogenetically and phylogenetically derives from the sensory rhinencephalic apparatus. The 'inner fundamental layer' contains effector (motor) elements. With advancing evolution, the two fundamental layers become intermingled. According to Jakob, sensory, motor, and associative elements exist in all cortical areas. Thus, he attributes a certain homogeneity to the cortex as an organ, and contradicts the theories of Flechsig and Cajal on association and memory centres.

At the same meeting in 1911, a discussion was held on the nature of hypnosis and amnesia. Following presentations by Bernheim and Claparède, Jakob mentioned an observation upon himself during a time he had been ill with typhoid fever in 1905, while in Argentina: the first 4 weeks had passed normally, and then a relapse came; from all that occurred afterwards, he remembered nothing, but argued that all dream life, on the other hand, was perfectly conscious to him [42].

Hemispheric Rotation around the Sylvian Pivot

Jakob explains the complex structure of the mammalian cortex through two separate events, hemispheric rotation and the formation of *Urwindungen* ('sagittal pregyri') [18].

The concept of hemispheric rotation around the sylvian pivot in the sagittal and coronal planes [18] is an original idea of Jakob [43, 44]. The rotation, which begins around the insular area as an axis, can directly explain the emergence of the sylvian fissure and the configuration of the remaining cortical sulci, such as the calcarine and parieto-occipital sulcus, and the 'radial or rotation sulcus system'; it results in a maturation gradient that implies heterochrony in cortical differentiation.

Hemispheric rotation, according to Irsigler [44], is one of seven events, upon which morphogenesis rests in the context of evolutionary theory, i.e. the transition from extant allocortical (reptilian and paleomammalian) formations preserved throughout vertebrate phylogeny and considered to be the foundation of species-specific behaviours in the animal scale from reptiles to humans. The other six events are Edinger's allocortical-isocortical contiguity, Spatz's allocortical base folding, hemispheric lateralization, Spemann's morphogenetic induction and metamorphosis, Sperry's chemo-affinity, and cyto-architectonic/connectivity factors.

Formation of Pregyri and Presectors

Jakob suggested the development of four sagittal cortical *Urwindungen* ('pregyri'), laterally to Ammon's formation [18]. He designated them as (I) gyrus callosomarginalis (also gyrus fornicatus, limbicus or splenialis), where he places the visceral cortex, (II) the bodily axis – hind limb zone located between the splenial and ectomarginal sulcus, (III) the forelimb zone between ectomarginal and suprasylvian sulcus, and (IV) the facio-mandibulo-lingual zone between the suprasylvian and marginal sulcus. The formation of these 'segments' has its origin in the base of the marginal sulcus, the insular area of higher mammals.

As the final most important principle in the organization of the cerebral cortex Jakob considered the *Sektorenentwicklung* ('development of sectors'), already noted in the brains of lower vertebrates such as the edentates [18]. With this concept, he explained regional variations in cortical architectonics, which he ascribed to five *Ursektoren* ('presectors'), i.e. frontal, central, parietal, occipital and temporal, and a rich *Untersektorengliederung* ('subsector conformation'). The entire cortical mantle was viewed as a system of similarly constructed radiating sectors in a fan-shaped form (*fächerförmig*), with their tip oriented towards the insula, and their expansions towards the upper hemispheric edge.

Based on the pattern of projection and association fibre growth, he reckoned that the sectors, with no excep-

tion, possess centripetal virgate parts in their coronae (*Stabkranzanteilen*), with centrifugal segments appearing only in certain areas, consistently across species. All sectors are receptively active, serving simultaneously both projection and association functions. Jakob rejected the separation of the cerebral cortex into independent projection and association centres [18].

In accordance with the sector principle (*Sektorenprinzip*), Jakob made a provisional attempt to cyto-architectonically partition the human cortex into 5 frontal, 3 central, 3 parietal, 2 occipital and 5 temporal sectors. Apart from those, he separated Ammon's formation, the uncus, the splenial formation ('visceral cortex') and the insular cortex [17].

Anatomical Centres of Emotion

Jakob suggested that the supracallosal gyrus is associated with the 'visceral cortical centre' [17, 18]. Here exist and have their highest central location the feelings from the visceral organs, especially in association with food intake, digestion, defecation and the sexual organs, that is, functions directly connected with the preservation of the individual and mating.

In 1964, Orlando [45] exposed succinct arguments in favour of a chronological and conceptual priority of Jakob over neuro-anatomist James W. Papez (1883–1958) on the formulation of the anatomical basis of the visceral brain. In 1937, Papez [46] suggested an anatomical basis of emotion, abiding in the mammillary bodies, anterior nucleus of the thalamus, cingulate gyrus, hippocampus and their interconnections, in what became later known as 'circuit of Papez'.

It is a fact that, based on clinical and patho-anatomical data from senile dementia and general paresis cases, as well as experimental evidence from retrograde degeneration and comparative anatomical and phylo-ontogenetic studies, Jakob had arrived at the conclusion, already in 1911 [17, 18], that the superior limbic (supracallosal or cingulate) gyrus is linked to afferent pathways that convey visceral thoracic-abdominal-pelvic sensations of the body and subserve internal feelings related to emotion. He wrote that 'the limbic cortex [cingulate gyrus] constitutes the hitherto unknown visceral cortical centre' and pinpointed to the involvement of the mammillary peduncles, mammillary bodies, mammillothalamic bundle of Vicq D'Azyr, thalamus, and the triangular system of the hypothalamus. The splenial zone of the reptilian brain, which corresponds to the superior limbic gyrus from lower mammals to primates, conveys visceral sensations from the mammillary bodies via the mammillothalamic bundle to the anterior nucleus of the thalamus.

It seems therefore that Jakob preceded Papez by more than a quarter of a century on the existence of a visceral zone in the brain. In his 1913 monograph [22], Jakob concluded that 'from opossum to humans, hunger and love reside in the limbic cortex, and from there they emit categorical imperatives that form individual temperament and affection', conclusively coupling the temporal rhinencephalon with emotional and affective behaviour.

Cerebral Cortical Organization

The cyto-architecture of the mammalian cerebral cortex has been classically described as having an orthogonal organization. According to Colombo et al. [47], a vertical pattern or 'ensemble' was envisaged and schematically illustrated by Jakob and Onelli [22] in a concept similar to that later formalized by Lorente de Nó, Powell and Mountcastle, Hubel and Wiesel, Szentágothai, and Goldman and Nauta, currently recognized as a key characteristic of cortical organization [48].

Already in 1906, Jakob was teaching in his classes that cerebral microcircuits and the neuropil formed electrical interference models, which he described as reverberations, similar in certain ways to what are today termed 'holograms' [49, 50]. Again, Jakob's ideas may have anticipated by some six decades neurophysiological concepts brought forth at a much later time in the English bibliography [50–53]. These ideas touch upon the topic of consciousness, with concepts of brain dynamics related to oscillations. Jakob viewed psychic activity in the scope of an integrated structural-functional context, sustained by reverberating neural 'macrocircuits' and 'microcircuits' in an oscillatory coupling, which he termed 'representational atomicities' [54].

In the transition of physiological events to the integrative experience of consciousness, the integration of multiple levels of messages into a coherent picture in a process known as 'binding' [55] has been compared to the physical phenomenon of resonance, associated with 30- to 80-Hz oscillations or γ-waves, to which Crick and Koch [56] had at one time attributed the neural basis of consciousness. Synchronous neural oscillations occurring globally throughout the brain might lead us towards a theory of cognition to explain how conscious awareness arises from neural events [57].

Recapitulation

Jakob is said to have established, in his time, one of the most important neurobiological laboratories in the world [8]. He dedicated his life to the investigation of unknown areas of nervous system biology. His initial work touched upon zoology, comparative and pathological neuro-anatomy and histophysiology with a special emphasis on the natural history of the cerebral cortex.

By the age of 46, that is, until his permanent move to Argentina, Jakob had steadily built a record of solid neurological works and concepts that echoed in the standard literature at the time. His knowledge and understanding of the nervous system helped him approach concepts of evolution, development and function in tandem and in a comprehensive context.

Jakob's work is extraordinary in quality, quantity and diversity; his ways of probing into cerebral themes may as well justify the intensifier 'ingenious'. That proclamation in the present article's title is rooted in the celebrated *Cytoarchitektonik* by von Economo and Koskinas [9], who reserve the use of similar terms only on three occasions: Jakob's ingenious idea *(geniale Ansicht)*, Cajal's use of the Golgi method *(ganz glänzend* = 'totally brilliant'), and Meynert's association of the granularity of the area striata in the calcarine cortex with sensory function *(geniale Intuition* = 'ingenious intuition'). Moyano [5] also resorts to the word 'ingeniosas' to characterize two small books of satirical poems published by Jakob in Buenos Aires under the pseudonym *Dr. Aussenseiter* ('Dr. Outsider'): *Die Apotheose der Null* (1932) and *Die Apotheose des Unendlichen* (1944).

The culmination, during the latter part of Jakob's career, of topical ideas that are relevant to modern cognitive neuroscience and neurophilosophy is covered in the companion paper [58].

Acknowledgements

The authors gratefully acknowledge the courtesy of the Staatsbibliothek zu Berlin; the British Library; the Interlibrary Loan Department of the University of Macedonia Library; the Ruth Lilly Medical Library of Indiana University, Indianapolis; the Bernard Becker Medical Library of Washington University, St. Louis; the Library of Congress, and the National Library of Medicine of the United States.

References

1 Pedace EA: Contribución de la Escuela Neu-robiológica Argentina del Profesor Chr Jakob en el estudio del lóbulo frontal. Arch Neurocir 1949;6:464–466.
2 Plotkin MB: Freud in the Pampas: The Emergence and Development of a Psychoanalytic Culture in Argentina. Stanford, Stanford University Press, 2001, pp 16–17.
3 Bonnet EFP: Cristofredo Jakob. Precursor de la histopatologia forense en la República Argentina. Segundo Congreso Nacional de Historia de la Medicina Argentina, Córdoba, 1970, pp 68–72.
4 Carnese FR, Goicoechea AS, Cocilovo JA: Argentina; in Spencer F (ed): History of Physical Anthropology: An Encyclopedia, vol 1 (Garland Reference Library of Social Science, vol 677). New York, Garland Publishing, 1997, pp 101–107.
5 Moyano BA: Christfried Jakob (25/12/1866 – 6/5/1956). Acta Neuropsiquiátr Argent 1957;3:109–123.
6 López Pasquali L: Christfried Jakob. Su obra neurológica, su pensamiento psicológico y filosófico. Buenos Aires, López, 1965.
7 Orlando JC: Christofredo Jakob: su vida y obra. Buenos Aires, Editorial Mundi, 1966.
8 Meyer L: Cristofredo Jakob. A veinticinco años de su muerte. Acta Psiquiátr Psicol Am Lat 1981;27:13–14.
9 von Economo CF, Koskinas GN: Die Cytoarchitektonik der Hirnrinde des erwachsenen Menschen. Textband und Atlas. Wien, Springer, 1925.
10 Jakob C: Aortitis syphilitica. Inaugural-Dissertation zur Erlangung der medicinischen Doktorwürde vorgelegt der medizinischen Fakultät der Universität Erlangen im Juli 1890. Erlangen, Vollrath, 1891.
11 von Strümpell A: Lehrbuch der speziellen Pathologie und Therapie der inneren Krankheiten, ed 26 (rev by C Seyfarth). Leipzig, Vogel, 1927.
12 Jakob C: Atlas des gesunden und kranken Nervensystems nebst Grundriss der Anatomie, Pathologie und Therapie desselben. München, Lehmann, 1895.
13 Jakob C: Atlas des gesunden und kranken Nervensystems nebst Grundriss der Anatomie, Pathologie und Therapie desselben, ed 2. München, Lehmann, 1899.
14 Ochs S: Beading phenomena of mammalian myelinated nerve fibers. Science 1963;139:599–600.
15 Jakob C: Atlas der klinischen Untersuchungsmethoden nebst Grundriss der klinischen Diagnostik und der speziellen Pathologie und Therapie der inneren Krankheiten. München, Lehmann, 1897.
16 von Strümpell A, Jakob C: Wandtafeln für den neurologischen Unterricht. München, Lehmann, 1897.
17 Jakob C: Das Menschenhirn: Eine Studie über den Aufbau und die Bedeutung seiner grauen Kerne und Rinde. München, Lehmann, 1911.
18 Jakob C, Onelli C: Vom Tierhirn zum Menschenhirn: Vergleichende morphologische, histologische und biologische Studien zur Entwicklung der Grosshirnhemisphären und ihrer Rinde. München, Lehmann, 1911.
19 Gray PA Jr: The cortical lamination pattern of the opossum, *Didelphys virginiana*. J Comp Neurol 1924;37:221–263.
20 Jakob C: El hombre sin cerebro. Estudio anátomo-biopatológico sobre disgenesias y agenesias del sistema nervioso central del hombre; in Barabino SE, Besio Moreno N (eds): Congreso Científico Internacional Americano, vol 1: Relación general del funcionamiento del Congreso. Buenos Aires, Coni Hermanos, 1910, p 438.
21 Jakob C, Onelli C: Anatomía comparada del encéfalo de los mamíferos de la República Argentina, atlas y planchas; in Barabino SE, Besio Moreno N (eds): Congreso Científico Internacional Americano, vol 1: Relación general del funcionamiento del Congreso. Buenos Aires, Coni Hermanos, 1910, pp 455–456.
22 Jakob C, Onelli C: Atlas del cerebro de los mamíferos de la República Argentina: Estudios anatómicos, histológicos y biológicos comparados, sobre la evolución de los hemisferios y de la corteza cerebral. Buenos Aires, Guillermo Kraft, 1913.
23 Jakob C: Das Problem der drohenden Fleischteuerung und seine Lösung. Munch Med Wochenschr 1911;58:2197.
24 Jakob C: Besprechungen. Dtsch Z Nervenheilkd 1893;3:355–357; 1893;4:477–479; 1894;5:101–102; 1895;6:373–374; 1895;6:488–490; 1896;9:145–147.
25 Jakob C: Contribution à l'étude de la morphologie des cerveaux des Indiens. Rev Museo La Plata 1905;12:59–72.
26 Ariëns Kappers CU: The Evolution of the Nervous System in Invertebrates, Vertebrates and Man. Haarlem, De Erven F Bohn, 1929, p 270.
27 Jakob C: Acute alkoholische Neuritis bei einem fünfjährigen Kinde. Jahrb Kinderheilkd Physiol Erzhg (Leipz) 1893;36:210–214.
28 Jakob C: Ein anatomischer Beitrag zur Lehre von den combinirten Systemerkrankungen des Rückenmarkes. Dtsch Z Nervenheilkd 1894;5:115–126.
29 Jakob C: Ueber einen Fall von Hemiplegie und Hemianästhesie mit gekreuzter Oculomotoriuslähmung bei einseitiger Zerstörung des Thalamus opticus, des hintersten Theiles der Capsula interna, der vorderen Vierhügel- und Haubengegend, mit besonderer Berücksichtigung der secundären Degenerationen. Dtsch Z Nervenheilkd 1894;5:188–224.
30 Jakob C: Ein Beitrag zur Lehre vom Schleifenverlauf (obere, Rinden-, Thalamusschleife). Neurol Centralbl 1895;14:308–310.
31 Jakob C: Estudio sobre el estado de la leucocitosis en las enfermedades infecciosas. Rev Soc Méd Argent 1900;8:113–134.
32 Jakob C: Breves observaciones al artículo sobre el 'coup de chaleur' de los doctores Ayersa y Piñero. Rev Soc Méd Argent 1900;8:109–111.
33 Ayerza A, Piñero HG: 'Coup de chaleur'. Contribución a su estudio. Rev Soc Méd Argent 1899;7:417–453.
34 Jakob C: Sobre el desarrollo de la corteza cerebral. Rev Soc Méd Argent 1899;7:397–403.
35 Jakob C: ¿Existe ó no un centro de Broca? Sem Méd 1906;13:677–678.
36 Jakob C: Consideraciones anátomo-biológicas sobre los centros del lenguaje. Sem Méd 1906;13:733–737.
37 Jakob C: Sobre un caso de paraplejía espasmódica familiar progresiva con examen histopatológico completo. Rev Soc Méd Argent 1909;17:665–703.
38 DeLuca GC, Ebers GC, Esiri MM: The extent of axonal loss in the long tracts in hereditary spastic paraplegia. Neuropathol Appl Neurobiol 2004;30:576–584.
39 Jakob C: Über die Ubiquität der senso-motorischen Doppelfunktion der Hirnrinde als Grundlage einer neuen, biologischen Auffassung des corticalen Seelenorgans. J Psychol Neurol (Leipz) 1912;19:379–382.
40 Jakob C: Über die Ubiquität der senso-motorischen Doppelfunktion der Hirnrinde als Grundlage einer neuen biologischen Auffassung des kortikalen Seelenorgans. Munch Med Wochenschr 1912;59:466–468.
41 Jakob C: La psicología orgánica y su relación con la biología cortical. Z Ges Neurol Psychiatr 1914;9:804–805.
42 Jakob C: Im Anschlusse an die Ausführung des Herrn Claparède. Diskussionen über die Definition, psychologische Interpretation und therapeutischer Wert des Hypnotismus. J Psychol Neurol (Leipz) 1912;19:276–278.
43 Ranke O: Bücheranzeigen und Referate: Chr Jakob und Cl Onelli: Vom Tierhirn zum Menschenhirn; Chr Jakob: Das Menschenhirn. Munch Med Wochenschr 1911;58:2510–2512.
44 Irsigler FJ: Morphogenetic versus morphofunctional theory. Behav Brain Sci 1988;11:95–96.
45 Orlando JC: Sobre el cerebro visceral: documentación histórica de una prioridad científica. Rev Argent Neurol Psiquiatr 1964;18:197–201.
46 Papez JW: A proposed mechanism of emotion. Arch Neurol Psychiatry 1937;38:725–743.

[23]

47 Colombo JA, Fuchs E, Härtig W, Marotte LR, Puissant V: 'Rodent-like' and 'primate-like' types of astroglial architecture in the adult cerebral cortex of mammals: a comparative study. Anat Embryol 2000;201:111–120.

48 Reisin HD, Colombo JA: Considerations on the astroglial architecture and the columnar organization of the cerebral cortex. Cell Mol Neurobiol 2002;22:633–644.

49 Szirko M: Effects of relativistic motions in the brain and their physiological relevance; in Wautischer H (ed): Ontology of Consciousness: Percipient Action. Cambridge, MIT Press, 2005.

50 Georgiev DD: The nervous principle: active versus passive electric processes in neurons. Electroneurobiología 2004;12:169–230.

51 Pribram KH, Spinelli DN, Kamback MC: Electrocortical correlates of stimulus response and reinforcement. Science 1967;157:94–96.

52 Willshaw DJ, Buneman OP, Longuet-Higgins HC: Non-holographic associative memory. Nature 1969;222:960–962.

53 Westlake PR: The possibilities of neural holographic processes within the brain. Kybernetik 1970;7:129–153.

54 Ávila A, Crocco M: Sensing: A New Fundamental Action of Nature. Buenos Aires, Institute for Advanced Study, 1996, pp 22–26.

55 Singer W: Consciousness and the binding problem. Ann NY Acad Sci 2001;929:123–146.

56 Crick F, Koch C: Towards a neurobiological theory of consciousness. Semin Neurosci 1990;2:263–275.

57 Ward LM: Synchronous neural oscillations and cognitive processes. Trends Cogn Sci 2003;7:553–559.

58 Triarhou LC, del Cerro M: Semicentennial tribute to the ingenious neurobiologist Christfried Jakob (1866–1956). 2. Publications from the second Argentina period, 1913–1949. Eur Neurol 2006;56:189–198.

[24]

Historical Note

European Neurology

Eur Neurol 2006;56:189–198
DOI: 10.1159/000096425

Received: May 6, 2006
Accepted: July 12, 2006
Published online: October 19, 2006

Semicentennial Tribute to the Ingenious Neurobiologist Christfried Jakob (1866–1956)

2. Publications from the Second Argentina Period, 1913–1949

Lazaros C. Triarhou[a]   Manuel del Cerro[b]

[a]Economo-Koskinas Wing for Integrative and Evolutionary Neuroscience, Department of Educational and Social Policy, University of Macedonia, Thessaloniki, Greece; [b]Departments of Neurobiology and Anatomy, and Ophthalmology, University of Rochester, Rochester, N.Y., USA

Key Words
Christfried Jakob · Evolutionary neuroscience · Cognitive neuropsychology · Neurophilosophy · Triune brain · History of anatomic neuropathology

Abstract
Christofredo (also Christfried or Christian) Jakob is considered the father of neurology, neurobiology and forensic histopathology in Argentina, where he initially worked between 1899–1910 and then from 1913 onwards. He held professorships of neurobiology at the Faculty of Humanities and Educational Sciences of the University of La Plata and of anatomy and biology at the University of Buenos Aires, and established one of the most important neuropathological laboratories in South America. In the latter phase of his career, Jakob published important works on the pathological anatomy of neurological and neuropsychiatric disorders and formalized ideas on consciousness and neurophilosophy.

Copyright © 2006 S. Karger AG, Basel

Introduction

In the preceding article [1], we covered the activity of Christfried Jakob during his German years and the 'first Argentina period' of 1899–1910. In 1912, Jakob left Germany once again, to return to Argentina permanently. In 1913, he was appointed Chief of the Neuropathological Institute at the Hospicio Nacional de Alienadas, the mental asylum for women in the Federal Capital, and Professor and Director of the Institute of Biology at the Faculty of Philosophy and Letters of the University of Buenos Aires [2].

In 1919, during the period of university reform begun in Argentina after World War I, Jakob was named Professor and Chairman of Descriptive Anatomy at the University of Buenos Aires Faculty of Medicine. He suggested the introduction of histology and embryology in forensic medicine. Apparently, these ideas were not received well by either students or faculty and the new administration forced him (perhaps for other unfathomable reasons as well) to resign the chair. He did so graciously, by declaring that 'students did not understand me, neither did professors' [3].

On March 1, 1922, Jakob was named Professor of Neurobiology at the Faculty of Humanities and Educa-

KARGER

Fax +41 61 306 12 34
E-Mail karger@karger.ch
www.karger.com

© 2006 S. Karger AG, Basel
0014–3022/06/0563–0189$23.50/0

Accessible online at:
www.karger.com/ene

Lazaros C. Triarhou, MD, PhD
University of Macedonia
Egnatia 156, Bldg. Z-312
GR-54006 Thessaloniki (Greece)
Tel. +30 2310 891 387, Fax +30 2310 891 388, E-Mail triarhou@uom.gr

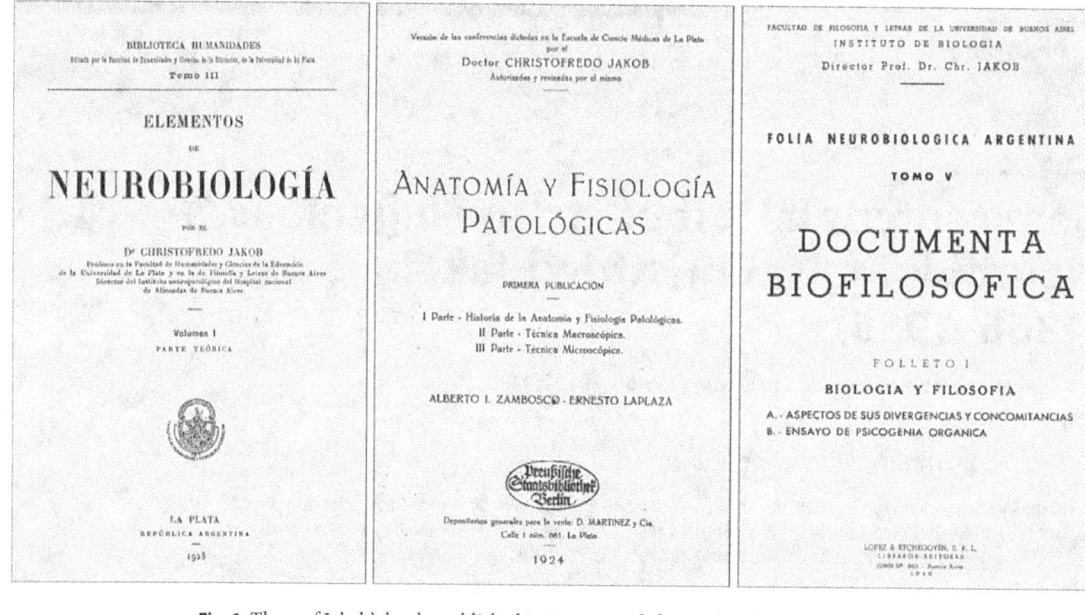

Fig. 1. Three of Jakob's books published in Argentina: left to right, *Elements of Neurobiology* (1923) [8], *Pathological Anatomy and Physiology* (1924) [9] and *Biophilosophical Documents,* being volume 5 of *Folia neurobiológica argentina* (1946) [13].

tional Sciences of the Universidad Nacional de La Plata, by order of the Secretary of Education and University President, Joaquín V. González; from 1921 to 1933, he held a joint appointment as Professor of Pathological Anatomy at the School of Medical Sciences of La Plata [4–6].

Jakob retired in 1945 [5], but kept his formal appointment in Buenos Aires as Chairman of Pathological Anatomy and continued to work in his laboratory at the Hospicio de Alienadas until 1954. He lived in a modest house in Belgrano (a residential neighbourhood in Buenos Aires city). Christofredo Jakob died in Buenos Aires on May 6, 1956. He was close to 90 years of age.

Unfortunately, his work was not always given the recognition it well deserved, even in Argentina. One of us (M.d.C.) recalls the degree to which don Christofredo's life and work were neglected by the majority of the Buenos Aires Medical Faculty of the late 1950s and early 1960s. In those rare occasions when Jakob was remembered, it was as if one referred to a great but remote individual, whose work was of doubtful relevance in the age of cellular and molecular neurobiology.

Newer Monographs

Some 20 monographs were published by Jakob in Spanish during the 'second Argentina period' of 1913–1949. Between 1915 and 1918, he wrote a two-volume treatise on general and special biology [7] for use in schools. The 1923 *Elements of Neurobiology* [8] (fig. 1) is a treatise on the nervous system that begins with an introduction to the triptych 'Cosmos – Life – Mind', continues with the history and methods of neurobiology, then moves on to developmental and comparative neurobiology (fig. 2), nervous histophysiology, organizational principles and principles of conduction, neural and psychic dynamics, and concludes with a chapter on neuropathology and psychopathology. In the preface, written in January 1923 on board the steamship *Cap Polonio* en route to the Tierra del Fuego, one finds Jakob's favourite quote, 'sun and brain are the creators of our worlds'. *Pathological Anatomy and Physiology* (in two volumes) [9] is based on lectures given during 1922 and 1923 at the School of Medical Sciences of La Plata (fig. 1) and covers the history of pathological anatomy and physiology, mac-

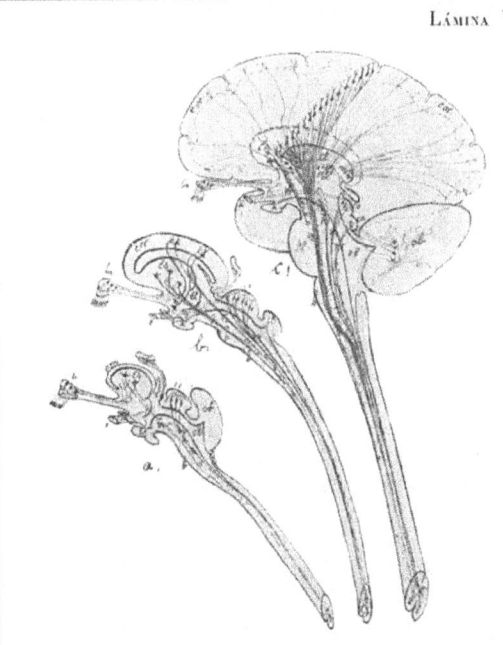

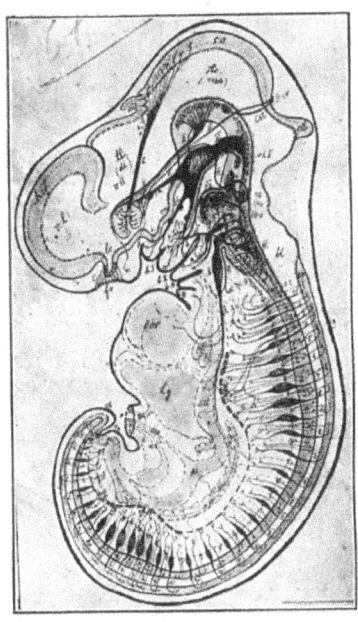

Fig. 2. Original drawings by Jakob, depicting cerebral phylogeny in the fish, reptilian and mammalian nervous system (left) and cerebrospinal ontogeny in a human embryo (right), from his book *Elements of Neurobiology* [8].

roscopical and microscopical techniques, and general pathophysiology.

Three volumes of class notes published by the Faculty of Humanities and Educational Sciences of the University of La Plata include 'Archiencephalon' (1932), 'Paleoencephalon' (1936) and 'Neoencephalon' (1937). Two additional seminar series and lecture notes were published in 1938, the 'Plan of the fundamental organization of the central nervous system of vertebrates' and 'The subcortical organization of the central nervous system of higher vertebrates: the paleoencephalon and its instinctive functions'. Some earlier lectures given at the Hospicio de las Mercedes and in the clinic of Dr. Ramos Mejía in Buenos Aires had been published in 1900 [10] and 1909 [11], respectively.

The *Folia neurobiológica argentina* comprise three richly illustrated atlas volumes [12] and five text volumes [13]. The atlas volumes cover the systematic and topographical anatomy of the human brain, its pathological anatomy in relation to the clinic, and its ontogeny and phylogeny; they total 1,200 pages, including over 1,000 figures plus 650 macrophotographic plates.

Atlas I is literally an atlas of anatomical tomography, a forerunner to modern computed tomography atlases of the head and neck, containing photographs in the coronal, horizontal and sagittal planes. Using a special technique, Jakob prepared his specimens with the brain in situ inside the skull (topographic), either opening windows of varying sizes or cutting the head in whole sections. Atlas II is an atlas of anatomical neuropathology

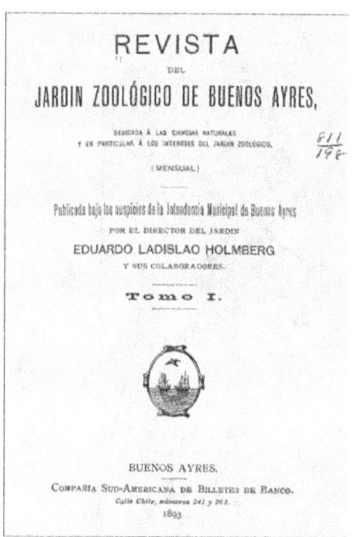

REVISTA

DEL

JARDIN ZOOLÓGICO DE BUENOS AYRES,

DEDICADA Á LAS CIENCIAS NATURALES
Y EN PARTICULAR Á LOS INTERESES DEL JARDIN ZOOLÓGICO.

(MENSUAL)

Publicada bajo los auspicios de la Intendencia Municipal de Buenos Ayres

POR EL DIRECTOR DEL JARDIN

EDUARDO LADISLAO HOLMBERG

Y SUS COLABORADORES.

Tomo I.

BUENOS AYRES.

COMPAÑÍA SUD-AMERICANA DE BILLETES DE BANCO.
Calle Chile, números 241 y 263.
1893

Fig. 3. Inaugural issue of the *Review of the Buenos Aires Zoological Garden,* where Christofredo Jakob published 15 articles between 1909 and 1921. Courtesy: Library of Congress.

with clinical correlations. Atlas III has three parts, on comparative neurobiology, systematic phylogeny and cerebral ontogeny, illustrated with specimens from 120 different animal species. Jakob concludes each of the three volumes with a summary on the historical course of cerebral anatomy, anatomic pathology, and phylo-ontogeny.

The five text volumes of the *Folia neurobiológica argentina* [13] cover in their 600 pages general neurobiology, the neurobiology of the edentate *pichiciego pampeano,* the pink fairy of the armadillo species also known as *ratoncito cascarudo* (*Chlamyphorus truncatus,* endangered from 1970), the reptilian *yacaré overo* or 'broadsnouted caiman' *(Caiman latirostris),* anatomo-clinical correlates of the frontal lobe, and issues in biological philosophy (fig. 1).

In the 1940s, Jakob et al. [14] produced three volumes on the human embryo, covering age determination and chronological development, problems of human embryology, gametogenesis, zygote formation, segmentation and gastrulation, and properties of the embryoblast.

Further Neuro-Anatomical and Neuropathological Studies

About two thirds of Jakob's 200 articles date from 1913 onwards. Beside works in anatomy, histology, embryology, phylogeny, pathology, philosophy and neuropsychology, there are 9 geography articles and 2 educational articles on teaching biological sciences and neuro-anatomy at schools and university [15, 16].

Jakob published 13 articles in the *Archives of Psychiatry and Criminology,* a journal founded and edited by José Ingenieros between 1902 and 1913 (later renamed to *Review of Criminology, Psychiatry and Legal Medicine*). These articles are [12, 17]: 'The frontal lobes and higher mental functions' (1906), 'The leptomeninges in mental disorders' (1909), 'Anencephalic monsters' (1910), 'Cerebral histology and psychology' (1911), 'Madness in animals' (1913), 'Motor aphasia and its localization' (1913), 'Psychology and its relation to cortical biology' (1913), 'The harmonic development of intelligence and the brain in the child' (1913), 'Organic psychology and its relation to cortical biology (1913), 'Heredity and psychopathology as dramatic factors in the work of Ibsen' (1929), 'Biological aspects of human typology and its application in Argentina' (1933), 'On the organic bases of memory' (1935), and 'Demonstration of lesions in the zone of language' (1936).

Between 1909 and 1921, Jakob published 15 papers in the monthly *Revista del Jardín Zoológico* (fig. 3). These articles include [5, 12]: 'Cerebrocranial autopsy of an elephant' (1910), 'The importance of comparative histoarchitecture for modern psychology' (1910), 'On the differential psychobiology between human and animal intelligence' (1913), 'An autopsied gibbon and chimpanzee' (1914), 'The language of animals' (1914), and 'An interesting teratological case' (1915).

In an extended article on the biological importance of the cerebellar system (1939) [18], Jakob reviewed 100 years of progress beginning with Purkinje's description in 1837 of the homonymous efferent cell, cerebellar phylogeny from fish through primates, cerebellar ontogeny in the human embryo, anatomical connections, and histophysiological correlations. He concluded the study by mentioning that the production of cognitive processes *(gnosiopoiesis)* uses peripheral information directly related to the thalamus and on to the parieto-occipito-temporal cortex; *praxiopoiesis* ('generation of actions') requires cerebellar information, which passes from the cerebellum and red nucleus to the thalamus and frontal cortex. Thus, these circuits become associated with 'gnosiopraxic' cortical systems to generate what Jakob calls 'ideopoietic dynamics'.

The year he retired, Jakob published an article [19] in which he described 'an unknown microganglion in the hypothalamic commissure' – in two cases doubled – located in the midline at the region of the mammillary bodies and posterior commissural zone, 'a phylogenetic memory in our brain'. The ganglion measured 0.30–0.45 mm in diameter, its cellular crown and neuropil centre with minimal vascularization being reminiscent of familiar ganglia in invertebrates. Jakob traced the embryological origin of the microganglion to the retromammillary sulcus, just rostral to the mammillary recess in the 3-month-old human embryo. That study is worth preserving, first as a piece of historical knowledge, and second, as potentially deserving some further exploration to determine whether anyone else has seen such a structure since Christofredo Jakob's description.

Further, he described the histology of two brains and a spinal cord in a calf born with two heads and a normal body in the Buenos Aires Zoo [20]. He was solicited for a forensic opinion based on histopathology on two historic occasions, a homicide in 1926 and a psychiatric case in 1930 [21].

Until the 1930s and even later, psychiatry in Argentina was articulated around what Nathan Hale calls the 'somatic style' [2], i.e. a view of mental illness pertaining to the body: the origin of psychiatric disorders could be traced to the morphology of the nervous system, and they were treated accordingly. Jakob had been influential in promoting such a notion for psychiatry.

His neuropathological articles include studies on olivopontocerebellar atrophy [22], pituitary tumours [23], intracranial fibro-chondro-osteoma [24], general paresis [25], cerebral arteriosclerosis [26], cortical changes in schizophrenia [27], forms of Schilder [28], Hallervorden-Spatz [29] and Pick [30] diseases, and the histophysiology and histopathology of the cerebello-hypothalamo-striatal and cerebello-hypothalamo-cortical systems in humans and apes, their diencephalic ontogeny and pathology in 162 cases that included chorea, parkinsonism, encephalitis lethargica, Wilson disease, and Luysian hemiballism [31].

Some additional neuropathological studies dealt with extrapyramidal syndromes [32, 33], the microscopical diagnosis of mental alienation [34], hemiplegia, hemiataxia and hemi-anaesthesia of cerebellar origin [35], infantile paralysis (acute poliomyelo-encephalitis) [36], issues in anatomic neuropathology [37], a family with combined mental retardation, visual and hearing impairment

and spastic quadriplegia [38], and dolorous spastic paraplegia from compression of the inferior dorsal medulla by a dural psammomatous endothelioma [39].

In 1942, Jakob published a case of myoclonus associated with a primary lesion of the deep cerebellar nuclei and hypertrophic degeneration of the contralateral inferior olive [40], and a case of secondary degeneration of the central tegmental bundle from a lesion of the cerebellar dentate nucleus [41]. He stated that the central tegmental bundle in humans contains both crossed and uncrossed axons from the dentate nuclei, which join pallidal axons in their trajectory through the brainstem. Jakob thus confirmed the cerebello-olivary fibres in the superior cerebellar peduncle, which had originally been described by Cajal, as a component of the central tegmental bundle, in which Jakob had also identified a dorsal and ventral part that 'continue their descending course spinalward to medullary centres' [42].

A distinct form of lower bilateral ('bibasal') cerebellar degeneration combined with dementia was described by Aranovich [43] as 'Jakob type', based on morphological criteria; it was identified in 15 of 31 cases of cerebellar atrophy in women over 50 years old. The lesion results from an abiotrophic process involving loss of Purkinje cells and originating in the fundus of the marginal sulcus and progresses through destruction of neighbouring lamellae and atrophy of subjacent white matter.

Ideas Pertinent to Cognitive Neuroscience and Neurophilosophy

Jakob viewed 'form as stabilized function and function as change of form; in other words, the vital energy of an organism is a single entity that will present itself as form in the latent state and as function in the kinetic state' [13]. He wrote that 'form, structure and function are inseparable, if not identical, and only scholastic science has managed to separate them … only a basis that is fundamentally biological, morphostructural and histophysiological at the same time, unified in an ample ontogenetic and phylogenetic context, can let us address in legitimate ways the serious questions of modern neuro- and psychobiopathology' [12]. Jakob always considered morphology in a functional context and formulated ideas on the integrative function of the brain. The neurobiology of the frontal lobes occupied him for more than four decades [13, 44, 45].

The following quote is relevant to the topical issue of free will in modern cognitive science: 'It is mnemonic

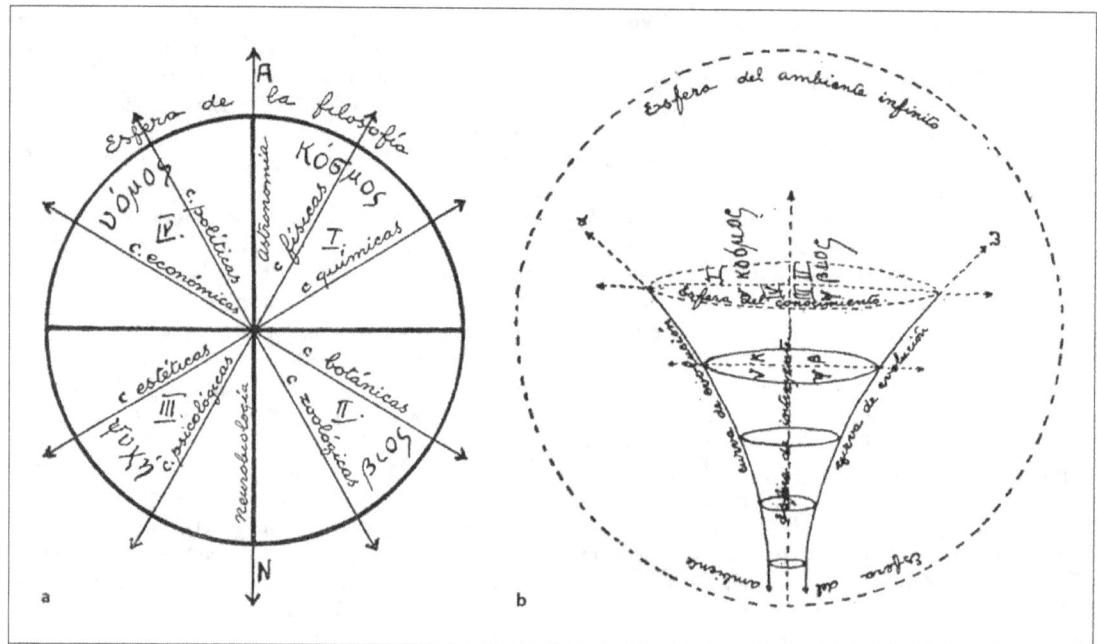

Fig. 4. a Jakob's drawing of the empirical sphere with the four quadrants (I–IV) of the sciences: cosmos (κόσμος), life (βίος), psyche (ψυχή) and order (νόμος). The four quadrants comprise (clockwise from upper-right to upper-left) the physical and chemical (I), botanical and zoological (II), psychobiological and aesthetical (III) and economic and political (IV) sciences. In the two opposite poles (A and N), astronomy and neurobiology invade, respectively, the ultramacrodynamic and the ultramicrodynamic. The sphere of philosophy is marked on the outside of the upper part of the circle. **b** Invasion of the sphere of infinite environment by that of progressive knowledge, delimited by the curve of evolution. From a 1945 article on the philosophical meaning of the human brain [53]. Courtesy: National Library of Medicine.

function that raises the cortical apparatus to its creative potential, its influence and dominant hierarchy in the psyche of the individual, and liberates it from non-salvable ties of reflex law and elaborated instinct; that amplified expectation of action that we call *volitional freedom* consists of the possibility of anticipating the result of a given situation and selecting among various possibilities the one best suited to the momentary constellation and its individual advantage' [6].

The philosophical writings of Jakob include an essay on the philosophy of nature according to Kant [46]; an analysis of heredity and psychopathology as dramatic factors in Ibsen [47]; the psychobiological views of Descartes over three centuries [48]; the significance of Cajal's work for neuropsychiatry [49] and biophilosophy [50] with special emphasis on the cognitive processes of *gno-siopoiesis* and *praxiopoiesis;* a conference on the religion of nature and human future [51], and a presentation of ideas on life and mental experience in relation to time at vegetative, phylogenetic and ontogenetic levels under the encompassing title 'From tropism to the general theory of relativity' [52].

As a basis for the construction of a future philosophy of the brain, i.e. the 'synthesis of neuropsychodynamic theories proven universally valid', useful for a biological psychology and psychiatry, Jakob focused on three concepts: (a) a central organization, (b) heredity and (c) the evocation and transformation of physical processes, through neurohistological, physiological and neurobiological, to psychological phenomena [53]. Considering that the utmost problem of science and philosophy converges in cerebral function, as the most immediate issues

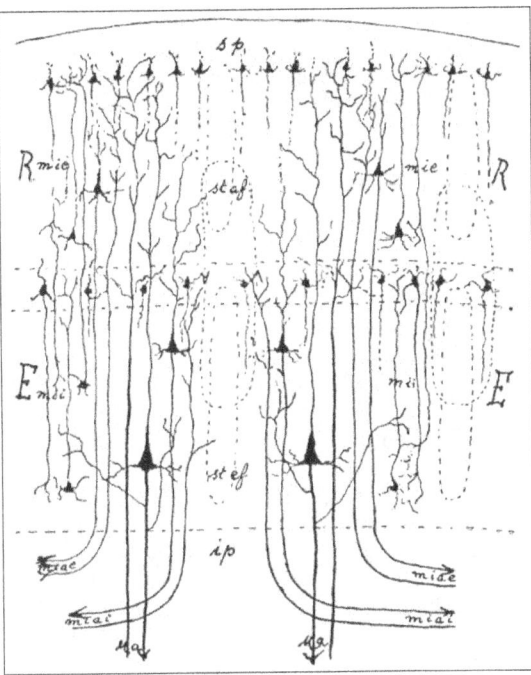

Fig. 5. Jakob's drawing of the microdynamic (mi) organization of the neocortex independently of macrodynamic (ma) events. From Jakob [53]. Courtesy: National Library of Medicine.

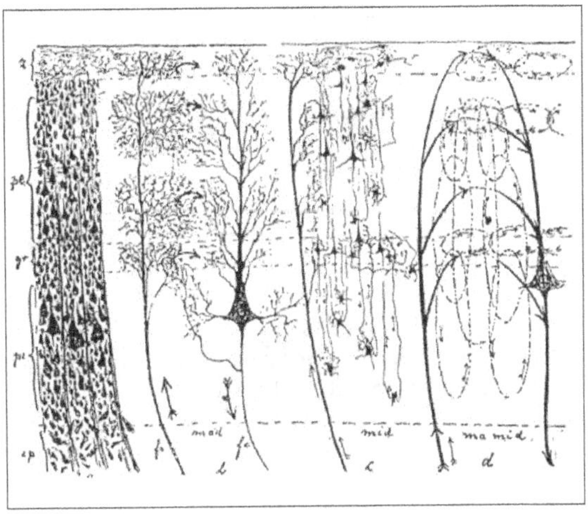

Fig. 6. Neocortical histotopography with its macrodynamic (ma) and microdynamic (mi) events according to Jakob [53, 54]. The scheme depicts the most probable trajectories of nervous current in the cortex: (a) the cortical layers; (b) termination of a thalamo-cortical fibre and a large motor neuron projecting its axon e.g. to the pes pedunculi; (c) an incoming afferent fibre (fa) in relation to small interconnecting cells, equivalent to the cortical 'microdynamic apparatus'; (d) probable circuits intercalated among afferent and efferent (fe) fibres in the various cortical layers. Z = Zonal (molecular) layer; pe = external pyramidal layer; gr = granule cell layer; pi = internal pyramidal layer. Courtesy: National Library of Medicine.

for the centuries to come he considered: (a) the laws and steps of the quest for cerebral phylogeny and ontogeny; (b) the micro-organization of neuroblasts and the activation of their functional derivatives, normal and pathological, and (c) the poly-energetic cosmo-bio-neuro-psychic transformation in the synthetic transformation of the outer (objective) and inner (subjective) environment (fig. 4). The actual problems in terms of his brain philosophy would then be: neurobiogenesis, neurodynamics and neuropsychogenesis. In his later works [13, 53, 54], Jakob returned to the microdynamic and macrodynamic concepts he had been teaching since the early 1900s, to explain the nature of memory and conscious activity (fig. 5, 6).

One last facet of Jakob's multifarious talent was music. On September 13, 1923, he gave a conference at the National College in La Plata on 'Musical biodynamism' [55], which he illustrated by performing piano works by Wagner and Chopin, as well as works for cello and piano by Grieg and Pergolese, accompanied by Professor Juan Chabra [5]. In 1926, he published an article on 'The spirit of music in pre- and post-Kantian philosophy' [56], in which he reviews fundamental musical elements from the ancient Greeks to the European classical composers and deals with the biology, physiology, psychology, pathology, aesthetics and paedagogy of music.

A 'Triune System of the Psyche' Based on Phylogeny

According to Jakob [12], in the phylogenetic scale, 'psychodynamic' or 'ontopsychic' nervous functions are preceded by 'plasmodynamic' activities, which encompass tropism and pulsatility. The 'neurodynamic' or 'phylopsychic' processes are subserved by three underlying hierarchical levels of the vertebrate central nervous system, designated as archineuronal, paleoneuronal and neoneuronal, and encompass *archikinesias* (reflex actions similar to invertebrates), *paleokinesias* (instinctual-automated reactions), and *neokinesias* (conscious motor reactions) [8, 57].

In the framework of the dynamic workings of the human cerebral cortex, *neokinesias* include *gnosias* (cognitive processes related to conscious orientation in one's environment), *praxias* (active individual intervention) and *symbolias* (ideative abstraction to facilitate interindividual communication, such as the sociogenetic processes on which human culture is based) [8, 58, 59].

In a certain way, Jakob's proposition on *archipsychic, paleopsychic* and *neopsychic* phenomena in 1923 [8] appears to have anticipated the 1973 'triune brain' concept of Paul MacLean [60] by half a century.

Concluding Remarks

Christofredo Jakob's pursuits spanned beyond neurobiology, neuropathology and comparative neuroanatomy into areas of geography, cognitive science and philosophy, and even cultural issues on Argentina's intellect in relation to the Romanic and Germanic traditions [61].

An inimitable conjunction of broad and firm educational foundations, affluent specimen resources, depth of scholarly focus, the good fortune of longevity and an enduring lucid mind must have conduced to the scientific output of Christofredo Jakob. Some of his contributions may still be useful sources of information for modern researchers in evolutionary and ontogenetic neuroscience, cognitive neuropsychology, and neurophilosophy. In 1939, he stated: 'In the end, in future centuries, we shall perhaps return to a genuine human micro-neuro-psychobiology, and thus *homo sapiens* will be able to aspire to a title which seems more like an irony at the moment, if one considers that we certainly ignore 99 percent of the actual functions of our neuronal elements, and this, despite three centuries of great efforts made by our teachers!' [12].

López Pasquali [62] describes his mentor Jakob as impressive in physical style, corpulent and plethoric, accessible and with humour; a temperament conferring untiring energy and optimism in a multitude and variety of interests; an extreme discipline, methodical character and solid scientific method, facilitating the expedient use of time and task completion. Gregorio Bermann [63] recalls Jakob's extraordinary vitality, until the end of his long life, his animated figure with the precise movements, his vivacious and sharp eyes, his pictorial language, with the foreign and admirable eloquence of the facts that filled his lectures. Besides biomedicine and philosophy, Jakob's diversions included music, poetry, mineralogy and journeys through unexplored territories of Argentina, Chile, Bolivia and Peru. He also taught biology at high schools, as he believed that the governing ideas of thought are generated in adolescence.

Jakob commanded a culture that included the classical Greek and Latin literatures. He begins his thesis with an original quote from Aretaeus of Cappadocia on the vena cava. In one of his later papers, we read: 'The most cultured people that have existed, the Hellenic, had beautified, *honoris causa,* in that art (music) the special name of *art of the Muses* and, symbolically, the great Plato had declared that philosophy is the science and art of truth like music' [56].

Christofredo Jakob's philosophical stance on life is exemplified in a *credo* he imparted for the man of the future [51]: 'I believe in the harmony of the Universe; in the triumph of Life; in the victory of the human spirit; in the mutual responsibility of individuals and nations; in an expanding universal confraternity and in a God of justice and love for all, without discrimination against races or religions, and who will be revealed to us, inside and through us, with the progressive humanization of the future, yet not with talk but with deeds'.

Acknowledgements

The authors gratefully acknowledge the courtesy of the Staatsbibliothek zu Berlin; the British Library; the Interlibrary Loan Department of the University of Macedonia Library; the Ruth Lilly Medical Library of Indiana University, Indianapolis; the Bernard Becker Medical Library of Washington University, St. Louis; the Library of Congress and the National Library of Medicine of the United States.

References

1 Triarhou LC, del Cerro M: Semicentennial tribute to the ingenious neurobiologist Christfried Jakob (1866–1956). 1. Works from Germany and the first Argentina period, 1891–1913. Eur Neurol 2006;56:176–188.

2 Plotkin MB: Freud in the Pampas: The Emergence and Development of a Psychoanalytic Culture in Argentina. Stanford, Stanford University Press, 2001, pp 16–17.

3 Strejilevich L, Quiroga LF: Christofredo Jakob (1866–1956): maestro de 40 generaciones. Alcmeon Rev Arg Clín Neuropsiquiátr 1999;30:201–204.

4 Carnese FR, Goicoechea AS, Cocilovo JA: Argentina; in Spencer F (ed): History of Physical Anthropology: An Encyclopedia, vol 1 (Garland Reference Library of Social Science, vol 677). New York, Garland Publishing, 1997, pp 101–107.

5 Orlando JC: Christofredo Jakob: su vida y obra. Buenos Aires, Editorial Mundi, 1966.

6 Piva JR, Virasoro CA: Christofredo Jakob: neurobiólogo, científico en diálogo filosófico. Com Mus Prov Cs Nat Florentino Ameghino 2004;9:1–18.

7 Jakob C: Tratado de biología general y especial para el uso de la enseñanza elemental, secundaria y superior en la República Argentina. Buenos Aires, Guillermo Kraft, 1915–1918, vol 1-2.

8 Jakob C: Elementos de neurobiología. La Plata, Facultad de Humanidades y Ciencias de la Educación de la Universidad Nacional de La Plata (Biblioteca Humanidades), 1923, vol 1: Parte teórica.

9 Jakob C: Anatomía y fisiología patológicas: versión de las conferencias dictadas en la Escuela de Ciencias Médicas de La Plata (primera y segunda publicación, comp por Alberto I Zambosco y Ernesto Laplaza, aut y rev por el autor). La Plata, Martínez y Cía, 1924.

10 Jakob C: Lecciones sobre anatomía y fisiología del sistema nervioso, en sus relaciones con la psiquiatría. Sem Méd 1900;7:325–327, 354–358, 363–366, 403–408, 439–444, 479–482, 589–590, 647–652.

11 Jakob C: Curso sobre enfermedades del sistema nervioso en relación con su anatomía patológica dictado en la clínica del Dr Ramos Mejía (recogidas por PM Barlaro). Buenos Aires, La Ciencia Médica, 1909.

12 Jakob C: El cerebro humano. Folia neurobiológica argentina, atlas I–III. Buenos Aires, Aniceto López, 1939–1941.

13 Jakob C: Folia neurobiológica argentina, tomos I–V. Buenos Aires, Aniceto López López y Etchegoyen, 1941–1946.

14 Jakob C, Jakob A, Pedace EA: El embrión humano, folletos I–III. Buenos Aires, Aniceto López, 1942–1945.

15 Jakob C: Sobre la enseñanza de las ciencias biológicas en la escuela primaria y secundaria. Rev Humanidades (La Plata) 1927;16: 159–172.

16 Jakob C: La enseñanza universitaria de la anatomía cerebral. Rev Asoc Méd Argent 1937;51:799 803.

17 Ingenieros J: Indice general de Archivos de Psiquiatría y Criminología, años 1902–1913. Buenos Aires, Talleres Gráficos de la Penitenciaría Nacional, 1914, pp 3–26.

18 Jakob C: El sistema cerebeloso y su significación biológica. Rev Asoc Méd Argent 1939; 53:198–209.

19 Jakob C: Un microganglio desconocido en la comisura hipotalámica. Rev Neurol Buenos Aires 1945;10:219–224.

20 Jakob C: Dos cerebros con una médula. ¿Es admisible una compensación funcionalmente válida en los casos de duplicidad cefálica? Prensa Méd Argent 1946;33:403–405.

21 Bonnet EFP: Christofredo Jakob. Precursor de la histopatología forense en la República Argentina. Segundo Congreso Nacional de Historia de la Medicina Argentina, Córdoba, 1970, pp 68–72.

22 Jakob C, Beretervide JJ, Caballero E: Atrofía olivopontocerebelosa familiar. Prensa Méd Argent 1934;21:1997–2017.

23 Jakob C: Contribución a la histogénesis de las neoplasías de la hipófisis. Rev Neurol Buenos Aires 1936;1:99–125.

24 Jakob C, Pedace EA: Fibro-condro-osteoma primitivo del cerebro. Rev Asoc Méd Argent 1933;47:2435–2442.

25 Jakob C: Estudios terapéutico-experimentales sobre un tratamiento anti-tóxico de la parálisis general progresiva. Prensa Méd Argent 1916;2:353.

26 Jakob C, Moyano BA: La anatomía patológica de la arterioesclerosis cerebral. Rev Asoc Méd Argent 1938;52:244–254.

27 Jakob C, Pedace EA: Estudio anatomopatológico de la esquizofrenia. Rev Asoc Méd Argent 1938;52:326–334.

28 Jakob C, González T: Leucoencefalosis centrolobar simétrica progresiva familiar (enfermedad de Schilder). Arch Argent Neurol 1936;14:51–76.

29 Jakob C, Montanaro JC: Síndrome palidal por esclerosis amarilla simétrica del globulus palidus: una forma especial en el adulto de la enfermedad de Hallervorden Spatz. Rev Asoc Méd Argent 1939;53:111–124.

30 Jakob C: La demencia progresiva; un análisis neurobiológico de la enfermedad de Pick. Rev Neurol Buenos Aires 1946;11:81–94.

31 Jakob C: Histofisiología normal y patológica del sistema estrío-hipotalámico. Sem Méd 1931;38:129–151.

32 Jakob C: Síndrome de hemibalismo coreiforme cruzado por hemorrhagia en el núcleo hipotalámico. Arch Argent Neurol 1928;2: 1–15.

33 Jakob C, Montanaro JC: Encefaloesclerosis lobar bi-rolándica primitiva. Síndrome extrapiramidal predominante. Rev Otoneurooftalmol Cir Neurol 1928;2:224–248.

34 Jakob C, Delpiano J, Novaro V: Diagnóstico de alienación mental por el examen microscópico del cerebro. Prensa Méd Argent 1935;22:43–45.

35 Jakob C: Hemiplejía, hemiataxia y hemianestesia homolateral de origen cerebeloso. Arch Argent Neurol 1929;4:13–31.

36 Jakob C, Moyano BA: Sobre la anatomía patológica de la parálisis infantil (poliomieloencefalitis aguda). Rev Asoc Méd Argent 1936;50:1454–1481.

37 Jakob C, Pedace EA, Moyano BA: Los problemas actuales de la anatomía patológica del sistema nervioso. Rev Neurol Buenos Aires 1938;3:79–92.

38 Jakob C, Scaravelli A: A propósito de un caso de ocho hermanos con idiocia, sordomudez y cuadriplejía espasmódica familiar. Rev Neurol Buenos Aires 1940;5:283–299.

39 Jakob C, Prini I, Riedel C, Thénon J: Paraplejía dolorosa espástica por compresión de médula dorsal inferior por endotelioma psamomatoso dural (operación, mejoría). Sem Méd 1940;47:1387–1394.

40 Jakob C, Montanaro JC: Mioclonías óculo-laringo-faringo-velopalatinas en el síndrome bulbo-protuberancial; estudio anatomoclínico. Rev Neurol Buenos Aires 1942;7:85–128.

41 Jakob C: La sistematización del haz central de la calota como vía neoneuronal cerebelosa aferente olivobulbar. Rev Neurol Buenos Aires 1942;7:1–24.

42 Bebin J: The central tegmental bundle: an anatomical and experimental study in the monkey. J Comp Neurol 1956;105:287–332.

43 Aranovich J: La atrofia cerebelosa marginal bibasal de Chr Jakob. Rev Asoc Méd Argent 1937;51:115–122.

44 Jakob C: Estudios biológicos sobre los lóbulos frontales cerebrales. Sem Méd 1906;13: 1375–1381.

45 Jakob C, Pedace EA: La mision del lóbulo frontal frente a una cuantificación sintética de sus elementos productores. Arch Neurocir (Buenos Aires) 1949;6:467–474.

46 Jakob C: La filosofía de la naturaleza según Kant. Rev Filos (Buenos Aires) 1925;11:45–57.

47 Jakob C: La herencia y psicopatología como factores dramáticos en la obra de Ibsen. Rev Criminol Psiquiatr Med Leg 1929;16:306–316.

48 Jakob C: La psicología de Descartes a través de tres siglos. An Inst Psicol (Buenos Aires) 1938;2:297–327.

49 Jakob C: Santiago Ramón y Cajal. La significación de su obra científica para la neuropsiquiatría. Sem Méd 1935;42:529–536.

[33]

50 Jakob C: El significado de la obra de Ramón y Cajal en la filosofía de lo orgánico. Rev Humanidades (La Plata) 1938;26:237–256.

51 Jakob C: La religión de la naturaleza y el porvenir del hombre. Rev Humanidades (La Plata) 1930;22:107–119.

52 Jakob C: Del tropismo a la teoría general de la relatividad. Rev Humanidades (La Plata) 1922;3:45–58.

53 Jakob C: El cerebro humano: su significación filosófica. Rev Neurol Buenos Aires 1945;10: 89–110.

54 Jakob C: Sobre las bases orgánicas de la memoria. Rev Criminol Psiquiatr Med Leg 1935;22:84–114.

55 Jakob C: Biodinamismo musical: conferencia y concierto. Bol Univ (La Plata) 1923;7: 332.

56 Jakob C: El espíritu de la música en la filosofía pre y postkantiana. Rev Humanidades (La Plata) 1926;13:119–132.

57 Jakob C: La filogenia de las kinesias: sobre su organización y dinamismo evolutivo. An Inst Psicol (Buenos Aires) 1935;1:109–127.

58 Jakob C: La teoría actual de las gnosias y praxias como factores fundamentales en el dinamismo de la corteza cerebral. Crón Méd (Lima) 1921;38:17–24.

59 Crocco M: ¡Alma e' reptil! Los contenidos mentales de los reptiles y su procedencia filética. Electroneurobiología 2004;12:1–72.

60 MacLean PD: A Triune Concept of the Brain and Behaviour. Toronto, University of Toronto Press, 1973.

61 Jakob C: Das intellektuelle Argentinien und seine Beziehungen zur romanischen und germanischen Kultur. Berlin, Süd- und Mittelamerika Verlag GmbH, 1912, pp 3–24.

62 López Pasquali L: Christfried Jakob. Su obra neurológica, su pensamiento psicológico y filosófico. Buenos Aires, López, 1965.

63 Bermann G: El profesor Christofredo Jakob – In memoriam (Baviera, 1866 – Argentina, 1956). Rev Méd Córdoba 1957;45:45–47.

[34]

Journal of the History of the Neurosciences, 22:366–382, 2013
Copyright © Taylor & Francis Group, LLC
ISSN: 0964-704X print / 1744-5213 online
DOI: 10.1080/0964704X.2012.762830

An Avant-Garde Professorship of Neurobiology in Education: Christofredo Jakob (1866–1956) and the 1920s Lead of the National University of La Plata, Argentina

ZOE D. THÉODORIDOU,[1] ATHANASIOS KOUTSOKLENIS,[1]
MANUEL DEL CERRO,[2] AND LAZAROS C. TRIARHOU[1]

[1]Neuroscience Wing, Department of Educational and Social Policy, University of
Macedonia, Thessaloniki, Greece
[2]Departments of Neurobiology & Anatomy, Ophthalmology, and Neurology,
University of Rochester, Rochester, NY, USA

*The interdisciplinary trend in "Mind, Brain, and Education" has witnessed dynamic
international growth in recent years. Yet, it remains little known that the National
University of La Plata in Argentina probably holds the historical precedent as the
world's first institution of higher education that formally included neurobiology in the
curriculum of an educational department, having done so as early as 1922. The responsibility of teaching neurobiology to educators was assigned to Professor Christofredo
Jakob (1866–1956). In the present article, we highlight Jakob's emphasis on interdisciplinarity and, in particular, on the neuroscientific foundations of education, including
special education.*

Keywords Christfried Jakob, cerebral onto-phylogeny, history of neuroeducation,
neurophilosophy

Introduction

Some of the earliest attempts at applying neurobiological findings to education can be
traced to the work of the neurologist Henry Herbert Donaldson (1857–1938) and the educator Reuben Post Halleck (1859–1936; Théodoridou & Triarhou, 2009). The new tools
of biology and cognitive science have generated vast possibilities for this field, enabling
the integration of diverse disciplines that study human learning and development (Fischer
et al., 2007). Terms used interchangeably to denote this new branch of knowledge include
"neuroeducation" (Battro & Cardinali, 1996), "neurolearning" (Petitto & Dunbar, 2004),

The authors gratefully acknowledge Dr. Daniel S. Margulies of the Max Planck Institute for
Human Cognitive and Brain Sciences in Leipzig for his invaluable help, the anonymous review-
ers for their constructive criticism that led to an improved manuscript, and the courtesy of the staff
at Biblioteca de Humanidades de la Universidad Nacional de La Plata, Academia de Medicina de
Buenos Aires, Ibero-Amerikanisches Institut Preussischer Kulturbesitz zu Berlin, British Library,
Library of Congress, Smithsonian Institution Libraries-National Zoological Park Library, National
Library of Medicine of the United States, Ruth Lilly Medical Library of Indiana University, and
Bernard Becker Medical Library of Washington University for bibliographic assistance.
Address correspondence to Lazaros C. Triarhou, University of Macedonia, Egnatia 156,
Building Z-312, Thessaloniki, GR 54006, Greece. E-mail: triarhou@uom.gr

"nurturing the brain" (Ito, 2004), "developing the brain" (Koizumi, 2004), "mind, brain, and education" (Fisher et al., 2007), "educational neuroscience" (Geake, 2005; Szűks & Goswami, 2007), and "pedagogical neuroscience" (Fawcett & Nicolson, 2007).

The dynamic growth of "neuroeducation" and the opportunities for interdisciplinarity are evidenced by (a) the establishment of academic programs and departments, such as the "Centre for Neuroscience in Education" at the University of Cambridge, the "Mind, Brain, and Education Program" at the Harvard Graduate School of Education, and Dartmouth's "Center for Cognitive and Educational Neuroscience," (b) the emergence of the "International Mind, Brain and Education Society" and the launch of its official journal in 2007, as well as the launch of another new journal in 2012, the *Trends in Neuroscience and Education*, (c) the organization of congresses, meetings, and seminars at both the international (European Association for Research on Learning and Instruction, Zürich, 2010; International Mind, Brain and Education Society, Philadelphia, 2009, San Diego, 2011) and local levels (Collaborative Frameworks for Neuroscience and Education, UK, 2005–2006), and (d) the attention received in the mainstream press and by the lay public (see also Beauchamp & Beauchamp, 2012).

There are additional examples of the worldwide impact of the neuroeducation movement. In the United Kingdom, the Teaching and Learning Research Programme administered by the Economic and Social Research Council (TLRP-ESRC) organized a seminar series on Neuroscience and Education which brought together national and international education and science experts to discuss how these two areas may collaborate. At its conclusion, in June 2006, over 400 teachers, educational researchers, psychologists, and neuroscientists had participated in the series. In Germany, the Federal Ministry of Education and Research found it reasonable to concentrate on the latest line of research in educational neuroscience to improve education (Stern, 2006). In Finland, a national network on neuroscience and education hosted its activities in the University of Helsinki, beginning in 2009, under the theme "The Brain, Learning and Education Network."

Still, it remains a little known fact that the National University of La Plata in Argentina holds a historical precedent as most likely the world's first institution of higher education that formally introduced and included neurobiology in the curriculum of an education department. It was Christfried Jakob (1866–1956; Figure 1), a Bavarian-born neuropathologist, who promoted the initiative to teach brain structure and function to students of the Faculty of Philosophy and Letters as early as 1922.

Prior to 2006, a literature search in the PubMed database would yield no returns on Jakob, save an article by Meyer in Spanish (Théodoridou, 2011). Since 2005, a systematic effort began by one of us (LCT) and his colleagues to retrieve and revive Jakob's writings; this ongoing project has aimed at placing Jakob's ideas in a modern neuroscientific perspective (Théodoridou & Triarhou, 2011, 2012a, 2012b; Triarhou, 2008a, 2008b, 2009, 2010a, 2010b; Triarhou & del Cerro, 2006b, 2006c, 2007; Tsapkini, Vivas, & Triarhou, 2008; Vivas, Tsapkini, & Triarhou, 2007). In the present study, we review Jakob's contributions to education, an endeavor that he pursued throughout his career.

Jakob's Professional Life

Jakob (his forename castillianized to "Christofredo"), the founder of neuropathology in Argentina, adopted that country as his home and lived there from 1899 until his death in 1956, with only a brief visit to Europe from 1910 to 1912 (Moyano, 1957; Fumagalli & Saredo, 2005). He was recognized for his neuroanatomical work, particularly for the systematization of brain sectioning and the application of the Weigert method to the study of

Figure 1. Bust of Christofredo Jakob at the Pathology Laboratory, Moyano Psychiatric Hospital, Buenos Aires. Credit: http://www.flickr.com/photos/theodor_meynert/4695367482.

myelinated fiber tracts (Allegri, 2008). He authored 30 books and 200 articles on developmental, evolutionary, anatomical and pathological neurobiology (Figure 2; Triarhou & del Cerro, 2006c). His approach was largely based on studying species of Argentinian fauna (Papini, 1978, 1988). For example, his theory on the phylogenetic origin of the neocortex (Jakob, 1945) issued from his histological studies on *Amphisbaena*, a small apod reptile (Papini, 1988). Jakob is credited with combining phylogenetic, embryological, and functional approaches in a quest to explain the role of the cerebral cortex in cognition and behavior (Triarhou, 2008b). From such an attempt ensued his theory of a dual developmental-evolutionary origin and ubiquitous sensory-motor function of the cerebral cortex (Jakob, 1911, 1912a, 1912b; Triarhou, 2010b). He further studied and highlighted the fields of neurophilosophy, affective neuroscience, and educational neuroscience (Triarhou, 2008a; Théodoridou, 2011). He did so based on his collective academic, clinical, and research experience from multiple vocational frameworks, including universities (National University of Buenos Aires and National University of La Plata), hospitals (Las Mercedes Hospital in Buenos Aires and National Psychiatric Hospital for Women in the Federal Capital), and research laboratories (Laboratory of the Psychiatric and Neurological Clinic of Las Mercedes Hospital, Neuropathological Institute at the National Psychiatric Hospital for Women in the Federal Capital, and Institute of Biology at the Faculty of Philosophy and Letters of the National University of La Plata; Triarhou & del Cerro, 2006b, 2007).

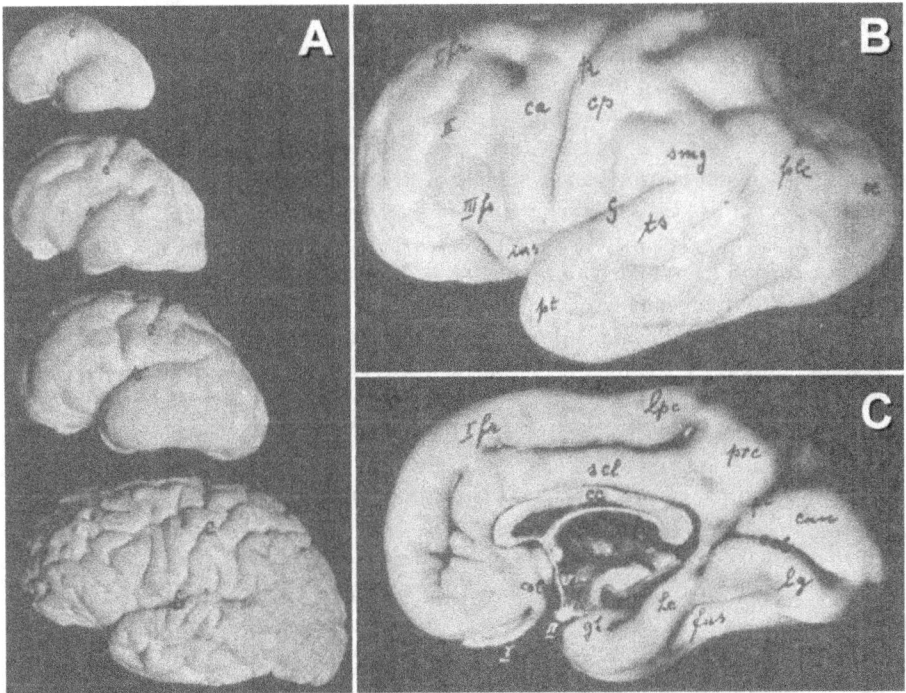

Figure 2. One of Jakob's main interests was developmental neurobiology, human and comparative. (a) Lateral view of left cerebral hemisphere in human embryos at 4, 5, 6, and 8 months of gestation, showing the progressive gyration (Jakob, 1941, p. 428). (b) Lateral view of left cerebral hemisphere in a fetus at a gestation age of 5½ months (Jakob, 1941, p. 433). (c) Midsagittal facies of right cerebral hemisphere in a human fetus at 5 months of gestation (Jakob, 1941, p. 433). From the private archive of L. C. Triarhou. Copying, redistribution, or retransmission without the author's express written permission is prohibited.

His work can be roughly divided into three periods, each having a slightly different focus. Jakob shared the "early period" of his work (1890–1912) between Germany and Argentina, carrying out anatomical studies (Théodoridou & Triarhou, 2012a). During his "middle period" (1913–1935), he developed his psychobiological ideas (Théodoridou & Triarhou, 2011). In the "late period" (1936–1949), he formulated a neurobiophilosophical synthesis (Théodoridou & Triarhou, 2012b).

Overall, Jakob realized the essence of the current definition of the mind, brain, and education convergence, that is, "the integration of disciplines that investigate human learning and development bringing together education, biology, and cognitive science" (Fischer et al., 2007, p. 1) as his life's paradigm. In 1899, at 33 years of age, Jakob had already attained renown after publishing a handbook of neuroanatomy and neuropathology that was translated in multiple languages (Triarhou & del Cerro, 2006b). In that year, he accepted an offer by Domingo Cabred (1859–1929), the Argentinian Professor of Psychiatry, to take over the organization of the Laboratory of the Psychiatric and Neurological Clinic of the Hospital of Las Mercedes at the National University of Buenos Aires (Orlando, 1966). The prospect of having 300 brains available for pathological study on an annual basis was a key factor that influenced his decision to leave Germany for South America (López Pasquali, 1965). Thus, he moved to Argentina, having signed a three-year contract.

During that time, he struggled to transform his innovative ideas into action. Already in 1906, encouraged by the visionary Minister of Public Education, Joaquín V. González

(1863–1923),[1] Jakob asked for permission from the University of Buenos Aires to introduce a new course under the title "The Nervous System and its Relationship to Education" (Orlando, 1966). However, he met with resistance, and it took several years for such an initiative to be effected. (It is not surprising that even in the past, the idea that neuroscience is or should be an important aspect of educating educators often sounded foreign to pedagogues or psychologists, who, not knowing a neuron from a glial cell, had little interest in neurobiology and avoided any serious study of the brain like the plague. The attitude that neuroscience was irrelevant for educators, other than as a subject matter to be taught by someone else, and the predilection exclusively for behavioral and psychological theories and practices are fortunately waning.)

The "Universidad Provincial de La Plata" was inaugurated on April 18, 1897 under the Administration of Dr. Guillermo A. Udaondo (1859–1922), Governor of Buenos Aires, with Dr. Dardo Rocha (1838–1921) serving as Rector.[2] The University of La Plata became nationalized by Act 4609 of Congress and by Provincial Law on September 29, 1905. When González was appointed its President the following year, he integrated several municipal scientific institutions into the university and brought substantial change by placing an emphasis on experimental and natural science methods (González, 1905). For example, he introduced modern academic physics to the country (Glick, 1996).

Jakob held faculty positions at the University of Buenos Aires from 1913 to 1944 and at the University of La Plata from 1922 to 1933 (Papini, 1988). In 1912, the Faculty of Philosophy and Letters of the University of Buenos Aires created a Professorship of Biology in an attempt to enhance the scientific status of the School (Nazar Anchorena et al., 1927; Orlando, 1966).[3] Jakob was then appointed Professor and assumed the task of building a solid biological basis for psychological and philosophical studies (Talak, 2008). The reverberation of the success of Ramón y Cajal in the Hispanic world and his Nobel Prize award must have played a part in the emphasis placed by the University of La Plata on the brain sciences. For instance, in 1906, the *Archives of Pedagogy* (official journal of the Faculty of Education) published in the inaugural volume two papers by Cajal on the

---

[1]González was admitted into the *Real Academia Española*, the Royal authority on the Spanish language, in 1906, and was elected to the Argentine Senate in 1916 (while still President of the University). Retiring from the latter in 1918, he returned to the University of Buenos Aires, where he taught Constitutional Law, Public Law, and a course in the History of Foreign Relations of Argentina. He contributed regularly to *La Nación* as a columnist and translated Rabindranath Tagore's *One Hundred Poems of Kabir*. He joined the International Law Association in 1919, and advocated on behalf of the League of Nations, as well as U.S. President Woodrow Wilson's efforts towards its ratification by a recalcitrant Senate. His most controversial work, *Patria y Democracia* (Fatherland and Democracy), was published in 1920 and delved into regional and political tensions in Argentina. González's efforts on behalf of the League of Nations helped lead to his nomination to the International Court of Justice at The Hague, becoming a member in 1921. González died in Buenos Aires in 1923; he was 60. *Fábulas Nativas*, a work of cultural anthropology, was published posthumously in 1924, and González left a bibliography of over a thousand works, including 50 books on a variety of academic subjects.

[2]Rocha was an Argentine naval officer, lawyer, and politician best known as the founder of the city of La Plata and of the University of La Plata. La Plata was planned by the architect Pedro Benoit in a regular pattern of diagonals and precisely placed squares. His success in La Plata led the Governor to seek his party's nomination for the presidency in 1886. Rocha was a well-known, well-connected, and persuasive candidate who had secured his place among Argentina's paramount "Generation of 1880" but lost the nomination to Miguel Juárez Celman, the Governor of the Province of Córdoba and President Roca's son-in-law.

[3]The Anchorenas were one of the most traditional, socially prominent, and richest families of Argentina. There was a saying: "Rich as the Anchorenas."

morphology and connections of nerve cells (Ramón y Cajal, 1906a, 1906b). The inclusion of highly technical neuroanatomical papers in a purely pedagogical journal is in itself a remarkable act. The following year, on December 14, 1907, the Science Museum of the University of La Plata bestowed Cajal the title of "Honorary Academician" (Ramón y Cajal, 1988, p. 628).

In 1913, Jakob published the "Atlas of the Brain of Mammals of Argentina" in collaboration with Clemente Onelli (1864–1924), the director of the Buenos Aires Zoo. Jakob and Onelli (1913) intended to establish the biological basis of mental phenomena, underscoring that psychology would either use the comparative method in studying nervous function and structure, in order to shed light on psychological phenomena, or it would be limited to the descriptive method (Papini, 1988). Eventually, in 1922, the National University of La Plata appointed Jakob as Professor of Neurobiology at the Faculty of Humanities and Educational Sciences (Triarhou & del Cerro, 2006c), rendering him one of the first academic teachers of neurosciences in a department of education (Figures 3 and 4). In fact, his course "Anatomy and Physiology of the Nervous System" (Figure 5) was taught in freshman year to the future teaching workforce of the country (Gotthelf, 1969).

Jakob's insight into the evolutionary basis of behavior, founded on his research in comparative developmental neuroanatomy, became known and cited. In one of the early textbooks of biological psychology (translated in French and German and prefaced by Wilhelm Ostwald, Nobel laureate in chemistry), the philosopher-psychiatrist José Ingenieros (1877–1925) credits Jakob and the palaeontologist Florentino Ameghino, along with Ramón y Cajal, van Gehuchten, Golgi, and von Lenhossék, for their fundamental discoveries in the anatomy and phylogeny of the nervous system (Ingenieros, 1922; Triarhou & del Cerro, 2006a). In the postscript to his autobiography, the great Ramón y Cajal (1988, p. 602) acknowledges Jakob (1922) among the foreign scientists of prestige who contributed to the two-volume *Festschrift* on the occasion of his retirement at the age of 70, including the Vogts, Loeb, Herrick, Sherrington, von Monakow, Marie, Houssay, and numerous others. Finally, in their monumental *Cytoarchitectonics*, von Economo and Koskinas (1925) argue that future research on the human cerebral cortex will have to be based on the work of three of the most important cortical neuroanatomists, that is, Theodor Kaes, Ramón y Cajal, and Christfried Jakob, whose ideas on cortical phylo-ontogeny they call ingenious.

In the capacity of Professor of Neurobiology in the School of Humanities and Educational Sciences, Jakob epitomized the neuroeducational idea through his teaching and research activities, discussed next.

Teaching Neurobiology to Educators

In considering the fervent growth of Educational Neuroscience, we note that the attempt to bridge neuroscience with the humanities goes back more than a century. The books of Donaldson (1895) and Halleck (1896) were published at the same time as Jakob's first atlas of the normal and pathological nervous system (1895). We do not have any evidence that Jakob was aware of Halleck's or Donaldson's books. Nevertheless, some years later he laid a new stone in the formation of Educational Neuroscience.

Throughout his career, Jakob defended the dissemination of neuroscientific knowledge into the humanities. He dealt with that topic in his *Documenta Biofilosófica* (Biophilosophical Documents; Jakob, 1946). He developed his rationale as follows (Théodoridou & Triarhou, 2012a): First of all, life sciences form a justified basis for an

FACULTAD DE HUMANIDADES Y CIENCIAS
DE LA EDUCACIÓN

Decano

Doctor Ricardo Levene.

Vicedecano

Doctor Juan José Nájera.

Consejeros académicos titulares

Señor Rafael Alberto Arrieta.
Doctor Alfredo D. Calcagno.
Señor Rómulo D. Carbia.
Doctor Juan Chiabra.
Profesor Pascual Guaglianone.
Doctor Enrique Mouchet.

Consejeros académicos suplentes

Doctor Tomás D. Casares.
Profesor Mateo Heras.
Profesor Francisco Legarra.
Doctor Enrique Loedel Palumbo.
Doctor Juan Millé Giménez.
Abogado Alberto Palcos.

Representantes de los estudiantes al Consejo académico

Señor José Armando Seco.
Señorita Araceli Saborido Gómez.

Secretario

Profesor Carlos Heras.

Profesores titulares

Historia de la filosofía: Doctor Alejandro Korn.
Lógica: Doctor Alfredo Franceschi.
Psicología: Doctor Enrique Mouchet.
Biología y anatomía y fisiología del sistema nervioso: Doctor Christo-
   fredo Jakob.
Higiene escolar: Ingeniero Antonio Restagnio.
Psicopedagogía: Doctor Alfredo D. Calcagno.
Legislación escolar argentina y comparada: Doctor Juan E. Cassani.
Ética: Doctor Coriolano Alberini
Didáctica general: Profesor José Rezzano.
Introducción a la filosofía: Doctor Coriolano Alberini.
Prehistoria argentina y americana: Doctor Luis María Torres.
Historia argentina: Doctor Ricardo Levene.
Historia de la civilización antigua: Profesor Pascual Guaglianone.
Historia de la civilización moderna: Profesor José A. Oría.
Introducción a los estudios históricos argentinos y americanos: Señor
   Rómulo D. Carbia.
Sociología: Doctor Ricardo Levene.
Composición y gramática: Señor Arturo Marasso.
Literatura castellana: Señor Arturo Marasso.
Literatura de la Europa meridional: Señor Rafael Alberto Arrieta.
Literatura de la Europa septentrional: Señor Rafael Alberto Arrieta.
Literatura argentina y de la América española: Doctor Arturo Cap-
   devila.
Literatura griega y latina: Doctor Leopoldo Longhi.
Latín, primer curso: Doctor Juan Chiabra.
Latín, segundo curso: Doctor Juan Chiabra.
Griego, primer curso: Doctor Leopoldo Longhi.

Figure 3. Roster of administrative officers and members of the Faculty of Humanities and Educational Sciences, National University of La Plata (Nazar Anchorena et al., 1927, pp. 20–21). The faculty roster includes some remarkable names. The Dean, Doctor (probably Law Doctor) Ricardo Levene, professor of Argentinian History and Sociology, was the author of the most respected and widely read *Historia de la República Argentina* available in the midtwentieth century. Alejandro Korn (1860–1936) was an Argentine physician, psychiatrist, philosopher, reformist, and politician. For 18 years, he was the director of the psychiatry hospital in Melchor Romero (a locality of La Plata in the province of Buenos Aires), named for the city. He was the first university official in Latin America to be elected, thanks to the students' vote. He is considered the pioneer of Argentine philosophy. Along with Florentino Ameghino, Juan Vucentich, Almafuerte, and Carlos Spegazzini, he is considered to be one of the five wise men of La Plata. He was still remembered with respect in the university days of one of the authors (MdC) as a man of extraordinary culture and liberal ideals. Rafael Alberto Arrieta was one of the best known and respected writers around 1945–1964 (MdC). Leopoldo Lugones: If Arrieta was the man of letters that saw the present and dreamed the future, Lugones was the man that saw the present and wanted to live in the past. A respected poet, we reluctantly have to admit. His son Leopoldito, as he was sarcastically called, became the chief torturer of the Argentinian police, or so it was said. Apparently he wanted to restore "la Argentina de la Cruz y de la Espada." From the private archive of L. C. Triarhou. Copying, redistribution, or retransmission without the author's express written permission is prohibited.

objective, rational, and scientific development of philosophical orientations (Jakob, 1946). Thus, the scientific field that studies nervous structure and function is indispensable for psychology and its related sciences. Further, knowledge of the evolution of the human brain in correlation with cognitive development, as well as brain alterations and their sequelae on memory, behavior, language, and other abstract processes, forms the natural foundation of a conscious learning science, as the creation and preservation of higher cognitive functions (intellect, volition, and emotions), instincts and reflexes depend on human cerebral organization. Thus, Jakob called for a learning science aware of the biological mechanisms that underpin learning.

Figure 4. The National University of La Plata, Argentina in 1926. *Upper left:* The School of Humanities and Educational Sciences. Central building of the University. *Upper right:* Jakob's Laboratory of the Nervous System at the School of Humanities and Educational Sciences. *Lower:* Jakob's Laboratory of Pathological Anatomy and Physiology at the School of Medical Sciences viewed from two different angles (Nazar Anchorena et al., 1927, pp. 297, 309, 350). From the private archive of L. C. Triarhou. Copying, redistribution, or retransmission without the author's express written permission is prohibited.

In a commentary on "The Function of Biology in a Faculty of Philosophy and Letters," Jakob (1942) argued that future teachers and professors, aware of their highest mission, ought to know the fundamental facts of brain development as well as the physiological capacity of its mentality and disorders, because development and its dynamics form the anatomical-physiological substrate of their instructional efforts.

Jakob presented these ideas in a monograph on the frontal lobe, which he characterized "rather as a plan for future research and not as an essay with solutions" (Jakob, 1943b, p. 9). Recognizing the "humanizing" role of the frontal lobe, he described the meaning of its dynamics for science and philosophy (Théodoridou & Triarhou, 2012b).

The fact that the human frontal cortex covers about 30% of the total cortical surface generated the hope that unravelling its function might eventually explain human behavior (Raichle, 2002). Jakob (1906) viewed the major part of the frontal lobe as a central station with multiplier and combinatorial characteristics, constantly receiving stimuli from all the motility organs via multiple pathways. The role of the frontal lobe in integrating information from multiple brain areas supports its crucial involvement in learning, comprehension, and reasoning (Baddeley, 2002). According to Fuster (2006), actions related to human behavior, reasoning, and language are organized by means of interactions between prefrontal and posterior networks at the top of the "perception-action cycle." Jakob placed in the frontal lobe the centers of experiential accumulation that results from personal intervention, elaborated progressively for elemental and higher human skills and stimulated by affective states.

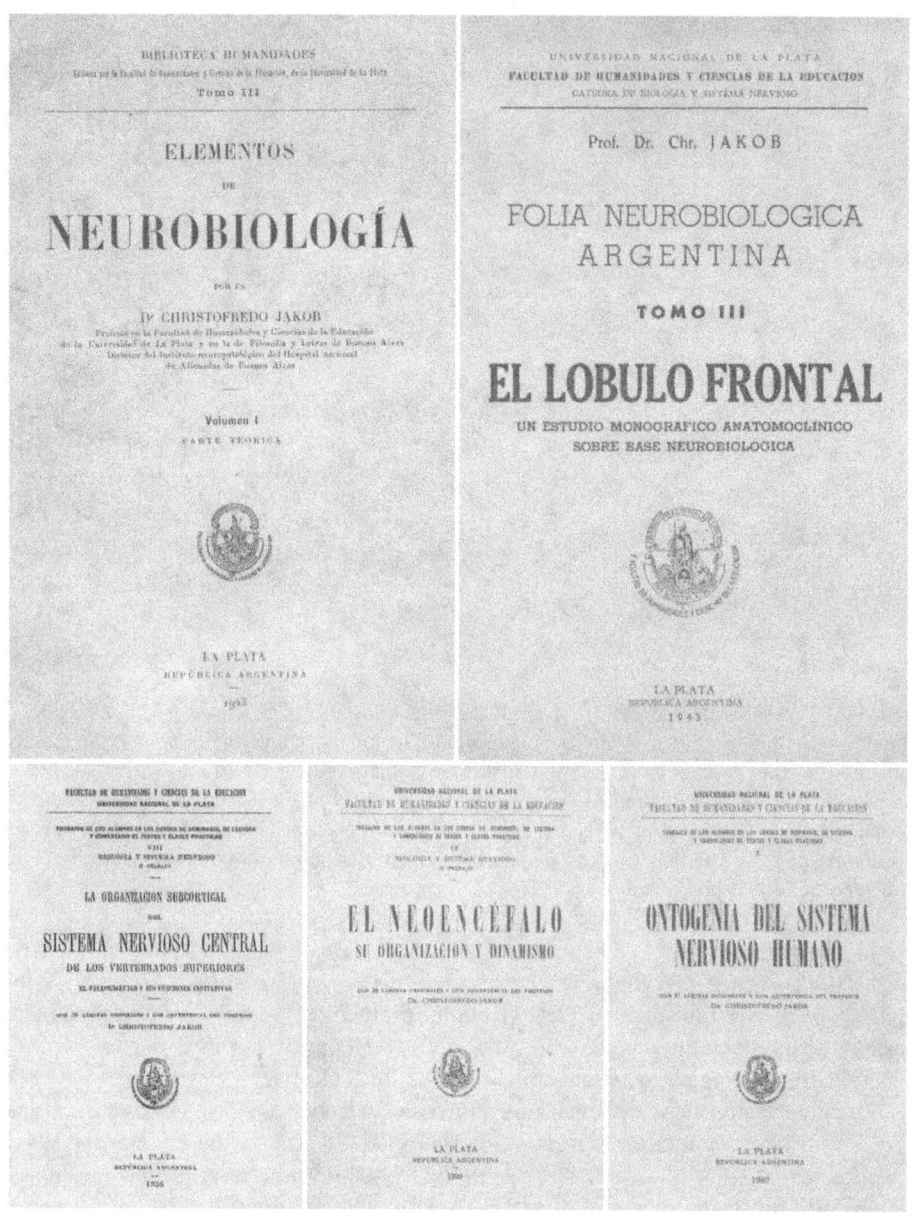

Figure 5. Jakob's neurobiology textbooks, *upper row*, published under the auspices of the School of Education at La Plata: the "Elements of Neurobiology," *left*, and "The Frontal Lobe," *right* (Jakob, 1923, 1943b). Jakob's booklets on the nervous system, lower row, based on his lectures at the School of Education in La Plata: "The Subcortical Organization of the Central Nervous System," "The Neoencephalon, its Organization and Dynamics," and "Ontogeny of the Human Nervous System" (Jakob, 1936, 1939, 1940). From the private archive of L. C. Triarhou. Copying, redistribution, or retransmission without the author's express written permission is prohibited.

In particular, Jakob explained the relation of the frontal lobe to education as follows: "The development of a social intelligence that will reinforce individual inclinations and will put emphasis on the active engagement of the student in the formation of concrete knowledge issues from the importance of frontal lobe functions" (Jakob, 1943b, p. 140).

As a result of the progress in the brain sciences in recent decades, traditional philosophical questions have been steered in new directions (Churchland, 2008), inviting a broad and divergent body of scientists to work together. Following philosophy, education has been enriched with new information stemming from the neurosciences.

For example, imaging studies pinpoint to neural systems that are responsible for the acquisition of reading skills and further support the idea that we can remedy inefficiencies in those systems through intervention; behavioral outcomes are accompanied by neural changes in the expected areas (Goswami, 2006). Concerning arithmetic, the finding that the brain has a preferred mode of representation bears directly on the teaching of mathematics: It suggests that teachers should build on this spatial system when teaching ordinality and place value. Other aspects of education that have been informed by neuroscience include second language learning, lifelong learning, and early learning (Blakemore & Frith, 2005).

However, a fair and fruitful dialogue among the sciences presupposes a minimum amount of knowledge and familiarity. The current demand for educators' literacy in neuroscience seems very attuned with Jakob's arguments (Ansari & Coch, 2006).

In a survey of teachers, almost 90% thought that knowledge of the brain was important, or very important, in designing educational programs (Pickering & Howard-Jones, 2007). In 1999, the Teaching and Learning Research Programme (TLRP) commissioned Blakemore and Frith to review neuroscientific findings that might be of relevance to educators. At the same time, a project on "Learning Sciences and Brain Research" was launched by the Centre for Educational Research and Innovation (CERI) at the Organisation for Economic Cooperation and Development (OECD). Major research and funding institutions from all over the world took part in that attempt: the Sackler Institute (United States), the University of Granada (Spain), and the RIKEN Brain Science Institute (Japan); the National Science Foundation (United States), the Lifelong Learning Foundation (United Kingdom), and the City of Granada (Spain); and INSERM (France) (OECD, 2001). The first phase of the project (1999–2002) brought together international scientists to review the potential implications of brain research for policy makers. The second phase (2002–2006) focused its activities on three areas: literacy, numeracy, and lifelong learning (Howard-Jones & Economic and Social Research Council, 2007).

Brain Topics, Psychological Theories, and Educational Implications

The potential cross-fertilization of neuroscience and education and its inherent limitations are one of the main points of discussion in the educational neuroscience literature (Beauchamp & Beauchamp, 2012). The production of "usable knowledge" (Christodoulou, Daley, & Katzir, 2009, p. 65) has proven a demanding task for researchers, as it entails an innate danger of misapplication, that is, oversimplification, misunderstanding, or generalization of scientific data (Beauchamp & Beauchamp, 2012).

Jakob's elaborations form a justified basis for educational implications, given that they cover multiple aspects of learning, both horizontally (evo-devo processes, normal and pathological conditions, structure, and function) and vertically (in-depth analyses and formulation of original theories). In addition, one of the most influential schools of educational thought, that is, the Piagetian, is based on a cognitive theory with biological roots.

Piaget (1964) argued that learning is subordinated to development. In a similar line of thinking, Jakob held the firm belief that cognitive and socioemotional development go hand in hand with cerebral development in a course he termed "psychogenesis" (Jakob, 1913,

1919, 1921, 1935). Thus, he explored its evolution and development in his neurodynamic postulate (for reviews see Théodoridou & Triarhou, 2011, 2012b).

Jakob's theory on cognitive development could have impact on the formulation of learning theories and practices, as the quest for a mutual understanding among cognitive scientists, neuroscientists, and educators is at the epicenter of current research.

In Jakob's neurodynamic theory, every system is an arc or circuit, composed of long ramifications or afferent and efferent pathways, "macrodynamics" of charge and discharge, and of a center or an inserted formation of increasing complexity, comprising cells and short fibers that constitute the "microdynamics" or "associative and commissural systems" (López Pasquali, 1965). The first macrodynamic circuit (reflexes) is only capable of responding in an instant and invariable form. The second macrodynamic circuit (instincts) preserves the information and mounts it up through discharges. In the third macrodynamic circuit, any entering or exiting element becomes registered and furthermore interacts interfocally. Such a system subserves the emergence of "psychism," that is, "the neurobiophylactic complex of neuroenergetic reception, assimilation and reaction, which regulates the organism's vital necessities against variable factors in the external and internal milieu" (Jakob, 1939, p. 8). Psychogenesis crescents in the "neopsychic" stage. Then, three kinds of neurocognitive processes occur: (a) *gnoses*, which secure the conscious orientation in one's environment; (b) *praxes*, which underlie active individual intervention; and (c) *symbolisms*, which subserve the communication of abstract ideas by means of human language (Jakob, 1919, 1921, 1935).

Treating Developmental Disorders

One of the most important of Jakob's contributions to education is his effort to establish a biological treatment theory for developmental disorders (Théodoridou, 2011). Jakob (1913) viewed mental retardation as the result of a degenerative psychogenetic process that necessitates a biological treatment. Thereby, the introduction of principles for a biological classification with practical and functional value would be meaningful. Jakob condemned the existing classifications as insufficient and ineffective for the formulation of both psychological and educational intervention.

Still, he acknowledged the contribution of existing classifications (clinical, anatomo-pathological, and educational) to psychology, medicine, and education, respectively. However, he stressed the importance of a functional connection between these aspects on the grounds of the dynamic nature of mental retardation. He noticed that each of those aspects and their interactions may influence prognosis in a child with mental retardation, for example, the time of onset of a disease, the extent of a lesion, individual differences, and the amount and quality of educational opportunities.

On that basis, Jakob (1913) divided normal cognitive and socioemotional development into the following psychogenetic stages (see also, Théodoridou, 2011):

1. "Psychobiomolecular stage," characterized by the irritability of the protoplasm.
2. "Psychoneuromolecular stage," when elementary nervous organization is not differentiated, although there is a nervous irritability.
3. "Elementary psychoreflexive stage," signalled by the differentiation of the reflexive, the nuclear, the spinal, and the bulbar systems.
4. "Complex psychoreflexive stage," when the successive maturation of instincts, impulses, and subcortical arcs is realized, along with the differentiation of afferent and efferent arcs.

5. "Crepuscular stage," when the reflexes and functions such as respiration, sucking, some movements, and mimicry dominate as primary, diffuse, and preconscious cortical perceptions.
6. "Stage of provisory psychological fixations":
   a. "Stage of elementary temporary fixations," when the superior instincts, affects, active mimicry and combined voluntary perceptions and actions first appear;
   b. "Stage of complex temporary fixations," when elements of articulate language, concrete ideas, orientation in the environment in terms of space and time as well as affective and voluntary actions emerge.
7. "Stage of definitive psychoenergetic associations":
   a. "Stage of permanent concrete associations," when an egocentric realism, the formation of personality, actions of affective inhibition, elementary consciousness, as well as the perception of time, space, and causality are observed.
   b. "Stage of elementary abstract associations," when the conscious personality is formed with elements of self-critisism and egoism; judgment, reasoning, and acts of intellectual inhibition are also evident in this stage of infantile analytic empiricism.
8. "Puberty," when the synthesis of ideas is possible, and aesthetic and ethic tendencies are evident.
9. And finally, "second puberty," when thought becomes rational and speculative.

Jakob's designation of egocentric realism precedes once again a Piagetian concept (see also, Théodoridou & Triarhou, 2011, 2012b). Piaget (1926, 1932) introduced the concept of egocentrism in his early writings (Light, 1983). The roots of the concept of egocentrism can be traced back to Freud's influence on Piaget, in particular on Freud's concepts of the "primary process" (i.e., the mode of functioning in service of the immediate gratification of needs) and the "secondary process" (i.e., the regulation and control of needs to attend to the demands of reality; Kesserling & Müller, 2011). Piaget (1920) initially distinguished between autistic (i.e., symbolic) and logical, scientific thought. Piaget's (1920) notion of autistic thought is derived from Bleuler and is much different from the contemporary use of this term as a designator of a particular developmental disorder (Kesserling & Müller, 2011). The pleasure principle dictates autistic thinking that is "personal, incommunicable, confused, undirected, indifferent to truth, rich in visual and symbolic schemas, and above all, unconscious of itself and by the affective factors by which it was guided" (Piaget, 1928, pp. 204–205). Later, Piaget introduced the concept of egocentrism as an intermediate level between these modes of thought. However, Piaget's study of his own infants led to a revision of the concept of egocentrism, which from the mid-1930s was conceptualized as a phenomenon that recurs at the beginning of different developmental stages.

Jakob (1913) viewed developmental disorders as energetic and dynamic conditions defined by multiple components. Thus, he stressed the importance of elucidating the internal and external causes of developmental cognitive arrest and their complex consequences on cerebral growth. He maintained that such knowledge would help scientists to contain the extent and severity of the factors that compromise the learning powers of the brain.

Apart from their biological characteristics, in his 1913 article, Jakob further provided the psychological profile that corresponds to each stage. He maintained that "degenerative psychogenesis" runs through the same stages as normal psychogenesis. He further made suggestions for suitable interventions, according to the stage that the child with disabilities falls into. Finally, Jakob discussed the concept of *patopedagogía* ("pathopedagogy") long before the fields of special and remedial education were formally introduced (Théodoridou, 2011).

In 1940, Jakob, in collaboration with Antonio Scaravelli, published a study of eight siblings from Tupungato with familial mental retardation, deafness, and spastic quadriplegia (Jakob & Scaravelli, 1940). Maintaining his interest in disability, Jakob dealt with heredity factors that cause pathological characteristics both from a neurological and from a socio-anthropological viewpoint.

Until the end of the eighteenth century, charity rather than education served as the underlying guidance for any special provisions for disabled children (Winzer, 1993). Afterwards, some sparse attempts were made for a methodologically sound, science-informed special education by pioneers such as the French physician-educator Jean Marc Gaspard Itard (1774–1838), the French psychologist Eduard Seguin (1812–1880), the Italian physician-educator Maria Montessori (1870–1952), the Belgian teacher and psychologist Jean-Ovide Decroly (1871–1932), and the Soviet defectologists Aleksandr M. Shcherbina (1874–1934), Lev S. Vygotsky (1896–1934), and Ivan A. Sokolyanskii (1898–1960).

The demand for a neuroscience-informed special education (Goswami, 2004) ensued from the unification of the mind, brain, and education sciences under modern attempts defined as neuroeducation (Battro, Fischer, & Léna, 2008) and educational neuroscience (Petitto & Dunbar, 2004; Szűks & Goswami, 2007). Learning and education can be viewed as a new field of the natural sciences with the entire human lifespan as its subject, including various problems such as fetal environment, childcare, language acquisition, general and special education, as well as rehabilitation (Koizumi, 2004). In this line of thought, Ito (2004) suggested that research should aim at providing new knowledge about the pathogenesis of developmental disorders on the solid basis of neuroscience. New knowledge should aid the appropriate assessment and treatment of patients, based on an accurate identification of individual-specific deficiencies, and environmental factors that might prevent children from behaving appropriately. Therefore, it would be helpful in solving problems rooted in antisocial behaviors of students. Further advantages of the adoption of a biological perspective include the timely diagnosis of special educational needs, the monitoring and comparison of the effects of different kinds of educational input on learning, and an increased understanding of individual differences in learning, and the best ways to customize input to the learner (Goswami, 2004).

Conclusion

A polymath, Jakob contributed original ideas to diverse aspects of education and pedagogy. Jakob's organizational skills have been considered exceptional (Papini, 1978). Throughout his career, Jakob became heavily involved in the organization of services, laboratories, clinics, and academic departments (Jakob, 1916, 1937, 1943a) and is credited as the father of Argentinian neurobiology and neurology (Orlando, 1966; Triarhou & del Cerro, 2007), having established an intellectual lineage of distinguished researchers and clinicians that included José Ingenieros, Braulio Moyano, and many others.

With his deep understanding of human brain function, he shed light on cognitive development and consciously aimed at the enhancement of learning theories and practices. Jakob further promoted the right of handicapped children to an appropriate education, paving the path for the grounding of special education on a scientific basis (Jakob, 1913). Impressively, he put forth the idea of teaching a course on "The Nervous System and Its Relationship to Education" as early as 1906. Although such a ground-breaking idea did not find fertile soil right away, eventually, in 1922, he became one of the first academics to formally teach neurobiology in a School of Education, at the National University of La Plata in Argentina.

Thus, Jakob introduced fundamentals of neuroeducation many decades before the discipline was formalized. At the same time, the National University of La Plata should be credited for its pioneering administrative decisions on educational and research policies that rendered such a neuroeducational connection possible. Ninety years later, Jakob's innovative thinking may still open up new horizons for a neuroscience-informed learning science and interdisciplinarity.

References

Allegri RF (2008): The pioneers of clinical neurology in South America. *Journal of the Neurological Sciences* 271: 29–33.

Ansari D, Coch D (2006): Bridges over troubled waters: Education and cognitive neuroscience. *Trends in Cognitive Sciences* 10: 146–151.

Baddeley AD (2002): Fractionating the central executive. In: Stuss DT, Knight RT, eds., *Principles of Frontal Lobe Function*. Oxford, Oxford University Press, pp. 246–260.

Battro AM, Cardinali DP (1996): Más cerebro en la educación. *La Nación*. [Online]. Retrieved from http://www.byd.com.ar/cereln.pdf

Battro AM, Fischer KW, Léna P (2008): *The Educated Brain: Essays in Neuroeducation*. Cambridge, Cambridge University Press.

Beauchamp M, Beauchamp C (2012): Understanding the neuroscience and education connection: Themes emerging from a review of the literature. In: Della Sala S, Anderson M, eds., *Neuroscience in Education: The Good, the Bad, and the Ugly*. Oxford, Oxford University Press, pp. 13–30.

Blakemore S-J, Frith U (2005): *The Learning Brain: Lessons for Education*. Malden, Massachusetts, Blackwell.

Christodoulou JA, Daley SG, Katzir T (2009): Researching the practice, practicing the research, and promoting responsible policy: Usable knowledge in mind, brain, and education. *Mind, Brain, and Education* 3: 65–67.

Churchland P (2008): The impact of neuroscience on philosophy. *Neuron* 60: 409–411.

Donaldson HH (1895): *The Growth of the Brain: A Study of the Nervous System in Relation to Education*. London, Walter Scott, Ltd.

Fawcett AJ, Nicolson RI (2007): Dyslexia, learning, and pedagogical neuroscience. *Developmental Medicine and Child Neurology* 49: 306–311.

Fischer KW, Daniel DB, Immordino-Yang MH, Stern E, Battro A, Koizumi H (2007): Why mind, brain, and education? Why now? *Mind, Brain, and Education* 1: 1–2.

Fumagalli A, Saredo G (2005): Christofredo Jakob (1866–1956) and the birth of neurology in Argentina. *Journal of the Neurological Sciences* 238(Suppl. 1): 160–161.

Fuster J (2006): The cognit: A network model of cortical representation. *International Journal of Psychophysiology* 60: 125–132.

Geake J (2005): Educational neuroscience and neuroscientific education: In search of a mutual middle-way. *Research Intelligence (Warborough)* 92: 10–13.

Glick TF (1996): Science in twentieth century Latin America. In: Betthel L, ed., *Ideas and Ideologies in Twentieth Century Latin America*. New York, Cambridge University Press, pp. 287–359.

González JV (1905): *La Universidad Nacional de La Plata: Memoria sobre su Fundación*. Buenos Aires, Talleres Gráficos de la Penitenciaría Nacional.

Goswami U (2004): Neuroscience, education and special education. *British Journal of Special Education* 31: 175–183.

Goswami U (2006): Neuroscience and education: From research to practice? *Nature Reviews Neuroscience* 7: 406–411.

Gotthelf R (1969): Historia de la psicología en la Argentina, segunda parte. *Revista Latinoamericana de Psicología (Bogotá)* 1: 183–198.

Halleck RP (1896): *The Education of the Central Nervous System: A Study of Foundations, Especially of Sensory and Motor Training*. New York, Macmillan Company.

Howard-Jones P, Economic and Social Research Council (2007): *Neuroscience and Education: Issues and Opportunities*. London, TLRP.

Ingenieros J (1922): *Prinzipien der biologischen Psychologie*. Leipzig, F. Meiner.

Ito M (2004): "Nurturing the brain" as an emerging research field involving child neurology. *Brain and Development* 26: 429–433.

Jakob C (1895): *Atlas des gesunden und kranken Nervensystems nebst Grundriss der Anatomie, Pathologie und Therapie desselben*. München, J. F. Lehmann.

Jakob C (1906): Estudios biológicos sobre los lóbulos frontales cerebrales. *La Semana Médica* 13: 1375–1381.

Jakob C (1911): *Das Menschenhirn: Eine Studie über den Aufbau und die Bedeutung seiner grauen Kerne und Rinde*. München, J. F. Lehmann.

Jakob C (1912a): Über die Ubiquität der senso-motorischen Doppelfunktion der Hirnrinde als Grundlage einer neuen, biologischen Auffassung des corticalen Seelenorgans. *Journal für Psychologie und Neurologie (Leipzig)* 19: 379–382.

Jakob C (1912b): Ueber die Ubiquität der senso-motorischen Doppelfunktion der Hirnrinde als Grundlage einer neuen biologischen Auffassung des kortikalen Seelenorgans. *Münchener Medizinische Wochenschrift* 59: 466–468.

Jakob C (1913): La psicopatogenia de los niños retardados: Psicogenesis degenerativa y su tratamiento biológico. *Revista de la Asociación Médica Argentina* 21: 1003–1016.

Jakob C (1916): Encuesta sobre en plan de estudios de medicina y la formación del profesorado universitario. *Revista del Círculo Médico Argentino y Centro de Estudiantes de Medicina* 16: 1–2.

Jakob C (1919): La teoría actual de las 'Gnósias y Práxias' como factores fundamentales en el dinamismo cortical. *Revista del Círculo Médico Argentino y Centro de Estudiantes de Medicina (Buenos Aries)* 19: 1266–1275.

Jakob C (1921): La teoría actual de las gnosias y praxias como factores fundamentales en el dinamismo de la corteza cerebral. *La Crónica Médica (Lima)* 38: 17–24.

Jakob C (1922): Sobre tumores teratogénicos del cerebro (A propósito de un teratoma del conducto de Sylvio). In: Junta para el homenaje a Cajal, eds., *Libro en honor de D. S. Ramón y Cajal—Trabajos Originales de Sus Admiradores y Discípulos, Extranjeros y Nacionales*, tomo II. Madrid, Jiménez y Molina Impresores, pp. 415–431.

Jakob C (1935): La filogenia de las kinesias: Sobre su organización y dinamismo evolutivo. *Anales del Instituto de Psicología de la Facultad de Filosofía y Letras de la Universidad de Buenos Aires* 1: 109–127.

Jakob C (1937): La enseñanza universitaria de la anatomía cerebral. *Revista de la Asociación Médica Argentina* 51: 799–803.

Jakob C (1939): *El Neoencéfalo: Su Organización y Dinamismo*. Buenos Aires, Universidad Nacional de La Plata–Imprenta López.

Jakob C (1941): La función psicogenética de la corteza cerebral y su posible localización (Aspectos de la ontopsicogénesis humana). *Anales del Instituto de Psicología de la Facultad de Filosofía y Letras de la Universidad de Buenos Aires* 3: 63–80.

Jakob C (1942): La función de la Biología en la Facultad de Filosofía y Letras. *Logos (Buenos Aires)* 1: 159–161.

Jakob C (1943a): Clinica–Laboratorio. *Revista de la Asociación Bioquímica Argentina* 9: 42–44.

Jakob C (1943b): *Folia Neurobiológica Argentina, III. El Lóbulo Frontal: Estudio Monográfico Anatomoclínico sobre Base Neurobiológica*. Buenos Aires, Aniceto López-López y Etchegoyen.

Jakob C (1945): *Folia Neurobiológica Argentina, IV. El Yacaré (Caimán latirostris) y el Origen del Neocortex: Estudios Neurobiológicos y Folklóricos del Reptil más Grande de la Argentina*. Buenos Aires, Aniceto López Editor.

Jakob C (1946): *Folia Neurobiológica Argentina, V. Documenta Biofilosófica*. Buenos Aires, López y Etchegoyen.

Jakob C, Onelli C (1913): *Atlas del Cerebro de los Mamíferos de la Republica Argentina: Estudios Anatómicos, Histológicos y Biológicos Comparados sobre la Evolución de los Hemisferios y de la Corteza cerebral*. Buenos Aires, Guillermo Kraft.

Jakob C, Scaravelli A (1940): A propósito de un caso de ocho hermanos con idiocia, sordomudez y cuadriplejía espasmódica familiar. *Revista Neurológica de Buenos Aires* 5: 283–299.

Koizumi H (2004): The concept of "developing the brain": A new natural science for learning and education. *Brain and Development* 26: 434–441.

Kesserling T, Müller U (2011): The concept of egocentrism in the context of Piaget's theory. *New Ideas in Psychology* 29: 327–345.

Light P (1983): Piaget and egocentrism: A perspective on recent developmental research. *Early Child Development and Care* 12: 7–18.

López Pasquali L (1965): Christfried Jakob. *Su Obra Neurológica, Su Pensamiento Psicológico y Filosótico.* Buenos Aires, López.

Moyano BA (1957): Christfried Jakob, 25/12/1866–6/5/1956. *Acta Neuropsiquiátrica Argentina* 3: 109–123.

Nazar Anchorena BA, Amaral SM, Alegre PJ (1927): *La Universidad Nacional de La Plata en el Año 1926: Publicación Oficial.* La Plata & Buenos Aires, Casa J. Peuser, Ltda.

OECD (Organisation for Economic Co-operation and Development) (2001): *Preliminary Synthesis of the Third High Level Forum on Learning and Sciences and Brain Research: Potential Implications for Education Policies and Practices.* Retrieved from http://www.oecd.org/edu/ceri/15302896.pdf.

Orlando JC (1966): *Christofredo Jakob: Su Vida y Obra.* Buenos Aires, Editorial Mundi.

Papini MR (1978): La psicología experimental argentina durante el período 1930–1955. *Revista Latinoamericana de Psicología (Bogotá)* 10: 227–258.

Papini MR (1988): Influence of evolutionary biology in the early development of experimental psychology in Argentina (1891–1930). *International Journal of Experimental Psychology* 2: 131–138.

Petitto LA, Dunbar K (2004): New findings from educational neuroscience on bilingual brains, scientific brains, and the educated mind. *Conference on Building Usable Knowledge in Mind, Brain, and Education,* 6–8 October 2004, Cambridge, MA.

Piaget J (1920): La psychanalyse dans ses rapports avec la psychologie de l'enfant. *Bulletin Mensuel de la Société Alfred Binet* 20: 18–34 & 41–58.

Piaget J (1926): *The Language and Thought of the Child.* London, Routledge & Kegan Paul.

Piaget J (1928): *Judgment and Reasoning in the Child.* New York, Harcourt, Brace and Company.

Piaget J (1932): *The Moral Judgment of the Child.* London, Routledge & Kegan Paul.

Piaget J (1964): Cognitive development in children: Development and learning. *Journal of Research in Science Teaching* 2: 176–186.

Pickering SJ, Howard-Jones P (2007): Educators' views on the role of neuroscience in education: Findings from a study of UK and international perspectives. *Mind, Brain, and Education* 1: 109–113.

Raichle ME (2002): Foreword. In: Stuss DT, Knight RT, eds., *Principles of Frontal Lobe Function.* Oxford, Oxford University Press, pp. vii–ix.

Ramón y Cajal S (1906a): Inducciones fisiológicas de la morfología y conexiones de las neuronas. *Archivos de Pedagogía y Ciencias Afines (La Plata)* 1: 216 236.

Ramón y Cajal S (1906b): Morfología de la célula nerviosa. *Archivos de Pedagogía y Ciencias Afines (La Plata)* 1: 92–106.

Ramón y Cajal S (1988): *Recollections of My Life* (translated by E. H. Craigie and J. Cano). Birmingham, Alabama, The Classics of Neurology and Neurosurgery Library–Gryphon Editions.

Stern E (2006): *Educational Research and Neurosciences—Expectations, Evidence and Research Prospects.* Berlin, Bundesministerium für Bildung und Forschung.

Szűcs D, Goswami U (2007): Educational neuroscience: Defining a new discipline for the study of mental representations. *Mind, Brain, and Education* 1: 114–127.

Talak AM (2008): Christofredo Jakob: La tradición neurobiológica en la primera psicología en Argentina. *VI Encuentro de Filosofía e Historia de la Ciencia del Cono Sur Montevideo,* 27–30 May 2008, Montevideo, Uruguay.

Théodoridou ZD (2011): *Christfried Jakob on the Cerebral Cortex: Neurobiological, Neurophilosophical and Neuroeducational Concepts* (Unpublished Doctoral Dissertation). Thessaloniki, Greece, University of Macedonia.

Théodoridou ZD, Triarhou LC (2009): Fin-de-siècle advances in neuroeducation: Henry Herbert Donaldson and Reuben Post Halleck. *Mind, Brain, and Education* 3: 117–127.

Théodoridou ZD, Triarhou LC (2011): Christfried Jakob's 1921 theory of the gnoses and praxes as fundamental factors in cerebral cortical dynamics. *Integrative Psychological and Behavioral Science* 45: 247–262.

Théodoridou ZD, Triarhou LC (2012a): Challenging the supremacy of the frontal lobe: Early views (1906–1909) of Christfried Jakob on the human cerebral cortex. *Cortex* 48: 15–25.

Théodoridou ZD, Triarhou LC (2012b): Christfried Jakob's late views (1930–1949) on the psychogenetic function of the cerebral cortex and its localization: Culmination of the neurophilosophical thought of a keen brain observer. *Brain and Cognition* 78: 179–188.

Triarhou LC (2008a): Centenary of Christfried Jakob's discovery of the visceral brain: An unheeded precedence in affective neuroscience. *Neuroscience and Biobehavioral Reviews* 32: 984–1000.

Triarhou LC (2008b): The books of Christofredo Jakob: Lasting treasures of evolutionary neuroscience. *Society for Neuroscience Abstracts* 38: 221.16.

Triarhou LC (2009): Tripartite concepts of mind and brain, with special emphasis on the neuroevolutionary postulates of Christfried Jakob and Paul MacLean. In: Weingarten SP, Penat HO, eds., *Cognitive Psychology Research Developments*. Hauppauge, New York, Nova Science Publishers, pp. 183–208.

Triarhou LC (2010a): Final publications of Christfried Jakob: On the frontal lobe and the limbic region. In: Flynn CE, Callaghan BR, eds., *Neuroanatomy Research Advances*. Hauppauge, New York, Nova Science Publishers, pp. 165–169.

Triarhou LC (2010b): Revisiting Christfried Jakob's concept of the dual onto-phylogenetic origin and ubiquitous function of the cerebral cortex: A century of progress. *Brain Structure and Function* 214: 319–338.

Triarhou LC, del Cerro M (2006a): An early work [1910–1913] in *Biological Psychology* by pioneer psychiatrist, criminologist and philosopher José Ingenieros, M.D. (1877–1925) of Buenos Aries. *Biological Psychology* 72: 1–14.

Triarhou LC, del Cerro M (2006b): Semicentennial tribute to the ingenious neurobiologist Christfried Jakob (1866–1956): 1. Works from Germany and the first Argentina period, 1891–1913. *European Neurology* 56: 176–188.

Triarhou LC, del Cerro M (2006c): Semicentennial tribute to the ingenious neurobiologist Christfried Jakob (1866–1956): 2. Publications from the second Argentina period, 1913–1949. *European Neurology* 56: 189–198.

Triarhou LC, del Cerro M (2007): Christfried Jakob (1866–1956). *Journal of Neurology* 254: 124–125.

Tsapkini K, Vivas AB, Triarhou LC (2008): "Does Broca's area exist?"—Christofredo Jakob's 1906 response to Pierre Marie's holistic stance. *Brain and Language* 105: 211–219.

Vivas AB, Tsapkini K, Triarhou LC (2007): Anatomo-biological considerations on the centers of language: An Argentinian contribution to the 1906 Paris debate on aphasia. *Brain and Development* 29: 455–461.

von Economo C, Koskinas GN (1925): *Die Cytoarchitektonik der Hirnrinde des erwachsenen Menschen: Textband und Atlas*. Wien, J. Springer.

Winzer MA (1993): *The History of Special Education: From Isolation to Integration*. Washington, DC, Gallaudet University Press.

Available online at www.sciencedirect.com

ScienceDirect

Brain and Language 105 (2008) 211–219

Brain and Language

www.elsevier.com/locate/b&l

'Does Broca's area exist?'
Christofredo Jakob's 1906 response to Pierre Marie's holistic stance

Kyrana Tsapkini [a,b], Ana B. Vivas [c], Lazaros C. Triarhou [d,*]

[a] Department of Psychology, Aristotelian University, Thessaloniki 54124, Greece
[b] Department of Cognitive Science, Johns Hopkins University, Baltimore, MD 21218, USA
[c] Department of Psychology, City Liberal Studies, Affiliated Institution of University of Sheffield, Thessaloniki 54624, Greece
[d] Economo-Koskinas Wing for Integrative and Evolutionary Neuroscience, Department of Educational and Social Policy,
University of Macedonia, 156 Egnatia Avenue, Building Z-312, Thessaloniki 54006, Greece

Accepted 13 July 2007
Available online 17 August 2007

Abstract

In 1906, Pierre Marie triggered a heated controversy and an exchange of articles with Jules Déjerine over the localization of language functions in the human brain. The debate spread internationally. One of the timeliest responses, that appeared in print 1 month after Marie's paper, came from Christofredo Jakob, a Bavarian-born neuropathologist working in Buenos Aires. The present study comprises an English translation of Jakob's 1906 paper and a discussion of Jakob's ideas on the localizationist–holistic approach regarding the role of Broca's area. This issue is still at the core of scientific debate in the light of current neuropsychological and neuroimaging findings. © 2007 Elsevier Inc. All rights reserved.

Keywords: Aphasia; Christfried Jakob (1866–1956); Language areas; Localization of function

1. Introduction

Two of the most vivid debates in the history of aphasiology (Benson, 1979) took place when Broca (1861) presented his case to the Anthropological Society of Paris in 1861 and when Marie (1906a) instigated a sensational localizationist–antilocalizationist controversy in 1906 that witnessed one of the most memorable scientific battles of French neurology (Goetz, 2003). Marie's arguments triggered a cascade of articles comprising two classic papers by Déjerine (1906a, 1906b), a further two articles by Marie (1906b, 1906c), and numerous reports by their assistants (Lecours & Joanette, 1984; Lotmar & de Montet, 1906; Marie & Moutier, 1906a, 1906b, 1906c, 1906d; Moutier, 1906; Souques, 1906) and by authors abroad (Monakow, 1906). The debate culminated with designated discussions at a special joint meeting of the New York and the Phila-

delphia Neurological Societies (McCarthy & Mills, 1906), and a three-session debate (Klippel, 1908) at the Neurological Society in Paris 2 years later (Goetz, 2003; Lecours & Joanette, 1984; Pearce, 2004). The three French 1908 debates, held on June 11th (clinical facts), July 9th (cerebral anatomy), and July 23rd (physiological pathology), involved the most explicit historical quarrel in aphasiology.

The 1906 dispute began when Marie (1906a) provokingly assailed Broca's claim that the inferior frontal gyrus ('third frontal convolution') of the left cerebral hemisphere plays any role in speech and passionately opposed the localization of brain functions, a strong tendency at the dawn of the twentieth century. Marie defended a 'holistic', while Déjerine a 'localizationist' view in their exchange. Speech disorders had also been dichotomized into motor and sensory varieties (Benson, 1979), an idea also suggested by Wernicke (1874) in his doctoral thesis. 'Localizationist' views were expressed by Broca, Wernicke, Lichtheim, Charcot, Bastian, and Déjerine. 'Holistic' views of

---

* Corresponding author. Fax: +30 2310 891 388.
E-mail address: triarhou@uom.gr (L.C. Triarhou).

0093-934X/$ - see front matter © 2007 Elsevier Inc. All rights reserved.
doi:10.1016/j.bandl.2007.07.124

language functions were advocated by Hughlings Jackson, Friedrich Goltz, (then) neurologist Sigmund Freud and Marie.

In his 1891 monograph on aphasia, Freud (1983) raised doubt about a mechanistic doctrine of brain centers (Jellinger, 2006) and argued against strict localization, a view not welcomed at the time (Eling, 2006). Challenging Wernicke and even Meynert, Freud argued that one cannot separate aphasias resulting from lesions to language centers or to fasciculi that connect them. He proposed a functional approach to the speech apparatus and distinguished between areas of purely pathologic importance and areas underlying the physiology of language (Jacyna, 2005), following in the footsteps of Jackson; the latter had advocated, 17 years earlier, that 'to locate the damage which destroys speech, and to locate speech are two different things' (Jackson, 1874). One should nevertheless mention that Hughlings Jackson did not espouse a totally holistic view of language (Lorch, 2004); his epochal 1864 paper on aphasia and right hemiparesis in patients with embolic stroke supported Broca's localization, and his analysis does not fit into a rigid localization–holism dichotomy, but rather is part of his neurophysiological principles.

By broaching the subject of Broca's area in his controversial paper, Marie (1906a) forced the neurological community to re-examine some of its prior conceptions under a harder anatomical look, and re-kindled a healthy skepticism that underlies all scientific advancement (Goetz, 2003). The debate spread internationally (Editorial, 1906; McCarthy and Mills, 1906).

A prompt response (Jakob, 1906a) under the title '*Does Broca's area exist or not?*' (Fig. 1) came the month after Marie's first 1906 article from Christofredo (or Christfried) Jakob (1866–1956), a German-born neuropathologist working in Buenos Aires. Having completed his doctoral thesis under Friedrich Albert von Zenker and served as assistant to Adolf von Strümpell at the Erlangen Medical Clinic, Jakob spent most of his professional life in Argentina, where he is considered the father of Neurobiology and Forensic Histopathology (Meyer, 1981; Orlando, 1966; Pedace, 1949). Jakob held professorships of Neurobiology at the Faculty of Humanities and Educational Sciences of the University of La Plata and of Anatomy and Biology at the University of Buenos Aires, and established one of the most important neuropathological laboratories in South America. He has left an invaluable legacy of 200 articles and 30 monographs (Triarhou and del Cerro, 2006a,b, 2007). Economo and Koskinas (1925) consider Jakob one of the three most ingenious neuroanatomists of the early twentieth century, the other two being Santiago Ramón y Cajal (1852–1934) and Theodor Käs (1852–1913).

In attempting to reconstruct the historical foundations of neuropathology, we present—a century later—Jakob's topical views on aphasia in a modern context; his 1906 paper (Jakob, 1906a) antedates much better known articles by e.g., Déjerine and Marie from that landmark year

Año XIII.—N.º 26   BUENOS AIRES JUNIO 28 DE 1906   677

LA SEMANA MEDICA

DIRECCIÓN, REDACCIÓN, ADMINISTRACIÓN É IMPRENTA: CALLAO, 737
Unión Telefónica 276 (Juncal)

FISIOPATOLOGIA

¿Existe ó no un centro de Broca?

POR EL

Dr. Chr. JAKOB
Jefe del Laboratorio de Psiquiatría y Neurología
Hospicio de las Mercedes

En uno de los últimos números de *La Semaine Médicale*, P. Marie publica un estudio anátomo-clínico sobre las afasias, y llega á las siguientes conclusiones:

1.º Que en todas las afasias existe un fondo considerable de demencia, y más de lo que hasta ahora se ha creído;

2.º Que no existe sino una sola afasia: la sensorial, y su centro sería la primera temporal, los pliegues: supramarginal y curvo (centros de Wernicke y Déjèrine);

3.º Que no existe una afasia motriz de Broca, no teniendo nada que ver el pie de la tercera frontal con las funciones del lenguaje. Lo que comúnmente se designa bajo la denominación de afasia motriz, no sería, según él, sino una combinación de la afasia(sensorial)con la anartria, y localiza el centro de ésta en el nucleo lenticular.

Fig. 1. Title page of Christofredo Jakob's 1906 article (Jakob, 1906a).

(Table 1) regarding an issue still debated in contemporary neuroscience (Brais, 1992; Daffner, Schomer, Cosgrove, Rubin, & Mesulam, 1991; Donnan, Carey, & Saling, 1999; Grodzinsky, 2000; Lecours, Basso, Moraschini, & Nespoulous, 1984). A neuroanatomical sequel published by Jakob on the pre- and postnatal development of white matter tracts related to language based on Weigert-stained specimens (Jakob, 1906b) forms the subject of a separate article (Vivas, Tsapkini, & Triarhou, 2007).

2. Does Broca's area exist?[1]

In one of the last issues of *La Semaine Médicale*, Marie has published an anatomical–clinical study on aphasias, and has reached the following conclusions:

---

[1] An English translation of *¿Existe ó no un centro de Broca?* by Dr. Chr. Jakob, Chief of the Laboratory of Psychiatry and Neurology, Hospicio de las Mercedes, originally published in Spanish in *La Semana Médica*, Buenos Aires, vol. XIII, no. 26, pp. 677–678, June 28, 1906.

Table 1
Timeline of publications in the aftermath of Pierre Marie's first 1906 article

Author(s) and reference	Date published	Comment
Marie (1906a)	May 23	First paper of 'trilogy', third frontal gyrus plays no role in language
Jakob (1906a)	June 28	First paper of two, neuropsychological on Broca's area
Marie and Moutier (1906a)	July 6	Examination of the brain in a case of Broca's aphasia
Déjerine (1906a)	July 11	First paper of two, on sensory aphasia
Jakob (1906b)	July 12	Second paper of two, neuroanatomical on myelin development
Déjerine (1906b)	July 18	Second paper of two, on motor aphasia
Marie (1906b)	October 17	Second paper of 'trilogy', on subcortical (pure) aphasia
Souques (1906)	October 24	Defends Marie with motor aphasia without lesion of third frontal gyrus
Moutier (1906)	October 26	Autopsy of a case of Broca's aphasia
von Monakow (1906)	November 15	Article on aphasia and diaschisis
Marie and Moutier (1906b)	November 16	Case of softening of foot of third frontal gyrus without Broca's aphasia
Marie and Moutier (1906c)	November 23	New case of Broca's aphasia without lesion of third frontal gyrus
McCarthy and Mills (1906)	November 24	Discussion at New York & Philadelphia Neurological Societies
Marie (1906c)	November 28	Third paper of 'trilogy', historical critical essay of Broca's doctrine
Lotmar and de Montet (1906)	December 1	Mental function in Broca's aphasia from Déjerine's group
Marie and Moutier (1906d)	December 14	Case of lesion of foot of third frontal without language trouble
Editorial (1906)	December 29	British Medical Journal, Pierre Marie on aphasia

1. There is a considerable element of dementia in all aphasias, more so than what was thought until now.
2. There is only a single type of aphasia, sensory, and its center would be at the first temporal, the supramarginal and the angular gyrus (areas of Wernicke and Déjerine).
3. A motor Broca's aphasia does not exist, and the foot of the third frontal [gyrus] has nothing to do with language functions. What is commonly designated under the denomination of motor aphasia, according to him, is nothing other than a combination of (sensory) aphasia with anarthria, with its center located at the lenticular nucleus.

Such a way of thinking justifiably attracts attention, firstly, because it involves an author with great experience on the subject, and secondly, because it raises the old discussion between Trousseau and Broca, which seemed to have been definitively resolved after the work of Charcot, Wernicke, Déjerine, Monakow and their disciples.

If we investigate the reasons why Marie reached conclusions so different from the ones accepted by science up to this point, we see two. First, the observation in some cases of a lesion of the foot of the third frontal [gyrus] without a co-existing motor aphasia, a point already noticed by other investigators; and second, the opportunity to detect motor aphasia in a small number of cases as well, without the autopsy revealing a lesion in that region.

But these two observations are not totally new. Already in 1881, Exner and Goltz, and in 1883 von Gudden, published cases of lesions of the third frontal [gyrus] without motor aphasia. But from the discussions that took place in scientific congresses, it had been established according to the ideas suggested by Wernicke in 1893, Monakow and others, that such cases, although not rare, demonstrate nothing against the cited localization, since these authors do not take into account the *restitution* of cortical functions. In effect, both clinical and experimental physiology has demonstrated that when a cortical center is destroyed,

in both humans and animals, the vicinity of that area takes over its functions. For example, dogs and apes, experimentally rendered hemiplegic, recover after some months the usual habitual walk; in the same way, a person with hemiplegia, hemianesthesia, etc. may improve to the point of recovery, and talking precisely about motor aphasia, numerous cases in the literature demonstrate the great curability of this affection.

Already in 1898, Monakow used to say that the entire region of Broca could be found destroyed and even more of the cortex in the vicinity, and aphasia would disappear after a certain period of time. For that reason, Exner's and Goltz's cases might have previously been accompanied by an aphasia, which could have passed unnoticed; in the case of Marie, the same critique is possible, because he should have demonstrated that [the patient] never had it.

The second argument, that there are motor aphasias without lesions to Broca's area, was already known and discussed for more than 15 years.

Sachs (school of Wernicke) insists on the possibility that there can be a motor aphasia through an inhibition of the left hemisphere (trauma, vascular disorders), without a focus of softening being formed, possibly leading to the death of the patient before it appears.

In the case of a total lesion of Wernicke's area, motor aphasia is observed indirectly, because the auditory area does not transmit signals to the motor area through the association pathways; moreover, in deaf-mutes, where the function of [Wernicke's] area is inhibited, language is impossible as well. One may cite transient motor aphasia without focal lesions of the third frontal [gyrus] in patients with general paresis, aphasia of uremics, intoxication, nonfocal [lesions], etc.[2]

Insofar as Marie affirms his views in such a categorical manner, without entering into the explanations just men-

---

[2] We exclude the cases of localization in the right hemisphere of the left-handed.

tioned, we can now accept, and therefore conclude, that the fundamentals of his theory—rejecting the existence of Broca's area—are in need of new and more convincing and detailed arguments.

With respect to what he calls a demented state of the aphasics, the fact is that it is observed in the institutionalized, since those patients fall invariably in dementia because of the vegetative and routine they have in the hospitals; on the contrary, it is harder to appear in patients assisted by their families and occupied with their tasks; and then the literature documents many motor aphasic patients who improve much and can even be useful, without showing significant disturbances of intelligence. Marie himself cites the case of a cook with aphasia, who even carried out his role in public, but whose incapacity, after a period of time in an institution, became total.

Therefore, the appearance of the demented state in motor aphasias depends on re-education and hygiene, which in a family setting is associated with the better care, as compared to an institution.

In sensory aphasias with extensive foci a mental decadence is noticed from the outset; Monakow points attention to the fact that the disappearance of an area as important for mental life results in severe disorders in intellectual work. If to these we add the co-existence of a generalized arteriosclerosis, it is clear that dementia will increase by the new alterations, leading to an ever decreasing blood supply.

As to the opinion of Marie of identifying subcortical aphasia with anarthria, we might always accept that, besides the lenticular nucleus, lesions involve equally well the cortico-capsular pathways, in other words, the trajectory from the frontal radiations of the internal capsule; because, both clinical and physiological studies have shown that cortical centers in the opercula participate in articulatory functions, while nothing was established for the lenticular nucleus; cases are even cited of total destruction of the latter without speech being affected.

Finally, that author opposes making subdivisions in the area that he accepts as [the area] of language.

Marie views aphasias as only differing in intensity, and not categorized into verbal blindness and alexia, as it is accepted today. It is true that there are doubts regarding this point, but such doubts are not clarified further by Marie's explanations; moreover, if we analyze the cases of alexia published by Déjerine, we see that they contradict [Marie's] thesis.

In another publication we shall occupy ourselves with the anatomical details of this question.[3]

3. Discussion

The present study makes justice to Jakob's 1906 article defending the existence of Broca's area from an anato-

mo-functional point of view; Jakob questions Marie's arguments and joins the localizationist tradition—defended by Déjerine among others—that has prevailed in the last half of the twentieth century and in recent years. He was correct in separating aphasia from dementia and in claiming that there could be lesions in Broca's area which do not cause aphasia or which cause it in a transient manner. This might be produced by the compensating action of neighboring areas. On the other hand, lesions in the inferior frontal gyrus accompanied by damage to the neighboring white matter and subcortical regions like the basal ganglia can lead to Broca's aphasia. Thus, Jakob goes in the right direction when speaking of the importance of the lesion in fibers projecting to the internal capsule.

Jakob's article reveals a lucid argumentation, placing special emphasis on the embodiment of the clinico-anatomical method. A practising neuropathologist, Jakob considered morphology in a functional context and formulated ideas on the integrative function of the brain. In general, the Buenos Aires school of neurology was a follower of the anatomo-clinical tradition: What could not be demonstrated in the clinic and the autopsy table, or under the microscope, was out of the realm, with little wish from the medical establishment to venture into terra incognita (Manuel del Cerro, Personal Communication, February 10, 2006). Through the initiative of Domingo Cabred, Professor of Psychiatry in Buenos Aires, an attempt had begun in 1898 for ways to promote neuroanatomy in the investigation of nervous and mental disorders (Orlando, 1966). Jakob had been influential in promoting a notion for psychiatry whereby disorders could be traced to the morphology of the nervous system; until the 1930s and even later, psychiatry in Argentina was articulated around such a 'somatic style', i.e., a view of mental illness associated with the body rather than with a 'soul' (Plotkin, 2001).

In some respects (Brais, 1992), Marie's work fell within a wider movement of criticism against associationist neuropsychology (Hécaen and Angelergues, 1965; Ombredane, 1951), which was also backed by philosopher Henri Bergson (1896) in his influential book Matter and Memory. Marie (1906a) claimed that 'aphasia is one' and that Broca's aphasia is none other than Wernicke's aphasia complicated by anarthria, the only notable difference between the two being that in Wernicke's aphasia patients speak more or less poorly, whereas in Broca's aphasia they do not speak at all. Marie also propounded the extreme lumping of categories, suggesting that there was only one type of aphasia (i.e., sensory).

The localization debate occupied the best part of the twentieth century. One group that continued to question localization is epitomized by Hildred Schuell and her colleagues in Minnesota (Schuell et al., 1964). An in-depth narrative of the historical controversy on cortical localization with special emphasis on language functions, including cytoarchitectonic and neurophysiological correlates, can be found in standard texts such as Finger (1994).

---

[3] An English translation of the sequel paper of Jakob can be found in Vivas et al. (2007).

An advocate of localized function, Déjerine is credited with proposing the first disconnectionist account of pure alexia and the existence of a 'visual verbal center' in the angular gyrus, functioning to store the visual images of words. Consequently, lesions of the left angular gyrus are associated with 'alexia with agraphia', whereas 'pure alexia' results from white matter lesions of the left occipital lobe that disrupt the path between the left and right visual areas and such a visual verbal center (Déjerine, 1891, 1892).

A turning point, tremendously influential for the holistic–localizationist controversy, was the proposition of Norman Geschwind that was also based on a disconnectionist framework (Catani and ffytche, 2005; Catani et al., 2005; Geschwind, 1965a,b, 1976). According to such a framework, the disconnection syndromes represented deficits in higher cognitive functions, resulting from lesions of the white matter or of association cortex. For neuroscience research, Geschwind's ideas led to the formulation of distributed network and connectionist theories of brain function (Catani and ffytche, 2005). The modern neuroimaging techniques offer investigators an unprecedented opportunity to supplement the information previously collected through traditional neuroanatomical tract-tracing methodologies with in vivo data on the connectivity of cortical and subcortical areas in the human brain under normal and pathological conditions.

The classically known pathway that connects Broca's and Wernicke's areas is the arcuate fasciculus. An additional pathway connecting Broca's and Wernicke's areas indirectly, termed the 'Geschwind territory' by Catani et al. (2005), passes through the inferior parietal lobe; it contains a caudal component that connects the temporal with the parietal cortex, and a rostral component that connects the parietal with the frontal cortex. In a confluence of reasoning, the disconnection model of Geschwind, as well as the more recently prediction by Catani et al. (2005) that lesions affecting different territories may cause different types of language deficits, are compatible with the importance attributed by Jakob (1906a) to subcortical projection pathways in Broca's aphasia.

Marie (1906a,b) further expressed the view that phonological errors arise at the level where phonemes are selected and ordered, whereas phonetic errors arise at the level of articulatory programming (Romani et al., 2002).

In claiming that Broca's aphasia is a combination of sensory aphasia and anarthria (Brais, 1992), Marie accordingly pointed out that the anatomical lesion of Broca's aphasia involves Wernicke's area and the lenticular nucleus. Jakob comments, toward the end of his article, that the associated lesions may not be located in the lenticular nucleus, but rather, in the cortico-capsular pathways, in other words, in the trajectory of the radiations of the internal capsule to cortical areas of the frontal lobe. Along a similar line of reasoning with Jakob, and 2 years later, Augusta Klumpke-Déjerine contested the 'lenticular zone' hypothesis during the second of the 1908 Paris debates and gave a different dimension, by convincingly contending

that its anterodorsal part includes association fibers emanating from or projecting to Broca's area, questioning the remainder of the 'Marie quadrilateral' (Klippel, 1908; Lecours, 1999).

In his later *Elements of Neurobiology*, Jakob (1923) wrote: 'What is of interest here from such a cortical symptomatology is, above all, the complex topic of the aphasias (Fig. 2), the apraxias and the agnosias; they are always of cortical origin and they can never be produced by lesions of subcortical centers or pathways; their effects result from the dynamic workings of the cortex over time and, thus, their alteration disturbs the piling of mental dynamics; therefore, we should take a closer look at the complicated game of the memory deficiencies or the partial dementias.'

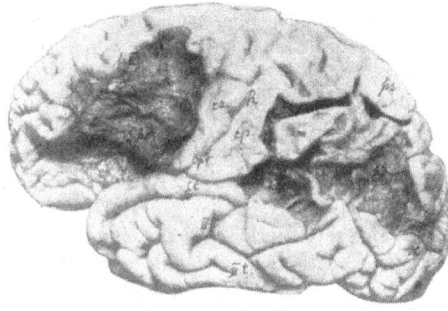

Fig. 49 a. — Caso de afasia sensomotora por reblandecimiento izquierdo cerebral (original) : zBr, zona de Broca; zW, zona de Wernicke; opr, opérculo rolándico; cn, cp, central anterior y posterior; R, Rolándica.

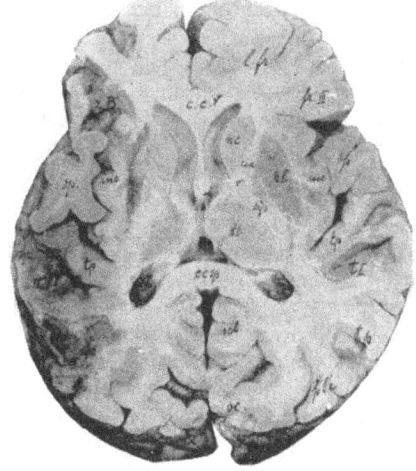

Fig. 49 b. — Corte horizontal de un caso de afasia sensomotora por reblandecimiento silviano izquierdo (original) : zBr, zona de Broca; zW, zona de Wernicke; opr, opérculo rolándico; ins, insula (compárese fig. 4o).

Fig. 2. Original pathological brain specimens of Christofredo Jakob with lesions affecting the speech areas, from his 1923 *Elements of Neurobiology* (Jakob, 1923). Upper field (Fig. 49a): Case of sensory-motor aphasia from left cerebral softening. Lower field (Fig. 49b): Horizontal section of a case of sensory-motor aphasia from softening of the left Sylvian fissure. Abbreviations: zBr, zone of Broca; zW, zone of Wernicke; opr, operculum of Rolando; cn, cp; precentral and postcentral gyrus; R, Rolandic sulcus; ins, insula.

Jakob and Monakow knew each other's work; Constantin von Monakow had in his possession an autographed copy of Jakob's 1899 *Atlas of the Normal and Pathological Nervous System* (Triarhou and del Cerro, 2006a). Jakob refers to von Monakow four times, to earlier works and to the second edition of *Brain Pathology* (Monakow, 1905), as the diaschisis paper (Monakow, 1906) was published in the latter half of 1906 (Table 1). Between 1902 and 1905, studying the localization of brain functions, Monakow developed the neuropsychological concept of 'diaschisis' to account for differences between acute transient severe symptoms after focal lesions in the brain and chronic residual, limited loss of function (Koehler and Jagella, 2002; Monakow, 1905, 1906), with particular relevance to aphasia. Diaschisis is a sudden-onset functional interruption, with points of impact distributed in places where fibers that originate from one area with a focal lesion finally enter a primarily undamaged area of gray matter. A difficult concept at the beginning, diaschisis gained importance with time in relation to recovery theories (Koehler and Jagella, 2002). Perhaps the recovery of function mentioned by Jakob in chronic clinical cases was meant in the context of diaschisis.

Jakob's rebuttal of Marie's view that Broca's area plays no role in language is erudite. Jakob makes it clear that damage to Broca's area without aphasia could be attributed to a restitution of language functions in other cortical areas; motor aphasia without damage to Broca's area can be explained by considering cortical connectivity more broadly. The speech disturbance resulting from lesions of Broca's area is a different condition from Broca's aphasia, which may result from damage far outside Broca's area; the more complex syndrome traditionally referred to as Broca's aphasia, including Broca's original case, is characterized by protracted mutism, verbal stereotypes, and agrammatism (Mohr et al., 1978). Most modern aphasiologists seem to agree that Broca's 1861 patient had a global aphasia that would not justify the diagnosis of 'Broca's aphasia' today (Eling, 1986).

The same issue was revived in the heated discussion that followed the advent of cognitive neuropsychology in the 1980s (Caramazza and Coltheart, 2006). The argument that there is a 'Broca's aphasia' without an underlying lesion of Broca's area and that there are Broca's area lesions without motor aphasia was made by advocates of the cognitive neuropsychology approach against strict neuroanatomical localization (Miceli and Caramazza, 1988), without necessarily advocating the holistic stance for brain function.

Having formulated intricate theories of language functions on the basis of neurological case studies, today's cognitive neuropsychology attempts to bridge such theories with cognitive neuroscience by means of neuroimaging techniques. Investigators use fMRI, PET, and transcranial magnetic stimulation (TMS) to advocate the important role of Broca's area in language functions and to specify such functions in detail. In the technical repertoire, neuro-

imaging (e.g., fMRI) may conceivably demonstrate localized function, and neurodynamical methods (e.g., EEG frequency analyses) are considered as more integrative measurements of summed nervous activity. Of course, using a specific methodology does not necessarily predetermine the backing of any theoretical view of brain function.

The question of the role of Broca's area remains open, even after a century of rigorous research. Modern imaging techniques allow scientists an unprecedented opportunity to visualize the brain in action. Although the involvement of Broca's area in language seems undisputed, there are multiple views on its exact role(s). Broca's area was found to mediate verbal inflection in both TMS (Shapiro et al., 2001) and fMRI (Tyler et al., 2004) studies. Embick et al. (2000) found that Broca's area is the site of syntactic processing, specifically, of 'syntactic working memory' (Fiebach et al., 2005). Broca's area is also thought of subserving language production, particularly phonological processing (Burton et al., 2001), of priming the motor response to heard speech even without speech production (Watkins and Paus, 2004; Wise et al., 1999), of being involved in semantic lexical processing (Klein et al., 2006), as well as three-dimensional mental rotation—i.e., non-musical visuospatial cognition—in orchestral musicians (Sluming et al., 2007). In considering the origins of human vocal skill, Passingham (1981) views the regulation of sequences of sounds—rather than the production of individual sounds—as the main role of Broca's area. There is a further cytoarchitectonic cortical differentiation, whereby Brodmann area 44 is likely involved in high-level aspects of programming speech production, whereas area 45 seems to be more involved in semantic aspects of language processing (Amunts et al., 2004; Grodzinsky and Amunts, 2006).

Besides the specific language functions that have been attributed to Broca's area through the use of neuroimaging techniques, there are claims that such functions may merely reflect higher-order cognitive processes. Thompson-Schill et al. (2005) claim that Broca's area, or rather the left inferior frontal gyrus, is responsible for lexical selection amongst competing candidates, whereas Rizzolatti and Craighero (2004) argue that the most important role that Broca's area played in an evolutionary sense was in action imitation, due to its location between motor areas and working memory areas. Evidence from primate studies suggests that more general cognitive demands 'prepared' Broca's area during phylogeny for the programming and sequencing of sounds, and eventually of phonemes and words (Petrides et al., 1993).

Another term with a controversial meaning, which has survived for over a century, is 'Broca's aphasia'. Jakob (1906a) writes that 'there are Broca's area lesions without Broca's aphasia and this can be attributed to the restitution of function by other areas of the brain.' The hypothesis that the right hemisphere (RH) takes over after left hemisphere (LH) damage has gained support. Points addressed in neuroimaging studies include (i) what happens when there is damage to the language areas of the LH and (ii) whether a RH homologous Broca's area exists. In a case with a Wernicke's area lesion, it was found that homologous RH areas seem to take over functionally (Musso et al., 1999; Weiller et al., 1995). After damage to Broca's area in the LH, other areas, such as the homologous RH Broca's and Wernicke's areas, as well as regions in the LH adjacent to Broca's area, seem to be more activated during tasks that normally require a functional involvement of Broca's area (e.g., the detection of syntactic anomalies) after rehabilitation, as Thompson (2000) notes. Such findings lend credence to Jakob's succinct point regarding the restitution of Broca's area functions by adjacent areas.

Regarding the general deterioration of intellectual function in aphasic patients and not merely the breakdown of language functions, Marie had been building such a concept since 1903 in a series of psychological studies co-authored with Vaschide and addressing mental life, immediate memory and ideative association (Brais, 1992). That is another point that Jakob rebuts, attributing the deterioration of cognitive function to the poor hygienic conditions of institutionalized patients as opposed to patients cared for at home. Jakob seems to be ahead of his time, since a current debate, with important implications for the rehabilitation from brain damage, pertains to the effect of enriched environment in compensatory function. Following the work of Schwartz (1964), several other researchers have shown that brain-damaged animals kept in an enriched environment—which offers opportunities for sensory, physical and social interactions—recover better than animals kept in bare cages. The benefit of an enriched environment has been correlated with increased survival of newly generated neurons, dendritic outgrowth and synaptogenesis (Will, Galani, Kelche, & Rosenzweig, 2004). Clearly, a century ago, as Jakob argues, homes might have offered a more stimulating environment compared to hospitals, where patients might have been kept isolated in their wards. Modern rehabilitation centers try, a century later, to instate the beneficial effects of a stimulating environment, like the one that Jakob mentions.

Jakob's rejection of the idea of a dementia–aphasia co-existence seems to have received support since. Severe language deficits in Alzheimer dementia are only found at later stages. Some demented patients show word production difficulties, especially in naming tools; those cases have been studied in the realm of semantic dementia that commences from a deterioration of parieto-temporal regions (Whatmough et al., 2003). The only dementia that has been associated with language disorders and begins as an aphasia without stroke is primary progressive aphasia (PPA); in that condition, patients may manifest problems with verbs, with characteristics of Broca's aphasia, e.g., syntactic difficulties and production problems (Hillis et al., 2002; Kertesz et al., 1994). Those cases quickly evolve into a more general impairment of cognitive functions. In Jakob's era it might have been difficult to make a differential diagnosis—especially in the absence of neuroimaging technology—in order to exclude a stroke that might confound a dementia of vascular origin with aphasia.

Acknowledgments

We thank Professor Harry A. Whitaker for reading an early version of this paper and generously offering his insightful input; the anonymous reviewers for their constructive comments that have led to a substantially improved manuscript; and the Library of the National Academy of Medicine of Buenos Aires for kindly furnishing us with a copy of Christofredo Jakob's original 1906 article. Supported in part by a basic science award from the Office of Intramural Research Funding Operations of the University of Macedonia to L.C.T.

References

Amunts, K., Weiss, P. H., Mohlberg, H., Pieperhoff, P., Gurd, J., Eickhoff, S., et al. (2004). Analysis of neural mechanisms underlying verbal fluency in cytoarchitectonically defined stereotactic space: The roles of Brodmann areas 44 and 45. Neuroimage, 22, 42–56.

Benson, D. F. (1979). Aphasia. In K. M. Heilman & E. Valenstein (Eds.), Clinical neuropsychology (pp. 22–58). New York, Oxford: Oxford University Press.

Bergson, H. (1896). Matière et mémoire: essai sur la relation du corps à l'esprit. Paris: Félix Alcan.

Brais, B. (1992). The third left frontal convolution plays no role in language: Pierre Marie and the Paris debate on aphasia (1906–1908). Neurology, 42, 690–695.

Broca, P. (1861). Perte de la parole, ramollissement chronique et destruction partielle du lobe antérieur gauche du cerveau. Bulletin de la Société d'Anthropologie (Paris), 2, 235–238.

Burton, M. W., Noll, D. C., & Small, S. L. (2001). The anatomy of auditory word processing: Individual variability. Brain and Language, 77, 119–131.

Caramazza, A., & Coltheart, M. (2006). Cognitive neuropsychology twenty years on. Cognitive Neuropsychology, 23, 3–12.

Catani, M., & ffytche, D. H. (2005). The rises and falls of disconnection syndromes. Brain, 128, 2224–2239.

Catani, M., Jones, D. K., & ffytche, D. H. (2005). Perisylvian language networks of the human brain. Annals of Neurology, 57, 8–16.

Daffner, K. R., Schomer, D. L., Cosgrove, G. R., Rubin, N., & Mesulam, M. M. (1991). Broca's aphasia following damage to Wernicke's area: For or against traditional aphasiology? Archives of Neurology, 48, 766–768.

Déjerine, J. (1891). Sur un cas de cécité verbale avec agraphie suivi d'autopsie. Comptes Rendus des Séances de la Société de Biologie et de ses Filiales (Paris), 43, 197–201.

Déjerine, J. (1892). Contibution a l'étude anatomo-pathologique et clinique des différentes variétés de cécité-verbale. Comptes Rendus des Séances de la Société de Biologie et de ses Filiales (Paris), 44, 61–90.

Déjerine, J. (1906a). L'aphasie sensorielle: sa localisation et sa physiologie pathologique. La Presse Médicale, 55, 437–439.

Déjerine, J. (1906b). L'aphasie motrice: sa localisation et sa physiologie pathologique. La Presse Médicale, 57, 453–457.

Donnan, G. A., Carey, L. M., & Saling, M. M. (1999). More (or less) on Broca. Lancet, 353, 1031–1032.

Economo, C. von, & Koskinas, G. N. (1925). Die Cytoarchitektonik der Hirnrinde des erwachsenen Menschen. Wien, Berlin: Verlag von Julius Springer.

Editorial. (1906). Pierre Marie on aphasia. British Medical Journal, 2, 1879–1880.

Eling, P. (1986). Speech and the left hemisphere: What Broca actually said. *Folia Phoniatrica (Basel), 38*, 13–15.

Eling, P. (2006). Neuroanniversaries 2006. *Journal of the History of the Neurosciences, 15*, 1–4.

Embick, D., Marantz, A., Miyashita, Y., O'Neil, W., & Sakai, K. L. (2000). A syntactic specialization for Broca's area. *Proceedings of the National Academy of Sciences of the United States of America, 97*, 6150–6154.

Fiebach, C. J., Schlesewsky, M., Lohmann, G., Cramon, D. Y. von, & Friederici, A. D. (2005). Revisiting the role of Broca's area in sentence processing: Syntactic integration versus syntactic working memory. *Human Brain Mapping, 24*, 79–91.

Finger, S. (1994). *Origins of neuroscience: A history of explorations into brain function.* New York, Oxford: Oxford University Press, [pp. 32–62, 374–385].

Freud, S. (1983). *Contribution à la conception des aphasies: une étude critique [1891] (trad. par C. van Reeth).* Paris: Presses Universitaires de France.

Geschwind, N. (1965a). Disconnexion syndromes in animals and man. Part I. *Brain, 88*, 237–294.

Geschwind, N. (1965b). Disconnexion syndromes in animals and man. Part II. *Brain, 88*, 585–644.

Geschwind, N. (1976) *Selected papers on language and the brain.* In R.S. Cohen & M.W. Wartofsky (Eds.) Boston studies in the philosophy of science (Vol. XVI, pp. 105–236), 2nd printing. Dordrecht, Boston: D. Reidel Publishing Company.

Goetz, C. G. (2003). Pierre Marie: Gifted intellect, poor timing and unchecked emotionality. *Journal of the History of the Neurosciences, 12*, 154–166.

Grodzinsky, Y. (2000). The neurology of syntax: Language use without Broca's area. *Behavioral and Brain Sciences, 23*, 1–71.

Grodzinsky, Y., & Amunts, K. (2006). *Broca's region: Mysteries, facts, ideas, and history.* New York, Oxford: Oxford University Press.

Hécaen, H., & Angelergues, R. (1965). *Pathologie du langage.* Paris: Librairie Larousse.

Hillis, A. E., Tuffiash, E., & Caramazza, A. (2002). Modality-specific deterioration in naming verbs in nonfluent primary progressive aphasia. *Journal of Cognitive Neuroscience, 14*, 1099–1108.

Jackson, J. H. (1874). On the nature of the duality of the brain. *Medical Press Circle New Series, 17*, 19–21, [pp. 41–44, 63–66].

Jacyna, S. (2005). Freud's critical study. *Cortex, 41*, 101–102.

Jakob, C. (1906a). Existe ó no un centro de Broca? *La Semana Médica (Buenos Aires), 13*, 677–678.

Jakob, C. (1906b). Consideraciones anátomo-biológicas sobre los centros del lenguaje. *La Semana Médica (Buenos Aires), 13*, 733–737.

Jakob, C. (1923). *Elementos de Neurobiología, volumen I: Parte Teórica* (pp. 219–234). (Biblioteca Humanidades, Tomo III). La Plata, República Argentina: Facultad de Humanidades y Ciencias de la Educación de la Universidad Nacional de La Plata.

Jellinger, K. A. (2006). A short history of neurosciences in Austria. *Journal of Neural Transmission, 113*, 271–282.

Kertesz, A., Hudson, L., Mackenzie, I. R., & Muñoz, D. G. (1994). The pathology and nosology of primary progressive aphasia. *Neurology, 44*, 2065–2072.

Klein, D., Zatorre, R. J., Chen, J. K., Milner, B., Crane, J., Belin, P., et al. (2006). Bilingual brain organization: A functional magnetic resonance adaptation study. *Neuroimage, 31*, 366–375.

Klippel, M. (1908). Discussions sur l'aphasie. *Revue Neurologique (Paris), 16*, 611–636, [pp. 974–1023, 1025–1047].

Koehler, P. J., & Jagella, C. (2002). Constantin von Monakow (1853–1930). *Journal of Neurology, 249*, 115–116.

Lecours, A. R. (1999). Aphasie: querelles. *Revue Neurologique (Paris), 155*, 833–847.

Lecours, A. R., Basso, A., Moraschini, S., & Nespoulous, J.-L. (1984). Where is the speech area, and who has seen it? In D. Caplan, A. R. Lecours, & A. Smith (Eds.), *Biological perspectives on language* (pp. 220–246). Cambridge, MA: MIT Press.

Lecours, A. R., & Joanette, Y. (1984). François Moutier or "from folds to folds". *Brain and Cognition, 3*, 198–230.

Lorch, M. P. (2004). The unknown source of John Hughlings Jackson's early interest in aphasia and epilepsy. *Cognitive and Behavioral Neurology, 17*, 124–132.

Lotmar, F., & de Montet, C. (1906). Examen de l'intelligence dans un cas d'Aphasie de Broca. *Revue Neurologique (Paris), 14*, 1063–1080.

Marie, P. (1906a). Revision de la question de l'aphasie: la troisième circonvolution frontale gauche ne joue aucun rôle spécial dans la fonction du langage. *La Semaine Médicale (Paris), 26*, 241–247.

Marie, P. (1906b). Revision de la question de l'aphasie: que faut-il penser des aphasie sous-corticales (aphasies pures)? *La Semaine Médicale (Paris), 26*, 493–500.

Marie, P. (1906c). Revision de la question de l'aphasie: l'aphasie de 1861 à 1866; essai de critique historique sur la genèse de la doctrine de Broca. *La Semaine Médicale (Paris), 26*, 565–571.

Marie, P., & Moutier, F. (1906a). Examen du cerveau d'un cas d'aphasie de Broca. *Bulletin et Mémoirs de la Société Médicale des Hôpitaux de Paris, 23*, 743–744.

Marie, P., & Moutier, F. (1906b). Sur un cas de ramollissement du pied de la troisième circonvolution frontale gauche chez un droitier, sans aphasie de Broca. *Bulletin et Mémoirs de la Société Médicale des Hôpitaux de Paris, 23*, 1152–1155.

Marie, P., & Moutier, F. (1906c). Nouveau cas d'aphasie de Broca sans lésion de la troisième frontale. *Bulletin et Mémoirs de la Société Médicale des Hôpitaux de Paris, 23*, 1180–1183.

Marie, P., & Moutier, F. (1906d). Nouveau cas de lésion corticale du pied de la troisième frontale gauche chez un droitier sans trouble du langage. *Bulletin et Mémoirs de la Société Médicale des Hôpitaux de Paris, 23*, 1295–1298.

McCarthy, D. J., & Mills, C. K. (1906). Discussion on aphasia, especially with reference to the views of Marie. *Journal of Nervous and Mental Disease, 34*, 459–473.

Meyer, L. (1981). Cristofredo Jakob: a veinticinco años de su muerte. *Acta Psiquiátrica y Psicológica de la América Latina, 27*, 13–14.

Miceli, G., & Caramazza, A. (1988). Dissociation of inflectional and derivational morphology. *Brain and Language, 35*, 24–65.

Mohr, J. P., Pessin, M. S., Finkelstein, S., Funkenstein, H. H., Duncan, G. W., & Davis, K. R. (1978). Broca aphasia: Pathologic and clinical. *Neurology, 28*, 311–324.

Monakow, C. von (1905). *Gehirnpathologie, zweite Aufl.* Wien: Hölder.

Monakow, C. von (1906). Aphasie und Diaschisis. *Neurologisches Centralblatt (Leipzig), 25*, 1026–1038.

Moutier, F. (1906). Examen nécropsique d'un cas d'aphasie de Broca. *Bulletin et Mémoirs de la Société Médicale des Hôpitaux de Paris, 23*, 1018–1020.

Musso, M., Weiller, C., Kiebel, S., Muller, S. P., Bulau, P., & Rijntjes, M. (1999). Training-induced brain plasticity in aphasia. *Brain, 122*, 1781–1790.

Ombredane, A. (1951). *L'aphasie et l'élaboration de la pensée explicite.* Paris: Presses Universitaires de France.

Orlando, J. C. (1966). *Christofredo Jakob: Su Vida y Obra.* Buenos Aires: Editorial Mundi.

Passingham, R. E. (1981). Broca's area and the origins of human vocal skill. *Philosophical Transaction of the Royal Society of London Series B (Biological Sciences), 292*, 167–175.

Pearce, J. M. S. (2004). A note on Pierre Marie (1853–1940). *Journal of Neurology Neurosurgery and Psychiatry, 75*, 1583.

Pedace, E. A. (1949). Contribución de la Escuela Neurobiológica Argentina del Profesor Chr. Jakob en el estudio del lóbulo frontal. *Archivos de Neurocirugía, 6*, 464–466.

Petrides, M., Alivisatos, B., Evans, A. C., & Meyer, E. (1993). Dissociation of human mid-dorsolateral from posterior dorsolateral frontal cortex in memory processing. *Proceedings of the National Academy of Sciences of the United States of America, 90*, 873–877.

Plotkin, M. B. (2001). *Freud in the Pampas: The emergence and development of a psychoanalytic culture in Argentina.* Stanford, CA: Stanford University Press.

Rizzolatti, G., & Craighero, L. (2004). The mirror-neuron system. *Annual Review of Neuroscience, 27*, 169–192.

Romani, C., Olson, A., Semenza, C., & Granà, A. (2002). Patterns of phonological errors as a function of a phonological versus an articulatory locus of impairment. *Cortex, 38*, 541–567.

Schuell, H., Jenkins, J., & Jimenez-Pabon, E. (1964). *Aphasia in adults: Diagnosis, prognosis and treatment*. New York: Harper & Row.

Schwartz, S. (1964). Effect of neonatal cortical lesions and early environmental factors on adult rat behavior. *Journal of Comparative and Physiological Psychology, 57*, 72–77.

Shapiro, K. A., Pascual-Leone, A., Mottaghy, F. M., Gangitano, M., & Caramazza, A. (2001). Grammatical distinctions in the left frontal cortex. *Journal of Cognitive Neuroscience, 13*, 713–720.

Sluming, V., Brooks, J., Howard, M., Downes, J. J., & Roberts, N. (2007). Broca's area supports enhanced visuospatial cognition in orchestral musicians. *Journal of Neuroscience, 27*, 3799–3806.

Souques, A. (1906). Aphasie motrice sans lésion de la troisième circonvolution frontale. *Bulletin et Mémoirs de la Société Médicale des Hôpitaux de Paris, 23*, 971–977.

Thompson, C. K. (2000). Neuroplasticity: Evidence from aphasia. *Journal of Communication Disorders, 33*, 357–366.

Thompson-Schill, S. L., Bedny, M., & Goldberg, R. F. (2005). The frontal lobes and the regulation of mental activity. *Current Opinion in Neurobiology, 15*, 219–224.

Triarhou, L. C., & del Cerro, M. (2006a). Semicentennial tribute to the ingenious neurobiologist Christfried Jakob (1866–1956). 1. Works from Germany and the first Argentina period, 1891–1913. *European Neurology, 56*, 176–188.

Triarhou, L. C., & del Cerro, M. (2006b). Semicentennial tribute to the ingenious neurobiologist Christfried Jakob (1866–1956). 2. Publica-tions from the second Argentina period, 1913–1949. *European Neurology, 56*, 189–198.

Triarhou, L. C., & del Cerro, M. (2007). Pioneers in Neurology: Christfried Jakob (1866–1956). *Journal of Neurology, 254*, 124–125.

Tyler, L. K., Bright, P., Fletcher, P., & Stamatakis, E. A. (2004). Neural processing of nouns and verbs: The role of inflectional morphology. *Neuropsychologia, 42*, 512–523.

Vivas, A. B., Tsapkini, K., & Triarhou, L. C. (2007). 'Anatomo-biological considerations on the centers of language': An Argentinian contribution to the 1906 Paris debate on aphasia. *Brain and Development*. doi:10.1016/j.braindev.2007.03.007.

Watkins, K., & Paus, T. (2004). Modulation of motor excitability during speech perception: The role of Broca's area. *Journal of Cognitive Neuroscience, 16*, 978–987.

Weiller, C., Isensee, C., Rijntjes, M., Huber, W., Muller, S., Bier, D., et al. (1995). Recovery from Wernicke's aphasia: A positron emission tomographic study. *Annals of Neurology, 37*, 723–732.

Wernicke, C. (1874). *Der Aphasische Symptomencomplex: Eine psychologische Studie auf Anatomischer Basis*. Breslau: Cohn und Weigert.

Whatmough, C., Chertkow, H., Murtha, S., Templeman, D., Babins, L., & Kelner, N. (2003). The semantic category effect increases with worsening anomia in Alzheimer's type dementia. *Brain and Language, 84*, 134–147.

Will, B., Galani, R., Kelche, C., & Rosenzweig, M. R. (2004). Recovery from brain injury in animals: Relative efficacy of environmental enrichment, physical exercise or formal training (1990–2002). *Progress in Neurobiology, 72*, 167–182.

Wise, R. J. S., Greene, J., Büchel, C., & Scott, S. K. (1999). Brain regions involved in articulation. *Lancet, 353*, 1057–1061.

Official Journal of
the Japanese Society
of Child Neurology

Brain & Development 29 (2007) 455–461

www.elsevier.com/locate/braindev

Review article

'Anatomo-biological considerations on the centers of language': An Argentinian contribution to the 1906 Paris debate on aphasia

Ana B. Vivas [a], Kyrana Tsapkini [b], Lazaros C. Triarhou [c,*]

[a] *Department of Psychology, City Liberal Studies, Affiliated Institution of University of Sheffield, Thessaloniki, Greece*
[b] *Department of Psychology, Aristotelian University, Thessaloniki, Greece*
[c] *Economo-Koskinas Wing for Integrative and Evolutionary Neuroscience, Department of Educational and Social Policy,*
University of Macedonia, 156 Egnatia Ave., Bldg. Z-312, Thessaloniki 54006, Greece

Received 25 September 2006; received in revised form 9 March 2007; accepted 18 March 2007

Abstract

In 1906, the year of the renowned holistic–localizationist controversy between neurologists Pierre Marie and Jules Déjèrine in Paris, Christfried Jakob, a protagonist researcher of the cerebral cortex at the time working in Argentina, published two relevant articles entitled 'Does Broca's area exist?' and 'Anatomo-biological considerations on the centers of language'. The two articles addressed neuropsychological and developmental aspects of language functions in normality and pathology with regard to the brain areas that subserve them. The present article provides an English translation of Jakob's second paper, on the embryonic and post-natal development of brain areas related to language. The information given and the views expressed may still shed, a century later, useful light on our understanding of brain–language relationships.
© 2007 Elsevier B.V. All rights reserved.

Keywords: Human brain development; Myelination; Christfried Jakob; Language areas

1. Introduction

The year 1906 was an *annus mirabilis* for the field of aphasiology [1–3]: it witnessed one of the most heated debates on the still lingering localizationist–holistic controversy regarding human brain function [4–6], triggered by Pierre Marie's sensational anti-localizationist stance [7].

Some of the timeliest responses [8,9] in that year came from Christfried Jakob (1866–1956), a neuropathologist working in Buenos Aires. Born and trained in Germany, Jakob spent most of his professional life in Argentina – where his name became 'castillianized' to Christofredo –

in affiliation with the Universities of Buenos Aires and La Plata [10,11]. Having made a name for himself in Europe through successful Atlases of human and comparative neuroanatomy [12–14], Jakob became the founder of the Argentinian school of neurobiology [10,11,15]. He authored 200 papers and 30 books in German and Spanish [16,17], and he studied the histophysiology of the frontal cortex extensively, both in the normal condition as well as in the aphasias, agnosias, and apraxias [18–26].

We previously rendered an English translation [27] of Jakob's first, 'neuropsychological' 1906 paper [8]. In the present article we present his developmental neuroanatomical sequel [9] (Fig. 1). In doing so, we aim at saving an important contribution from oblivion, which historical articles not written in English seem to be condemned to. The remarks made by Jakob still appear relevant today; in reading his paper, one realizes that even the

---

* Corresponding author. Address: University of Macedonia, 156 Egnatia Ave., Bldg. Z-312, Thessaloniki 54006, Greece. Tel.: +30 2310 891 387; fax: +30 2310 891 388.
E-mail address: triarhou@uom.gr (L.C. Triarhou).

0387-7604/$ - see front matter © 2007 Elsevier B.V. All rights reserved.
doi:10.1016/j.braindev.2007.03.007

Año XIII.—N.° 28   BUENOS AIRES JULIO 12 DE 1906   733

LA SEMANA MEDICA

DIRECCIÓN, REDACCIÓN, ADMINISTRACIÓN & IMPRENTA: CALLAO, 737
Unión Telefónica 276 (Juncal)

FISIOPATOLOGIA

Consideraciones anátomo-biológicas sobre los centros del lenguaje

POR EL

Dr. Chr. JAKOB

Jefe del Laboratorio de Psiquiatría y Neurología
Hospicio de las Mercedes

La doctrina que hoy predomina acerca de las funciones del lenguaje, no sólo se basa en la observación anátomo-clínica sino también en otra clase de estudios, que han confirmado los anteriores conocimientos y de una manera más demostrativa.

La anatomía y fisiología del desarrollo del sistema nervioso humano, la anatomía y fisiología comparadas, y la fisiología experimental nos han enseñado que los centros motrices y sensoriales, que guardan relación con el lenguaje están localizados en diferentes regiones

Fig. 1. Title page of Christfried Jakob's 1906 'developmental anatomical' article in *La Semana Médica* [9].

advent of sophisticated neuroimaging methods such as fMRI and diffusion tensor imaging (DTI) have not added much to ideas conceived 100 years ago on the basis of simple but concrete observations made with the Weigert stain and a light microscope.

2. Anatomo-biological considerations on the centers of language by Dr. Chr. Jakob[9]

The dominating doctrine today with regard to the functions of language is not based solely on anatomo-clinical observations, but also on other types of studies, which have confirmed previous knowledge in a more compelling way.

The anatomy and physiology of the development of the human nervous system, comparative anatomy and physiology, and experimental physiology have taught us that motor and sensory centers related to language are localized in different regions of the cortex.

I shall now continue to expose the results of investigations I conducted on the myelination of the centers of language.

The human cerebral hemisphere, after going through the vesicular stage, shows already in the fourth month a configuration that lets one grasp the future lobules, but its walls do still have a cellular structure, lacking the differentiation of fibers. In the fifth month, the formation of sulci and gyri begins and axis cylinders appear, but without a myelin sheath. At the beginning of this period, the vicinity of the Sylvian region becomes more and more prominent, to the point of transforming the cavity into a fissure through the growth of the opercula which gradually cover it. In the sixth month the frontal and Rolandic operculum are distinctly separated, just as the Ist from the IInd temporal [gyrus]. The walls have developed, thickening by the abundant formation of axis cylinders and the internal and external capsule become discernible, which are separated by the lenticular nucleus. If we obtain a section around the middle of this month, it strikes our attention that it is not possible to separate the future white matter from the cortical [matter], but on the contrary both form a homogenous mass of a red-grayish color. However, the same section at the level of the lenticulo-thalamic nucleus shows a differentiation between the two matters, because myelination of the rubro-thalamic-lenticular and luysio-lenticular pathways (ansa lenticularis) has already begun. At the beginning of the seventh month the lenticular nucleus appears already divided because the ansa has been myelinated, and the myelin lamellae have formed, whereas in the opercula, in the supramarginal [gyrus] and Ist temporal [gyrus] fibers with myelin are not distinguished. It cannot be doubted that in this period those myelinated centers fulfill functions, although we do not know how, insofar as in the cortex there is not but only one region that shows a similar development: the paracentral lobule (the center of movement of lower limbs), which already has connections with the thalamus.

From this period on, myelinated pathways develop in the cortex, whose knowledge we owe to Flechsig, and which are: the olfactory, whose advance operates from the periphery to the uncus of the hippocampus, and the optic, which is initiated in the nerve in order to be subsequently directed towards the occipital lobe. When these two pathways appear, the thalamo-rolandic pathway becomes more evident, filling in with myelinated fibers not only of the paracentral lobule but also the rest of both central [lobes] all the way to its opercular portion. At the beginning of the eighth month, it is noticed in the deep temporal auditory pathway, whose myelination takes place first in the cochlear, continues through the trapezoid body, the lateral fasciculus, the medial geniculate, and from here, passing through the capsule, to the mentioned gyrus.

Picture IV (Fig. 3) is a photographic reproduction, in natural size, of a horizontal section made across the left hemisphere in a fetus eight months of age, where we do not see any myelin in the frontal lobe, in the insula and superficial temporal [lobes] and very little in the occipital (calcarine region); however, white color stands out in the regions of the thalamus and lenticular nucleus, myelinated, and at the level of the internal capsule (posterior segment) the region corresponding to the thalamo-rolandic pathways, mixed with the pyramidal [pathway].

Beyond this central region of the capsule myelin is equally observed in fibers of the auditory [acoustic] radiations, which enter the deep temporal, but in the anterior segment of the internal capsule we only see the first outline of the thalamo-frontal pathway, which still has not gone across the extent of the corpus striatum.

From the considerations exposed above one may conclude that the first of the centers in relation to language, that is closer to its definitive state, is the auditory.

The mechanism by which the completion of myelination continues will be understood by analyzing diagrams I, II and III (Fig. 2), which represent two centers in three different periods of the process.

In number I we have two centers in their initial period: ca and co. To both arrive from their respective subcortical centers, csa and cso, the primary projection pathways.

In this period, myelinated fibers only occupy the axis [i.e., the central part] of the pathway, and soon thereafter, by the stowing of the sheath against the remaining fibers, [the pathway] grows little-by-little in width, finally transforming itself into a completely myelinated region.

These two centers (Fig. 2) have not established a relation between them yet. Consequently, a zone that Flechsig calls intermediate territory (im) remains empty. However, the myelinated portion in its center has been extended: first by more complete myelination of the pathway of projection vc, and second, by the appearance of other [pathways] with myelin fibers, which join together the different paths of each one of the centers and that are in the U (u) of Meynert, short fibers of association characteristic of each center. This period will be comprised between the ninth month and the third month of extrauterine life.

Just in the III scheme (Fig. 2), which corresponds to the fourth or fifth extrauterine month, besides the completion of each one of the described centers, advance the short fibers of association, invading the intermediary territory (im), and end up, finally, establishing contact between both regions.[1]

It results, thus, that from the centers of language, the auditory [center] becomes myelinated at the beginning of

the eighth month. In the ninth month the process invades the neighboring portion of the Ist temporal [gyrus], but at the same time the first myelinated fibers are observed in the foot of the III frontal (operculum frontale): pathway for the kinesthetic sensations of the labio-lingual regions and especially of the side situated rostral to the other opercula, which, as I already showed above, acquire their myelinated fibers somewhat earlier.

Now, if we compare photograph V, which reproduces a section of the right hemisphere of a fetus in term with the earlier, we notice myelination in the thalamus, in the posterior half of the posterior segment of the internal capsule (central pathways), optic radiations (very complete) and auditory pathway. In addition the external capsule has become myelinated – the anterior segment of the internal capsule has already gone ahead – and fibers are seen entering the frontal operculum, forming an arch. The remaining of the fibers of the frontal lobe lack a sheath, what is left from the III (capsular and orbital region), the temporal lobe with exception of the I and deep and the convexity of the occipito-parietal [cortex].

A neonate is then prepared to function in all the receptive-sensory pathways (primordial centers), but the sensations that arrive cannot have connections among them, because the intermediary association pathway has not been formed yet. Just when they appear these could exert an influence one upon the other (the auditory center on the visual and tactile, etc.); and this happens from the fourth month of life onwards, according to the findings of child physiology.

In picture VI (sixth postnatal month), the white color of already myelinated pathways and centers stands out in a more pronounced way, but intermediary territories (frontal, parietal, temporal lobes) are seen already invaded by myelinated fibers, and not only the opercular region of the III frontal [gyrus] but almost all the rest of its anterior portion (Fig. 3).

With this I conclude the *first period* of myelination, which begins during the sixth month of intrauterine life and ends by the fifth or sixth postnatal month. During the following months the *second period* continues: the one of myelination of short intracortical fibers (superficial and deep layers of association: of Baillarger, etc.), into details of which I shall not enter. To these, a *third period* will be added, ending right after puberty, and characterized by the formation of the sheath in the fine plexuses situated between the layers or bands.

In the newborn, all sensory pathways are already formed in their longitude but not in their latitude. Motor pathways that are not fully grown may not be excitable in the same manner as it [normally] happens with cortical centers. The myelinated pyramidal tract reaches up to the bulb; the remaining motor cortical pathways (fronto-pontine, temporo-pontine) do not reach the peduncle; from that, it is deduced that, in this

---

[1] In the intermediary territories the quantity of projection fibers is minor, due to the scarcer development of the thalamo-cortical pathways. With regard to the described association and direct "long pathways", according to my observations, they do not exist in sensory centers.

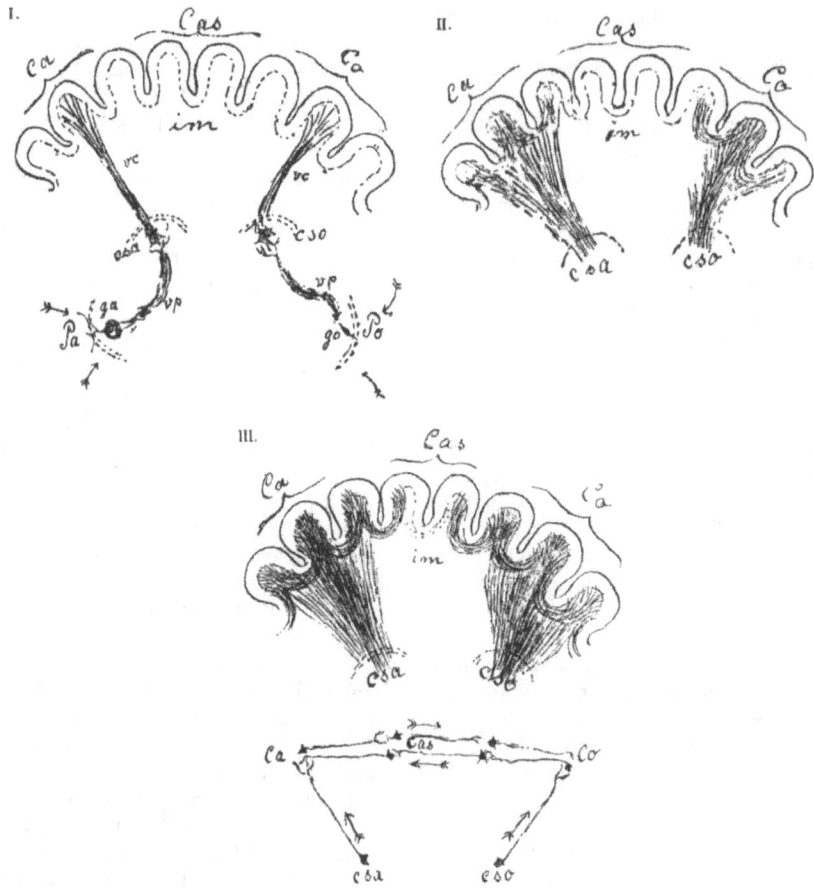

Fig. 2. Sketches by Jakob [9] of the developing fiber pathways (fields I, II, III). As the original figures do not have legends in the original article, the following information is deduced from the text description. Field I shows myelin pathway development before the eighth month of gestation; field II depicts the developmental stage between the ninth prenatal through the third postnatal month; field III represents an age of four to five months after birth. Abbreviations: *ca*, cortical center *a*; *co*, cortical center *o*; *im*, intermediate territory of Flechsig; *vc*, pathway of projection; *csa*, subcortical center *a*; *cso*, subcortical center *o*.

period of life, the motor cortical apparatus is less developed than the sensory. Thence all movements, outside reflexes, are ataxic. More complex movements have to be learned by the child, and in the way that this is achieved it advances in a correlative manner the myelination of the corresponding cortical territories.

In consequence, the cerebral cortex is exclusively the place where learned movements have their seat; the articulation of speech, which is the result of a series of complex and combined acts, is perfected in the child insofar as the center of these movements is completed. The newborn moves all the muscles (lips, tongue) it needs for the production of sound (shouts, etc.), but is unable to associate those movements for producing articulated speech.

In one-year old children, only the articulation of very few syllables is possible; already for this period the len-

ticular nucleus is completely developed, and if, according to Pierre Marie this would be one of the centers of the articulation of language, it cannot be understood why it does not possess a more developed relationship with the faculty of speech; because experimental physiology has taught us that the function of the opercula (centers of XII, X, etc.) is in direct relation to its development; and we find the demonstration of these facts in the child, whose aptitude for exteriorizing its ideas by means of speech increases at the same time as the texture of its cerebral cortex is completed.

What is necessary for the full development of word articulation is a constant excitation of the auditory center, which develops before the motor; deaf-mutes are an example that shows the importance of such a center.

In summary, we reach the conclusion that all movements which a child must learn in order to articulate

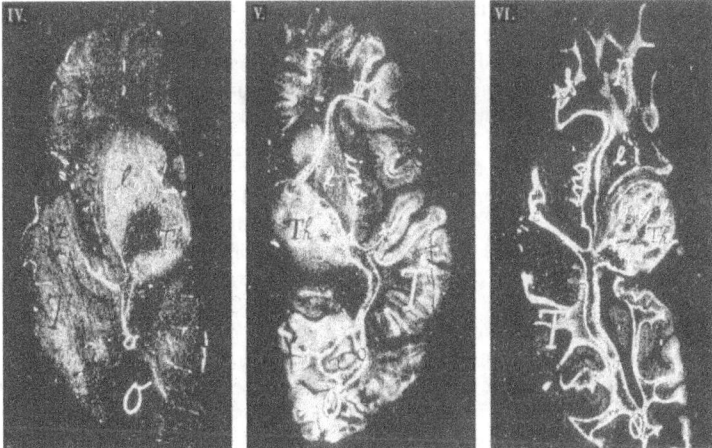

Fig. 3. Photographs of horizontal brain sections showing the development of myelin tracts (in white) at three different time-points. The brain of an 8-month-old fetus is shown in field IV, a fetus in term in V, and a six-month-old infant in VI. Fields IV and VI are horizontal sections of the the the left hemisphere, and field V of the right hemisphere. Abbreviations: *F*, frontal lobe; *T*, temporal lobe; *O*, occipital lobe; *IIIf*, third frontal gyrus; *l*, lenticular nucleus; *Th*, thalamus; *ins*, insula of Reil. From Jakob [9].

speech have their localization in the cerebral cortex and continue to develop in direct relation with it, and not with the lenticular nucleus, which is perfectly developed at a period during which the child is just beginning to learn articulation.

As far as the localization of "anarthria" is concerned in the pathways of the anterior segment of the internal capsule, to which many authors, outside P. Marie, refer, we must remember that they are not the centers of this function but pathways originating or ending up in the frontal cortex, where we must search their centers, and the lenticular-caudate pathways that pass through there, form a limited contingent and of functions still completely unknown.

Cases of aphasia that are under study will be published later.

3. Comment

In this 1906 article [9], Jakob provides an overview of the development of white matter pathways, largely based on the anatomical monitoring of myelination tracts associated with language-related areas, with their functional implications. His data feature the objectivity and longevity that characterize the anatomical approach, whereby observers would document mere facts, in contrast to current hypothesis-driven studies, in which interpretations may vary as a function of the paradigm adopted.

At the time when the article was published, the study of cortical myeloarchitectonics (the histological investigation of the cerebral cortex based on the anatomical patterns of myelinated fiber pathways in specimens stained by the Weigert method) was at its peak. In the late nineteenth century, Paul Flechsig, by studying the temporal order in which subcortical white matter fibers become myelinated, provided 'myelogenetic' maps of the human cerebral cortex with 45 areas numbered according to the sequence of myelin maturation, lower being earlier [28,29]. The field of myeloarchitectonics of the human cerebral cortex progressed between 1900 and 1906 mainly through the contributions of Cécile and Oskar Vogt in Berlin, who described over 200 myeloarchitectonic zones [30,31].

Bartels and Zeki have put it most succinctly [32]: The myeloarchitectonic maps of the cerebral cortex of Vogt and Vogt [33], as well as the cytoarchitectonic maps of Campbell [34], Brodmann [35], and von Economo and Koskinas [36], are, remarkably, still in use a century later. This may seem surprising given the advent of the modern neuroimaging techniques of infinitely greater sophistication. Perhaps one important clue to their success lies in the fact that they do not hypothesize about functions. These early brain researchers did not have to be overly concerned with attaching a function to each subdivision they delineated. Rather, they hoped that future functional and clinical studies would validate their subdivisions. This has turned out to be to a certain extent true; examples of architectural subdivisions reflecting to an unexpected degree functional differentiation are found in the striate, motor and parietal cortices.

Jakob argues that one may not assign articulatory functions to the lenticular nucleus, because children can only articulate a few syllables even with a fully developed lenticular nucleus. The concept underlying such an argumentation is simple and enduring [32]: it merely states that architectural differences are indicative of functional differences and that functional differences, conversely, necessitate architectural differences – the dominant theme in cortical studies. In modern terms, brain function is determined by neural hardware. Thus, by definition, differences in function require differences in hardware, visible in terms of cell types, connectivity, synaptic and molecular structures.

Regarding the contribution of the lenticular nucleus to speech production, Jakob challenges Marie and emphasizes that it is rather the striato-cortical pathways to the frontal lobe that subserve speech production. Jakob's conclusion was confirmed in the 1990s by neuroimaging and lesion studies. For example, using PET, Wise et al. [37] did find pallidal involvement in association with articulatory planning, but attributed it to the motor loops between cortical and subcortical regions. Similar results had been reported earlier in a patient with dysarthria that had actually resulted from white matter lesions, although it had been originally attributed to basal ganglia lesions in the left hemisphere [38].

In their influential paper on subcortical aphasia, Nadeau and Crosson [39], having reviewed 17 cases of striatocapsular infarction without language impairment and 33 cases with similar anatomical characteristics that were associated with language deficits, concluded that aphasia from subcortical damage does not exist. They proposed instead that subcortical aphasia is due to cortical damage that cannot be detected by neuroimaging methods or to a disruption of the fronto-striatal pathways or both. Finally, they also suggested that subcortical aphasia associated with thalamic lesions is due to destruction of the thalamo-cortical pathways that subserve the attentional gating system. Their work renewed the debate on subcortical aphasia [40,41], that had begun at the dawn of the 20th century, by integrating the evidence from numerous neurological cases.

One of the problems here is that language is a complex phenomenon involving both motor components as well as central higher-order abilities. For one to argue that subcortical aphasia does or does not exist depends on the definitions of aphasia [42,43]. Certainly there are motor difficulties impinging upon the organization and expression of language that may not be 'pure' aphasia, but that are generally included in such a category. For example, subcortical aphasias can affect language through hypoperfusion [44].

One of the difficulties in answering the question whether a pure subcortical aphasia exists is that typical functional imaging methods have not thus far permitted the functional study of white matter pathways. The newly developed technique of DTI [45,46], which can reveal the detailed anatomy of the white matter, promises to shed light on this long-standing debate. As current studies using DTI are largely focused on structural issues [46,47], Jakob's views on white matter tract myelination and its relation to language development will await future corroboration.

Most interesting is the developmental aspect of this early work, being probably preciser in characterizing the progression of myelination than modern imaging techniques. For example, current work on language development and imaging [48], using three-dimensional MRI, makes a distinction between the myelination of the sensory-motor network (at six months after birth) and that of the temporofrontal language network (at eighteen months). Covering a wider spectrum, Jakob firstly provides data on prenatal pathway myelination, and secondly dissociates the maturation of sensory and motor pathways (sensory becoming myelinated before motor) postnatally.

Acknowledgements

We gratefully acknowledge the Library of the National Academy of Medicine of Buenos Aires for kindly providing a copy of Christofredo Jakob's original 1906 article, as well as the constructive comments of the anonymous reviewers that have led to a substantially improved manuscript. Supported in part by a basic science award from the Office of Intramural Research Funding Operations of the University of Macedonia to L.C.T.

References

[1] Benson DF. Aphasia. In: Heilman KM, Valenstein E, editors. Clinical Neuropsychology. New York–Oxford: Oxford University Press; 1979. p. 22–58.
[2] Brais B. The third left frontal convolution plays no role in language: Pierre Marie and the Paris debate on aphasia (1906–1908). Neurology 1992;42:690–5.
[3] Goetz CG. Pierre Marie: gifted intellect, poor timing and unchecked emotionality. J Hist Neurosci 2003;12:154–66.
[4] Lecours AR, Basso A, Moraschini S, Nespoulous J-L. Where is the speech area, and who has seen it? In: Caplan D, Lecours AR, Smith A, editors. Biological perspectives on language. Cambridge, MA: MIT Press; 1984. p. 220–46.
[5] Daffner KR, Schomer DL, Cosgrove GR, Rubin N, Mesulam MM. Broca's aphasia following damage to Wernicke's area. For or against traditional aphasiology? Arch Neurol 1991;48:766–8.
[6] Grodzinsky Y. The neurology of syntax: language use without Broca's area. Behav Brain Sci 2000;23:1–71.
[7] Marie P. Revision de la question de l'aphasie: la troisième circonvolution frontale gauche ne joue aucun rôle spécial dans la fonction du langage. Semaine Méd (Paris) 1906;26:241–7.
[8] Jakob C. Existe ó no un centro de Broca? Semana Méd (Buenos Aires) 1906;13:677–8.
[9] Jakob C. Consideraciones anátomo-biológicas sobre los centros del lenguaje. Semana Méd (Buenos Aires) 1906;13:733–7.
[10] Orlando JC, Christofredo Jakob. Su Vida y Obra. Buenos Aires: Editorial Mundi; 1966.
[11] Meyer L, Christofredo Jakob. A veinticinco años de su muerte. Acta Psiquiát Psicol Amér Lat 1981;27:13–4.

[12] Jakob C. Atlas des Gesunden und Kranken Nervensystems nebst Grundriss der Anatomie, Pathologie und Therapie Desselben. II. Umgearbeitete Auflage. München: Verlag von J. Lehmann; 1899.

[13] Jakob C. Das Menschenhirn: Eine Studie über den Aufbau und die Bedeutung seiner Grauen Kerne und Rinde. I. Teil. Tafelwerk nebst Einführung in den Organisationsplan der Menschlichen Zentralnervensystems. München: J.F. Lehmann's Verlag; 1911.

[14] Jakob C, Onelli C. Vom Tierhirn zum Menschenhirn: Vergleichend Morphologische, Histologische und Biologische Studien zur Entwicklung der Grosshirnhemisphären und ihrer Rinde. I. Teil. Tafelwerk nebst Einführung in die Geschichte der Hirnrinde. München: J.F. Lehmann's Verlag; 1911.

[15] Pedace EA. Contribución de la Escuela Neurobiológica Argentina del Profesor Chr. Jakob en el estudio del lóbulo frontal. Arch Neurocirug 1949;6:464–6.

[16] Triarhou LC, del Cerro M. Semicentennial tribute to the ingenious neurobiologist Christfried Jakob (1866–1956). 1. Works from Germany and the first Argentina period, 1891–1913. Eur Neurol 2006;56:176–88.

[17] Triarhou LC, del Cerro M. Semicentennial tribute to the ingenious neurobiologist Christofredo Jakob (1866–1956). 2. Publications from the second Argentina period, 1913–1949. Eur Neurol 2006;56:189–98.

[18] Jakob C. Estudios biológicos sobre los lóbulos frontales cerebrales. Semana Méd (Buenos Aires) 1906;13:1375–81.

[19] Jakob C. Los lóbulos frontales y las funciones psíquicas superiores. Arch Psiquiat Criminol (Buenos Aires) 1906;5:678–99.

[20] Jakob C. La afasia motriz y su localización. Arch Psiquiat Criminol (Buenos Aires) 1913;12:79–85.

[21] Jakob C. La función psicogenética de la corteza cerebral y su posible localización. An Inst Psicol Fac Filos Letr Univ (Buenos Aires) 1941;3:63–80.

[22] Jakob C. El Lóbulo Frontal: Un Estudio Monográfico Anatomoclínico sobre Base Neurobiológica. Folia Neurobiologica Argentina. Tomo III. La Plata: Cátedra de Biología y Sistema Nervioso, Facultad de Humanidades y Ciencias de la Educación, Universidad Nacional; 1943. pp. 1–149.

[23] Jakob C. Ensayo de una histofisiopatología frontal. Rev Neurol (Buenos Aires) 1944;9:195–238.

[24] Jakob C, Pedace EA. La mision del lóbulo frontal frente a una cuantificación sintética de sus elementos productores. Arch Neurocirug 1949;6:467–74.

[25] Lores Arnaiz MR, Borrego Maturana F, Azzara S. Las ideas de Christofredo Jakob sobre mapa cortical y funciones superiores. Rev Hist Psicol 2002;23:9–36.

[26] Outes DL. A medio siglo de la muerte de Christofredo Jakob, 1956–2006: Fuentes de la concepción biológica de la doble corteza. Electroneurobiología 2006;14:3–28.

[27] Tsapkini K, Vivas AB, Triarhou LC. 'Does Broca's area exist?': Christofredo Jakob's 1906 response to Pierre Marie's holistic stance. Brain Lang 2007; [submitted].

[28] Flechsig PE. Gehirn und Seele. Zweite Auflage. Leipzig: Verlag von Veit und Comp.; 1896.

[29] von Bonin G. Essay on the Cerebral Cortex. Springfield, IL: Charles C Thomas Publisher; 1950.

[30] Vogt C. Étude sur la myélinisation des hémisphères cérébraux. Paris: Steinheil; 1900.

[31] Vogt O. Der Wert der myelogenetischen Felder der Großhirninde. Anat Anz (Jena) 1906;29:273–87.

[32] Bartels A, Zeki S. The chronoarchitecture of the cerebral cortex. Philos Trans R Soc Lond (Biol Sci) 2005;360:733–50.

[33] Vogt C, Vogt O. Allgemeinere Ergebnisse unserer Hirnforschung. J Psychol Neurol (Leipz) 1919;25:279–461.

[34] Campbell AW. Histological studies on the localisation of cerebral function. Cambridge: University Press; 1905.

[35] Brodmann K. Vergleichende Lokalisationslehre der Großhirninde in ihren Prinzipien dargestellt auf Grund des Zellenbaues. Leipzig: J.A. Barth; 1909.

[36] von Economo C, Koskinas GN. Die Cytoarchitektonik der Hirnrinde des Erwachsenen Menschen. Textband und Atlas mit 112 Mikrophotographischen Tafeln. Wien–Berlin: Verlag von Julius Springer; 1925.

[37] Wise RJS, Greene J, Büchel C, Scott SK. Brain regions involved in articulation. Lancet 1999;353:1057–61.

[38] Kim JS. Pure dysarthria, isolated facial paresis, or dysarthria-facial paresis syndrome. Stroke 1994;25:1994–8.

[39] Nadeau SE, Crosson B. Subcortical aphasia. Brain Lang 1997;58:355–402.

[40] Simard M, Panisset M. Commentary on 'subcortical aphasia' by Stephen E. Nadeau and Bruce Crosson. Brain Lang 1997;58:418–23.

[41] Wallesh C, Johannsen H, Bartels C, Hermann M. Mechanisms of and misconceptions about subcortical aphasia. Brain Lang 1997;58:403–9.

[42] Naeser MA, Alexander MP, Helm-Estabrooks N, Levine HL, Laughlin SA, Geschwind N. Aphasia with predominantly subcortical lesion sites: description of three capsular/putaminal aphasia syndromes. Arch Neurol 1982;39:2–14.

[43] Kuljic-Obradovic DC. Subcortical aphasia: three different language disorder syndromes? Eur J Neurol 2003;10:445–8.

[44] Hillis AE, Wityk RJ, Barker PB, Beauchamp NJ, Gailloud P, Murphy K, et al. Subcortical aphasia and neglect in acute stroke: the role of cortical hypoperfusion. Brain 2002;125:1094–104.

[45] Henry RG, Berman JI, Nagarajan SS, Mukherjee P, Berger MS. Subcortical pathways serving cortical language sites: initial experience with diffusion tensor imaging fiber tracking combined with intraoperative language mapping. Neuroimage 2004;21:616–22.

[46] Huang H, Zhang J, Wakana S, Zhang W, Ren T, Richards LJ, et al. White and gray matter development in human fetal, newborn and pediatric brains. Neuroimage 2006;33:27–38.

[47] Partridge SC, Mukherjee P, Henry RG, Miller SP, Berman JI, Jin H, et al. Diffusion tensor imaging: serial quantitation of white matter tract maturity in premature newborns. Neuroimage 2004;22:1302–14.

[48] Pujol J, Soriano-Mas C, Ortiz H, Sebastián-Gallés N, Losilla JM, Deus J. Myelination of language-related areas in the developing brain. Neurology 2006;66:339–43.

Neuroscience and Biobehavioral Reviews 32 (2008) 984–1000

Contents lists available at ScienceDirect

Neuroscience and Biobehavioral Reviews

journal homepage: www.elsevier.com/locate/neubiorev

Review

Centenary of Christfried Jakob's discovery of the visceral brain: An unheeded precedence in affective neuroscience

Lazaros C. Triarhou *

Economo-Koskinas Wing for Integrative and Evolutionary Neuroscience, Department of Educational and Social Policy, University of Macedonia, 156 Egnatia Avenue, Building Z-312, 54006 Thessaloniki, Greece

ARTICLE INFO

Article history:
Received 24 November 2007
Received in revised form 25 March 2008
Accepted 26 March 2008

Keywords:
Christofredo Jakob (1866–1956)
James W. Papez (1883–1958)
Cingulate gyrus
Mamillary body
Anterior nucleus of thalamus
Hippocampus
Anatomical basis of emotion
Visceral responses
Affect
Limbic system
Circuit of Papez

ABSTRACT

The benchmark discovery of the cingulate gyrus as a brain structure receiving stimuli from muscles and viscera (proprioception and interoception) is traced to a 1907/1908 article by neuropathologist Christfried Jakob. Further, the involvement of the mamillary bodies, anterior thalamic nucleus, cingulate cortex and hippocampus in the circuitry of the emotive brain (i.e. all elements of the 1937 'circuit of Papez') was published by Jakob in his 1911 and 1913 monographs on human and comparative neuroanatomy. In those works, Jakob also described the thalamocingulate projection, commonly attributed to a 1933 study by Le Gros Clark and Boggon, and introduced the term 'visceral brain', commonly attributed to a 1949 paper by MacLean. The present article includes the first English translations of Jakob's relevant passages, which incontrovertibly document his chronological priority in discovering the visceral brain and some of its key constituent elements.

© 2008 Elsevier Ltd. All rights reserved.

Contents

1. Introduction

A surge of research during the past three decades into the neural systems that underlie emotion, affect and mood has provided the basis for advancing the new discipline of affective

---

* Tel.: +30 2310 891 387; fax: +30 2310 891 388.
E-mail address: triarhou@uom.gr.

0149-7634/$ – see front matter © 2008 Elsevier Ltd. All rights reserved.
doi:10.1016/j.neubiorev.2008.03.013

neuroscience (Eccles, 1980; Davidson and Sutton, 1995; Berthoz et al., 2002; Derbyshire, 2003; LeDoux, 2003; Panksepp, 2003; Dalgleish, 2004).

Historical forerunners to the field of affective neuroscience from the neurobehavioral perspective, prior to the era of refined neuroanatomical knowledge, include Charles Darwin, William James, Siegmund von Exner, Sigmund Freud and Israel Waynbaum (Peper and Markowitsch, 2001; Dalgleish, 2004). The modern roots of that field of research are generally attributed to the physiological studies of Bard (1928), Cannon (1929) and Hess (1932), and the anatomical studies of Papez (1937a) and MacLean (1949).

In the neurophysiology domain, Woodsworth and Sherrington (1904) set the foundations by describing a pseudoaffective reflexive state in decerebrate cats. Bard (1928) and Cannon (1929) performed transection experiments in cats at the level of the diencephalon removing the dorsal thalamus and the cortex and documented the phenomenon of 'sham rage' and autonomic hyperactivity, a nondirected behavior devoid of any integrated cortical input that would impart a subjective emotional experience (Neylan, 1995). Thus, the older James–Lange theory, which held that emotions are initiated in motor and visceral activities that feed back to the neocortex before being felt, was replaced by the 'thalamic theory' of Cannon and Bard, according to which emotions are affective states translated in the thalamus into central and peripheral signals for 'fight or flight' (Marshall and Magoun, 1998). Subsequently, Walter Hess (1932) provided the first systematic study of the effects of electrically stimulating limbic and hypothalamic areas and evoking emotional behavior in conscious animals (Eccles, 1980); for his pioneering work on the diencephalic control of the activity of internal organs, Hess was awarded the 1949 Nobel Prize in Physiology or Medicine.

In 1937, the North American neuroanatomist James W. Papez (1883–1958) elaborated on a theory that had been proposed by Dana (1921) and other researchers and which held that the experience of emotion involved an interaction between the diencephalon and the cerebral cortex (Neylan, 1995). Papez proposed a circuit of rhinencephalic and limbic pathways as being the anatomical basis of emotional experience and its expression through visceral and instinctual activities, such as those involved in feeding, mating, mothering and aggression (Papez, 1937a; DeMyer, 1988). In his paper, Papez (1937a) proposed an anatomical corticothalamic mechanism of emotion, involving, in the medial facies of the human cerebral hemisphere, the hippocampus and its connection with the mamillary body through the fornix, and the connections of the mamillary body to the anterior thalamic nuclei and the cortex of the cingulate gyrus. In a companion paper published the same year, Papez (1937b) distinguished an afferent ('externalizing') and an efferent ('internalizing') visceral system, the latter comprising two major divisions, the muscular and the autonomic apparatus. Twenty years later, in the June 1957 meeting of the Society of Biological Psychiatry, Papez (1958) elaborated on the visceral brain (which he indiscriminately identifies with the rhinencephalon or the limbic lobe), expanding on the anatomical structures and connections subserving "the basic biologic needs of the individual", to comprise additional components of the olfactory, forebrain (including nucleus basalis, claustrum, and amygdala), hippocampal, and hypothalamo–hypophysial systems. Papez's original paper was reprinted some six decades later as a classic (Papez, 1995), selected as such on the basis of being "the most widely recognized citation in neuropsychiatry" (Neylan, 1995).

The 'circuit of Papez' is a loop of projections extending from the hippocampus via the fornix to the mamillary bodies of the caudal hypothalamus; then via the mamillothalamic tract to the anterior nucleus of the thalamus; via the superior thalamic peduncle (thalamocingulate fasciculus) to the cingulate gyrus (the dorsal limb of the fornicate gyrus); the cingulate gyrus connects via the cingulate fasciculus (cingulum) with the parahippocampal gyrus and the piriform area of the temporal lobe; and from the temporal lobe via the temporoammonic (perforant) tract of Cajal to the hippocampus (Ammon's horn) (Nauta and Feirtag, 1986; DeMyer, 1988). Fibers also project from the hypothalamus to the hippocampus, the amygdala and the mesencephalic reticular formation (Nauta and Feirtag, 1986); feed-in pathways to the circuit include projections from the septal and olfactory regions and amygdala (DeMyer, 1988).

A fact, little known in the English biomedical literature, is that the discovery of the 'visceral brain' can be traced to a 1907/1908 article published by neuropathologist Christfried Jakob (1866–1956) in Buenos Aires (Jakob, 1907/1908; Orlando, 1964). In that paper, entitled 'Localization of the soul and intelligence', one finds the first designation of the supracallosal (cingulate) gyrus as a brain structure that receives stimuli from muscles and viscera (proprioception and interoception in modern terms, respectively).

A native of Bavaria, Jakob obtained his medical degree from the University of Erlangen. He subsequently worked as assistant to Adolf von Strümpell (1853–1925) at the Erlangen Medical Clinic, and also practised medicine in Bamberg. Having made a name for himself in Europe through a successful atlas of neuroanatomy (Jakob, 1899), Jakob accepted an offer and moved to Buenos Aires in 1899. Save for an intermission between 1910 and 1912, he spent the rest of his life in Argentina, where 'Christofredo' Jakob is considered the founder of the national school of neurobiology (Moyano, 1957; Orlando, 1966; Meyer, 1981). Jakob's life and work have remained altogether unheeded in the English biomedical literature until recently (Triarhou and del Cerro, 2006a, b).

A number of authors in the Argentinian (Goldar, 1975, 1997; Meyer, 1981; Faccio, 1991; Crocco, 1994; Szirko, 1995; Piva and Virasoro, 2004; Pelliza, 2006) and in the Castillian literature (Barraquer-Bordas, 1976) have put forward arguments in terms of Jakob's precedence in proposing a cerebral circuitry that subserves emotion, and also in making the benchmark discovery of the 'visceral brain' and its association with the cingulate gyrus.

Jakob's priority in presenting for the first time a scientific exposition to explain the cerebral regulation of visceral phenomena was minutely documented in a 1964 article, published in Argentina by Jacinto Carlos Orlando (1921–2001), Chief of Neurology Service at the National Neuropsychiatric Hospital of Buenos Aires and an alumnus of Jakob's successor Braulio Moyano at the Laboratory of Pathological Anatomy (Orlando, 1964; Ure, 2001). Orlando (1964) exposed succinct arguments favoring the chronological and conceptual priority of Jakob over Papez regarding the formulation of the anatomical basis of the visceral brain. Jakob realized the first interpretation of the limbic or 'internal' brain as a viscero-emotional mechanism, including the connections of the anterior nucleus of the thalamus with the cingulate cortex in the neural circuit that subserves instinctive and emotional life. The mamillary body is the one that establishes the union of the hippocampal impulses with those that arrive at the diencephalon via the mamillary peduncle that carries visceral information, according to the old thesis of Jakob and Onelli (1911, 1913), which Goldar (1975) later backed with his own anatomical studies.

The entire circuitry encompassing the mamillary bodies, the anterior nucleus of the thalamus, the cingulate gyrus and the hippocampus, as well as the projection pathways that connect them, the components of the so-called 'circuit of Papez', was foreshadowed by Jakob decades before Papez, based on degeneration experiments in apes and dogs, and on human clinicopathological correlations in neurodegenerative and inflammatory

Fig. 1. Period leaflet (*left*) describing Jakob's two 1911 German monographs (Jakob and Onelli, 1911; Jakob, 1911). Frontispiece (*right*) of the monograph published two years later in Buenos Aires (Jakob and Onelli, 1913), with the author's inscribed dedication to Dr. Rodolfo Rivarola (1857–1942), Chairman of Ethics and Dean of the University of Buenos Aires School of Philosophy, and President of the National University of La Plata (private archive).

diseases such as senile and paralytic dementia (Jakob, 1911; Jakob and Onelli, 1911, 1913; Ranke, 1911; Siemerling, 1911). Thus, the "remarkable achievement" of Papez's paper, "allegedly written in just a few days" (Dalgleish, 2004), should actually be credited to Christfried Jakob.

Jakob discovered and published in 1911 the neurovisceral apparatus and its central connections, a most important cerebral macrocircuit that he taught in his classes, appearing in print for the first time in 1907/1908 (Jakob, 1907/1908). In 1909, he wrote that the limbic cortex represents an ano-vesico-genital center, inferred from human and comparative anatomical, embryological and neuropathological studies, progressive general paresis in particular (Jakob, 1909). The following year Jakob published two papers of relevance (Orlando, 1964). In one article, entitled 'The cortical cell in madness' (Jakob, 1910a), he wrote that "the great superior limbic (supracallosal) gyrus is connected via afferent pathways that derive from our body (visceral sensations: thoraco-abdomino-pelvic)" and mentions the division of the cerebral cortex into two regions, one lateral, receiving stimuli from the outside world, and one medial, at which arrive stimuli from visceral musculature and the sympathetic nervous system. In the second article entitled 'Motor aphasia and its localization', Jakob (1910b) reported on his Nissl studies of the thalamic nuclei in cases of aphasia, subsequently to his previous interest in Broca's area and the ontogeny of language-related pathways in the human brain (Jakob, 1906a, b; Vivas et al., 2007; Tsapkini et al., 2008).

In 1910, Jakob presented his ideas at the International Scientific Congress in Buenos Aires (Jakob and Onelli, 1910; Triarhou and del Cerro, 2006a), from which his three large atlases ensued, namely, *From Animal Brain to Human Brain* (Jakob and Onelli, 1911) subtitled 'Comparative morphological, histological and biological

studies on the development of the cerebral hemispheres and their cortex'; *The Human Brain* (Jakob, 1911) subtitled 'A study on the structure and meaning of gray matter nuclei and cerebral cortex of the human brain'; and the *Atlas of the Brain of the Mammals of Argentina* (Jakob and Onelli, 1913) subtitled 'Comparative anatomical, histological and biological studies on the evolution of the hemispheres and cerebral cortex' (Fig. 1).

The aim of the present article is to expose the relevant passages from Jakob's work, rendered into English – to my knowledge for the first time – based on the original German (Jakob, 1911; Jakob and Onelli, 1911) and Spanish (Jakob and Onelli, 1913) sources. The passages form the evidence that documents Jakob's priority in discovering the visceral brain and some of its key anatomical components. These publications have not been widely accessible for many decades. Often times, frequent citation of a work is taken as an 'objective' measure of influence in a field; such a criterion, nonetheless, discounts bibliographic omissions, especially with articles written in languages other than English or not readily accessible through the current digital databases or literature search engines. In the translation that follows, Jakob's terms are retained as they appear in the original, including the italics that he had used for emphasis. Editorial explanatory comments are positioned in brackets.

A note on anatomical terminology: Jakob depicts the cingulate gyrus under various terms, e.g. gyrus fornicatus (Fig. 2) in his 1899 German atlas (Jakob, 1899), callosomarginal, splenial and superior limbic gyrus in his 1911 and 1913 atlases (Jakob, 1911; Jakob and Onelli, 1911, 1913), gyrus cinguli in the 1926 Strümpell-Jakob *Icones Neurologicae* (Müller and Spatz, 1926), and supracallosal gyrus (Fig. 3) in his 1939 *Folia Neurobiológica* atlas (Jakob, 1939). He explains that "in the central portion of the median hemispheric

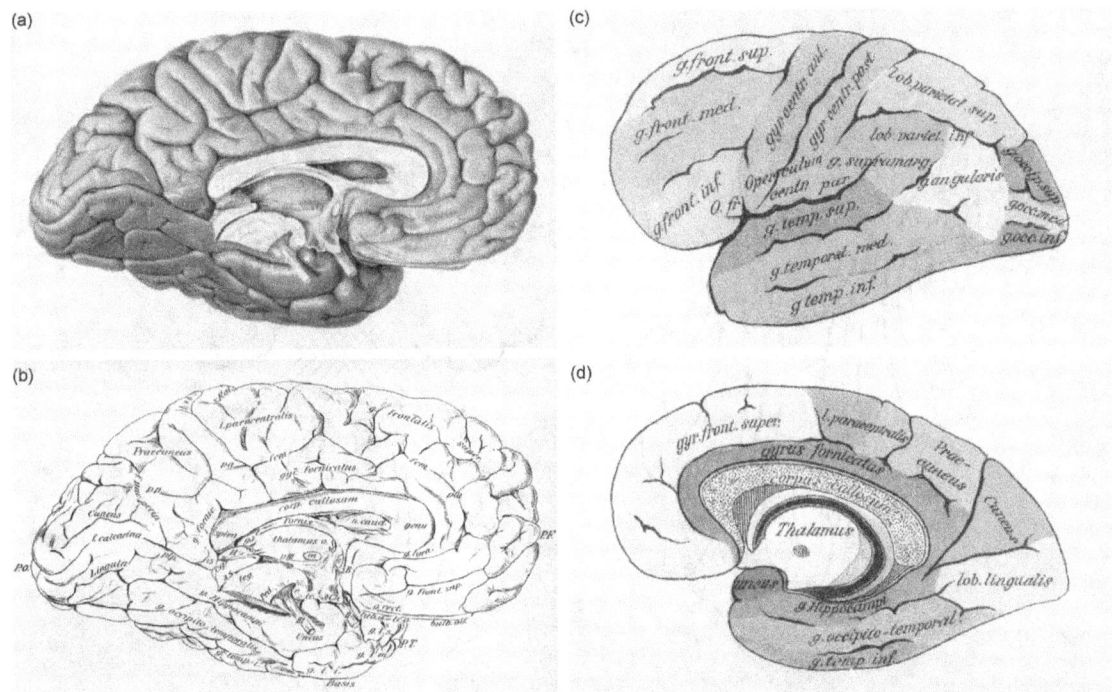

Fig. 2. Lithographs depicting the topographic anatomy of area structures in the median (a, b and d) and lateral (c) facies of the human brain, from plates 4 (a and b) and 21 (c and d) in Jakob's 1899 Atlas of the Nervous System (Jakob, 1899). According to Jakob, the 'lower marginal gyrus' corresponds to the hippocampus, and the 'upper marginal gyrus' to the fornicate gyrus, bounded above by the callosomarginal sulcus (*fcm*). *Abbreviations: g.*, gyrus; *l.*, lobe; *s.*, sulcus; *a.*, anterior, and *m.*, middle commissure; *c.*, mamillary body; *v.III*, third ventricle; *P.*, pole; *is.*, isthmus of fornicate gyrus; *pla.*, anterior, and *plp.*, posterior fold of fornicate gyrus; *pa.*, anterior, and *pp.*, posterior transition gyrus from fornicate gyrus to the precuneus.

surface the cortex ends above as the fornicate gyrus and below as the hippocampal gyrus" (Jakob, 1899). Moving from reptiles to higher mammals, the splenial zone is identified with the superior limbic gyrus (Jakob and Onelli, 1913; Orlando, 1964).

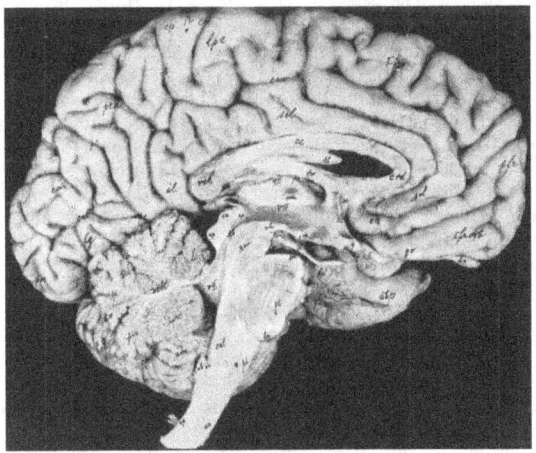

Fig. 3. Sagittal view of the median facies of the left side of a human brain denoting the topographic anatomy of area structures, from plate 33 in Jakob's Atlas of the Human Brain (Jakob, 1939). Photograph reduced to 45% of the original. *Abbreviations: scl*, supracallosal gyrus; *cc*, corpus callosum; *cal*, calcarine sulcus; *il*, limbic isthmus; *crd*, genu ('rodilla') of corpus callosum; *rcl*, splenium ('rodete') of corpus callosum; *tr*, fornix (cerebral trigonum); *cm*, callosomarginal sulcus (upper marking); *cm*, mamillary body (lower marking); *vIII*, third ventricle.

2. Evidence: Jakob's original passages

2.1. Organ projection (Jakob and Onelli, 1911)

Between the splenial sulcus and Ammon's horn lies the 'visceral cortex' (the oldest part of the neopallium); similar to Ammon's formation, it is located on the median surface. Between the splenial and ectomarginal sulci is the dome of the hemispheres, which contains the centers for the trunk, spinal column (axial formation) and hindlimbs. Thus, the remaining of the lateral hemispheric wall is left for the centers for the forelimb and the branchiogenous head organs (jaw, tongue, lips, etc.). Here, develop and appear later the suprasylvian (*ssl*) and possibly the ectosylvian (*ecsl*) sulci. Thus we see that the gyral partitions, initially appearing in the centroparietal part of the hemispheres, are to be fully understood essentially as a result of the projection of the body surface on the cortex of the neopallium, although one cannot deny that still other factors might be involved. Fig. 4 shows the *systemic correspondence* of such a projection, which clarifies in the most natural way the location of the cortical centers and their sagittal segmentation; the projection is also distributed in the sagittal direction (in the case of tetrapods, and exactly the same way in the case of humans as a subsequence of the axis alignment in the vertical direction). We may thus imagine that these longitudinally-coursing segmental zones of the cortical mantle have become differentiated segmentally: first Ammon's formation, then the visceral zone, the trunk-hindlimb zone, the forelimb zone, and finally the facio-mandibulo-lingual zone, while between each segment an intermission seems to occur; such a force seems to correspond morphologically to an arrest of the development of the cortical elements (sulcus floor), while the

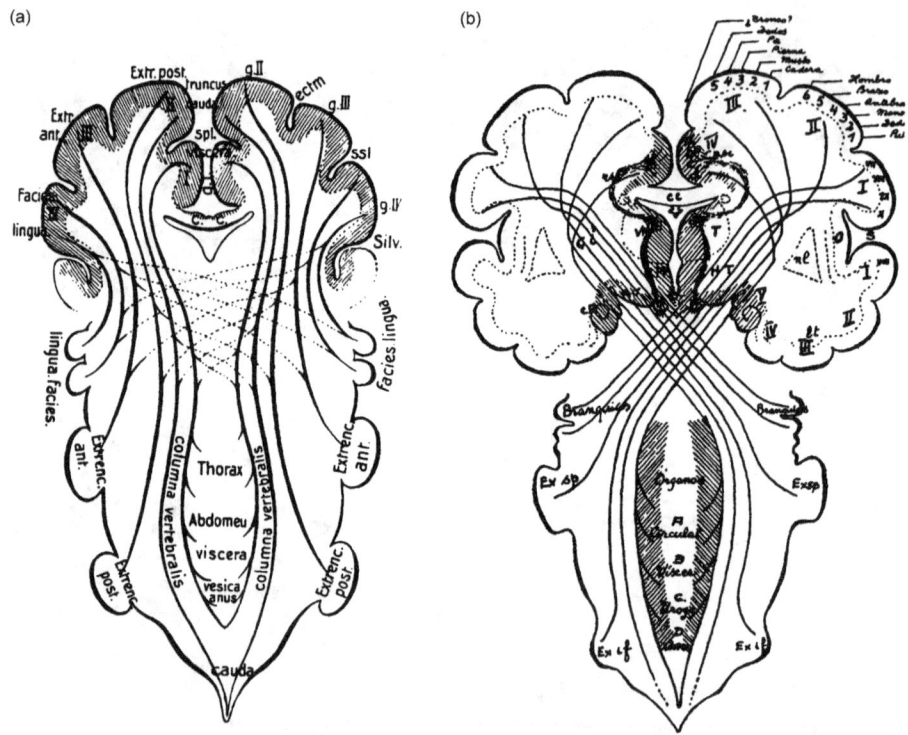

Fig. 4. (a) Scheme of the trunco-cortical projections from the viscera to the cingulate cortex, and from the extremities to the lateral cortical areas; originally Fig. 22 in Jakob and Onelli (1911), Fig. 51 in Jakob (1911) and unnumbered figure on page 21 in Jakob and Onelli (1913). (b) An updated version, originally Fig. 13 in Jakob's later article on the philosophical significance of the human brain (Jakob, 1945). Sympathetic system projections carrying endogenous signals occupy the medial aspect of the human cerebral hemisphere (oblique hatching) and somatic capsular projections (*ci*) carrying exogenous signals the lateral aspect contralaterally.

formation period should actually find its morphological expression in the protruding arch that depends on the cellular growth of the involved cortical part (thus the gyral formation). Even if the actual formation is not so schematic, it is in any event thought that there is certainly at least one of the formative forces important for the formation of the mechanism of the pre-gyrus system. This observation, once occurring in the central hemispheric part, should correlatively apply to the rest as well (frontal and temporal). Thus, we arrive at the acceptance of five sagittal *pre-gyri*, running from front to rear (Figs. 5 and 6); however, if we leave aside Ammon's formation as an earlier branched segment, we end up with the *four pre-gyri of the neopallium*, which begin from the oldest segment – the one that lies above the corpus callosum on the medial hemispheric wall, which I already described as 'visceral cortex' – and descend outwards. We thus discern that the '*visceral cortex*' exists in the callosomarginal gyrus (fornicate, limbic, splenial gyrus) or cortical pre-gyrus I (*g I*) between the corpus callosum (*cc*) together with Ammon's formation (= stria of Lancisi) and the splenial (*spl*) sulcus; as the *body axis–hindlimb zone* in pre-gyrus II (*g II*) between the splenial and ectomarginal (*ectm*) sulci; as the *forelimb zone* in pre-gyrus III (*g III*) between the ectomarginal and suprasylvian (*ssl*) sulci; and finally as the *face, mouth, tongue, etc. zones* in pre-gyrus IV (*g IV*) between the suprasylvian and marginal sulci (marginal or rhinal fissure). A cortical portion remains in all animal brains at the bottom of the marginal sulcus, which forms the *insular area* of higher mammals. From here (we may call it marginal cortex) all cortical segments begin their stepwise formation; that is why such insular cortex also forms morphologically the center of the hemispheric cortex; it represents

the transition between the cortex and the basal rhinencephalic zone.

2.2. Systems of projection (Jakob and Onelli, 1913)

Cortical elements necessitate for their adequate function a dual system of conducting fibers with lower nervous centers (Fig. 7), and by mediation of these with the periphery of the organism; in the afferent pathways enter sensitive, sensory and muscular excitations from the epidermis, from subcutaneous, muscular, periosteal and tendinous tissues, such as from mucosae, glands and serosae in a corticopetal direction towards the various segments and sectors of the cortex, in such a manner that each sector and segment are in a determined relationship with their corresponding peripheral organs, which thus constitute a true projection of the organism on the cortical mantle, and not, as generally thought, of its surface only, but equally well of all its internal regions. Therefore, *exogenous excitations* arrive from the external world (received by the organs of the senses and the epidermis, etc.) and, in a par, *endogenous excitations* of the internal world (of muscles and viscera) arrive from their respective pertinent zones, thus forming the so-called 'cortical and sensory centers'.

In an opposite direction to those afferent systems run the *corticofugal or motor projection* systems, and establish with them connections between determined cortical zones and the musculoskeletal, glandular and visceromotor systems.

By means of these efferent pathways of communication, the cortical zone projects, in the reverse of what is established further anterior, to the motor peripheral zones of the organism, from

(a)   (b)

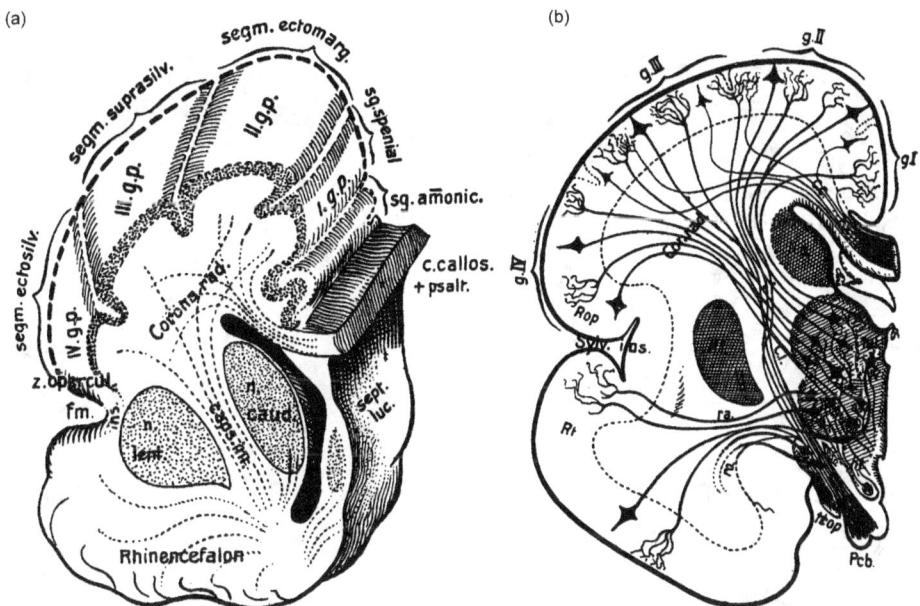

Fig. 5. Schematic drawing depicting the splenial segment (*sg. splenial*) of the cerebral cortex, the first pre-gyrus (*gI*) to be formed proceeding mediolaterally from the Ammonic segment (*sg. amonic.*). (a) Depiction of the segmentation of the cerebral hemisphere; originally Fig. 24b in Jakob and Onelli (1911), Fig. 26 in Jakob (1911), and unnumbered figure on page 17 in Jakob and Onelli (1913). (b) Projection pathways of the hemisphere of a mammal in a coronal section; originally Fig. 27 in Jakob and Onelli (1911) and unnumbered figure on page 20 in Jakob and Onelli (1913).

which they transmit their executive stimuli. In the phylogenetic evolution of these systems, there are fixed laws that establish the order of appearance of projection systems, e.g. their extent and location. The olfactory afferent and efferent pathways invariably develop and mature in the first place towards Ammon's segment (olfactory radiation, olfactorio-Ammonic and the efferent system called cerebral trigonum [fornix]); later develop the afferent and efferent visceral pathways (Fig. 4) that are subserved, above all, by the sympathetic system, projecting incoming stimuli of our internal organs to the splenial segment (primordial segment II). These two projection systems: (1) the *Ammonic* and (2) the *splenial*, already exist consistently in lowest cortical vertebrates (the gymnophiona, lower reptiles [actually amphibians, later corrected by Jakob (1918) to *Amphisbaena darwini*, a lower reptile (Outes, 2006)], whereas in higher vertebrates, little by little the following are added: (3) the *cutaneo-muscular* or *kinesthetic* system, (4) the *optic [visual]* system and (5) the *acoustic [auditory]* system (Figs. 5 and 6).

2.3. Splenial formation (Jakob, 1911)

This segment comprises the cortex lying directly upon the corpus callosum and outlining the fornicate gyrus (supracallosal, superior limbic), which towards the rear bends in the isthmus to the hippocampal cortex; the outermost edge of the hippocampal gyrus probably belongs to the splenial segment. The splenial cortex is the second oldest cortical formation of the hemispheric mantle after the Ammonic cortex. In the phylogenetic line it represents as interhemispheric cortex the 'visceral cortex', which is especially connected with the sympathetic system and the hypothalamus. '*Rhinencephalic cortex and visceral cortex*' are thus the oldest cortical zones, to which cutaneo-muscular, visual, auditory, etc. dual centers are further connected. The relationship of this zone to all viscera sensations (respiration, digestion, excretions, sexual

apparatus, etc.) is also maintained in humans and primates, as I am in a position to prove, based on my clinicoanatomical observations in humans and on my experimental observations in apes. This also fits closely in the general layout of the order of organ projection upon the cortex (Fig. 4). The cortex of the splenial formation is typically arranged in two layers, according to the scheme of a 'full cortex'. Outer and inner fundamental layers, granular layer and substrata are present; however, the inner ground layer clearly outweighs the outer here, as also larger pyramids are far more markedly and numerously structured in the inner layer than in the external fundamental stratum. Both ground layers become diminished towards the corpus callosum, their lower lamina disappearing; both contribute to the formation of the lateral induseum.

Fiber development is essentially in accord with that in the frontal cortex; the certain peculiarities (in the tangential stratum) need not be related here.

How do the pathways of origin and termination fare in this circumstance? Until now, all this was completely unclear; my personal studies place me in a position to offer a fully substantial clarification on the issue. Sensory pathways – transmitting from low centers and ending in the substrata of the outer layer and the granular layer of the splenial cortex – originate, just like in the case of all other cortical sectors, from the thalamus, and, as a matter of fact, mainly from its anterior nucleus. This nucleus receives, as we saw above, the transmitting bundle of Vicq d'Azyr from the mamillary bodies [mamillothalamic tract]. And in those terminate on their part the ascending sensory cap pathways, which are usually designated as the peduncles of the mamillary bodies. The general opinion until recently was that such pathways coursed downwards; however, I was able to prove already in 1902 (Jakob, 1902) that, in a case of gummatous infiltration of the upper pons and lower forebrain shank intermediate area, where both peduncles of the mamillary bodies were totally destroyed, the

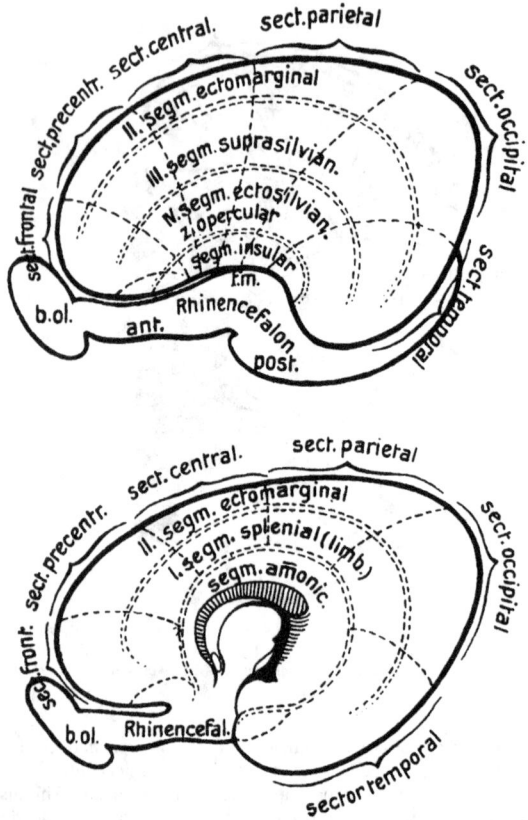

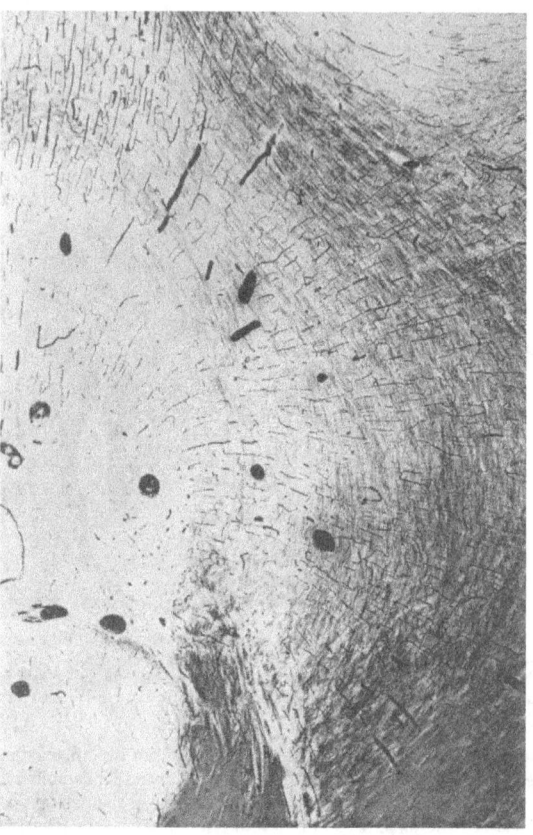

Fig. 6. General plan of the hemispheric sector formation; originally Fig. 24a in Jakob and Onelli (1911), Fig. 25 in Jakob (1911), and unnumbered figure on page 18 in Jakob and Onelli (1913).

Fig. 8. Histological photomicrograph showing the corona radiata in a newborn human brain at the level of the central sector. Lateral radiations to the Rolandic sulci, medial fasciculus paracaudatus (fronto-occipitalis) and its radiation (somewhat myelinated) towards the splenial sulcus. Magnification ×45; frame 1 of plate 62 in Jakob (1911).

cells of the mamillary bodies did not exhibit any retrograde change, although the peduncles appeared to have had undergone a complete secondary degeneration. The histological findings of Ramón y Cajal (1903) and von Kölliker (1902), obtained in animals,

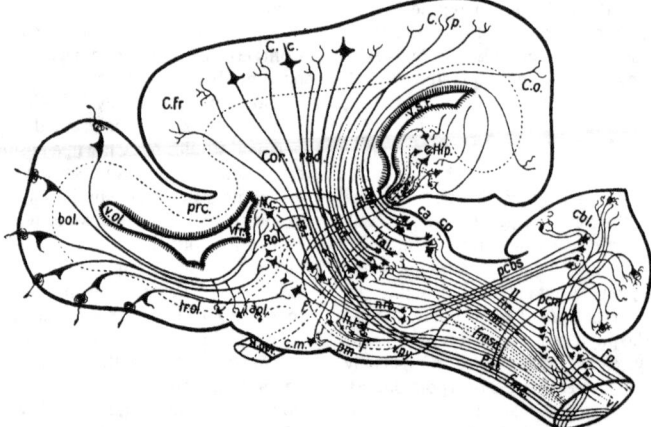

Fig. 7. Reciprocal projection pathways in the mammalian central nervous system (sagittal view); originally Fig. 26 in Jakob and Onelli (1911), and unnumbered figure on page 20 in Jakob and Onelli (1913).

on the origin of the bundle of Vicq d'Azyr, etc. are in perfect agreement as well. I then tested these conditions experimentally in the ape and obtained consistent results that all showed the entire system mamillary bodies – anterior nucleus of thalamus – splenial cortex, which incorporates the visceral sensations (mucosa and musculature) deriving from sympathetic spinal and bulbar column pathways, to be strengthened and relayed and forwarded. *The splenial cortex forms the hitherto unknown, visceral cortical center* (Fig. 4). My clinical observations in human cases (paralytic and senile dementia in particular) lead to the same conclusions, as also do phylogenetic, histological, experimental and clinical results, which support the existence and location of such a center.

The course of the thalamosplenial radiation [thalamocingulate fasciculus or superior thalamic peduncle], which emanates from the anterior nucleus of the thalamus, is peculiar. It forms a component of the fronto-occipital fasciculus (Fig. 8) falsely described as an association bundle by Déjerine (1895) (which is thus a bundle of the corona radiata); since the anterior nucleus is especially developed only in the anterior one-third of the thalamus, the part of the thalamocortical path destined for posterior splenial sections is bent in bundles caudally, and runs for a while backwards forming the fasciculus paracaudatus, to finally radiate into the distal segment of the fornicate gyrus. Besides association bundles, to which mainly belongs the cingulum, descending centrifugal pathways (spleniohypothalamic fasciculus) also emanate from the cortex; such pathways likewise descend from the fasciculus paracaudatus (ex-fronto-occipital) and end in hypothalamic nuclear areas (reticular nucleus, hypothalamic nucleus, etc.); their relationship with the sympathetic nuclei was already delineated above. It still remains to be determined whether any portion (motor pathways, etc.) runs differently, directly (via the fornix longus) to the spinal sympathetic columns.

2.4. Splenial zone (Jakob and Onelli, 1913)

All sectors studied until now derive from the portion of the cortical mantle that I have designated in the first part of this work as the *lateral formation*, and to such a formation belongs also the splenial zone, which represents the phylogenetically oldest region insofar as maturation of the lateral cortex is concerned. All sectors,

plus the splenial zone, form what authors call the *neopallium*. The splenial zone, as we have seen, lags in topography and chronology immediately behind the Ammonic region, and in my impression, it would be more correct to combine these two zones into the archipallium, and the other connected sectors as the neopallium, above all when compared with what I said earlier on the chronology of the phylogenetic evolution of the cortical centers; moreover, it is not necessary to attribute to questions of classification and nomenclature more importance than they deserve.

In the mammalian series, the splenial zone occupies a cortical region of great consistency in its outer and inner morphology. It is located in all of them in the median facies, forming, in mammals that lack a corpus callosum [*Lyencephala* according to Richard Owen (1857); for a description of the four subclasses of mammals suggested by Owen, cf. Striedter (2005)], the dorsal limiting cingular region of the median hemispheric side (dorsal marginal gyrus); in mammals that possess a corpus callosum (rodents, carnivores, etc.), it is located directly over that commissural system and it girdles it completely in the form of a semiarch. In higher mammals (lemurians, primates) and in humans, it is identified as the supracallosal or superior limbic gyrus (fornicate gyrus). It forms part of the frontal, central and parieto-occipital region, and it is subdivided in various subsectors that correspond to this topographic relationship.

The function of this vast cortical zone has been, until now, completely obscure. My personal studies (see on this issue the cited work *Das Menschenhirn* (Jakob, 1911)) remedy such a gap. The splenial zone receives, as I can state, its afferent pathways in its outer fundamental layer, from the anterior thalamic nucleus, and to that subcortical center arrive, through the mediation of the bundle of Vicq d'Azyr [mamillothalamic tract], the *visceral sensations at the mamillary ganglion*. All internal organs of the pelvis, abdomen and thorax emit the excitations of their mucosae, glands and muscular layers by means of this afferent system, common in the entire mammalian series all the way to humans, towards the splenial zone, and from its inner stratum emanate motor efferent pathways, which, through the mediation of the hypothalamus, are directed in a centrifugal fashion again to the visceral musculature and mucosa (Fig. 4). There exists, in addition, an associative apparatus (the cingulum) important among the

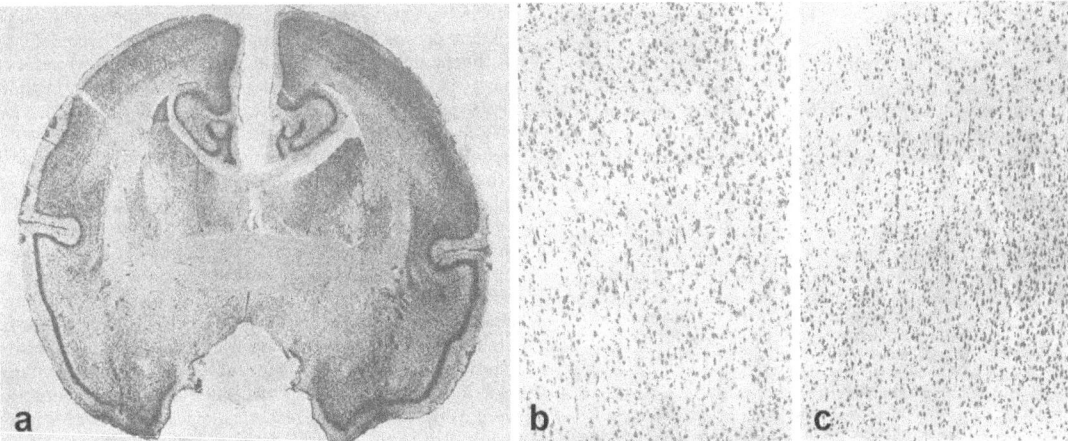

Fig. 9. Histological photomicrographs showing (a) the cerebral cortex of Didelphis ('comadreja') in a coronal section. Magnification ×22; frame 1 of plate 4 in Jakob and Onelli (1911, 1913). (b and c) The middle and anterior splenial cortical transition types (supracallosal gyrus) of the human brain. Magnification ×90; frames 2 and 3 of plate 86 in Jakob (1911).

various subsectors of this vast zone which, probably, extends towards its morphologic continuation: the inferior limbic gyrus or hippocampus with Ammon's formation.

The visceral center is furthermore in an associative relationship with all the other sectors. The visceral splenial center, according to its importance for the conservation of the individual, is already well developed in lower mammals and becomes differentiated in proportion to the general advancement of an organism in the [phylogenetic] series. Visceral sensations, in their different components, contribute with a most important contingent to the affective atmosphere and temperament of the individual. Here reside 'hunger and love', and from the splenial zone they emit their categorical imperatives[1] from the opossum to human (Fig. 9).

3. Discussion

The outlined text from Jakob's original sources covers the development, sectorization, formation and topography of projections, and the phylogenetic roots and correlates of the visceral brain. Jakob's methods comprised experimental operations in apes and dogs and histopathological analyses of human brains, using mainly the Weigert and Nissl stains to study myelination and retrograde degeneration, coupled with his deep understanding of the ontogeny and phylogeny of the nervous system.

Jakob's original contributions had been acknowledged by critics at the time of publication of his two major German atlases (Ranke, 1911; Siemerling, 1911). Ranke (1911) credits Jakob and Onelli (1911) for a far-sighted work that presumes the callosomarginal gyrus (=fornicate, limbic, splenial gyrus) as being the 'visceral cortex' and Jakob (1911) for placing special emphasis on the diencephalon, a detailed description of the thalamus and its cortical relations, and for presenting numerous new views on the morphology and physiology of this area and its connections with the cortex that depart from previous opinions (e.g. the connection between the mamillary bodies and the anterior nucleus of the thalamus and the fornicate gyrus as a central control for 'visceral stimuli', cf. 'first sagittal pre-segment'), based on his clinicopathological evidence and the experimental data from operations in apes and dogs (Ranke, 1911).

Siemerling (1911) credits Jakob and Onelli (1911) for the idea that the supracallosal gyrus is associated with a visceral cortical center. Here exist the feelings from visceral organs, especially insofar as they are associated with food intake, digestion, defecation and the sexual organs. Here have their highest central location functions directly connected with mating and the conservation of the individual (Siemerling, 1911).

Based on clinical and neuropathological observations in cases of senile and paralytic dementia, as well as experimental evidence from retrograde degeneration and comparative anatomical and phylo-ontogenetic studies, Jakob had concluded as early as 1911 (Jakob, 1911; Jakob and Onelli, 1911) that the superior limbic (supracallosal or cingulate) gyrus is linked to afferent pathways that convey visceral thoraco-abdomino-pelvic sensations of the

body and subserve the internal feelings related to emotion. He wrote that "the limbic cortex [cingulate gyrus] constitutes the hitherto unknown visceral cortical center" and pinpointed to the involvement of the mamillary peduncles, the mamillary bodies, the mamillothalamic bundle of Vicq d'Azyr (1786), the thalamus, the hypothalamus and the fornix. The splenial zone of the reptilian brain, which corresponds to the superior limbic gyrus in lower mammals to primates, conveys visceral sensations from the mamillary bodies via the mamillothalamic bundle to the anterior nucleus of the thalamus.

It is evident from the above texts that Jakob takes precedence for discovering the functional and evolutionary significance of the cingulate gyrus as a visceral center and for investigating some of the anatomical components of the mechanism of emotion and the thalamocingulate projection.

3.1. Functional and evolutionary significance of the cingulate gyrus as a visceral center

Papez (1937a) accepted a visceral emotional role for cingulate function. Herrick (1933) had proposed a forebrain regulator of affect based on cortical pathways strongly linked to olfactory centers. However, Jakob had propounded the termination of visceral signals to the cingulate gyrus as early as 1907/1908 (Jakob, 1907/1908). The credit for the suggestion that the cingulate cortex is a key substrate of the conscious experience of emotion (Papez, 1937a, b; Lane et al., 1998) and of the central representation of autonomic arousal (Critchley et al., 2000; Dalgleish, 2004) belongs to Jakob (1907/1908, 1911; Jakob and Onelli, 1911, 1913).

The notion of a visceral brain has meaning as an evolutionary concept (cf. Section 3.6, below). Somatic and visceral responses from the cingulate gyrus were reported 50 years ago (Showers and Crosby, 1958), but the visceral afferents and efferents of the anterior insular and cingulate cortices were identified more recently (Krushel and van der Kooy, 1988; Allen et al., 1991; Hanamori et al., 1998; Ito, 1998; Bagaev and Aleksandrov, 2006; Sikes et al., 2008). In the past decade, it has been reported that functional neuroimaging studies in healthy controls and in patients with neurological or psychiatric diseases have begun to provide a factual basis for the role of individual limbic and cortical regions as neural substrates of emotion (Lee et al., 2004).

The designation of the superior limbic lobe as a distinct functional entity concerned primarily with the regulation of the visceral organs and the elaboration of affective behavior, as well as the naming of the forebrain emotional circuits 'visceral brain' has been credited for half a century by specialists in the field (Fulton, 1953; LeDoux, 2003) to MacLean (1949) – who, like Jakob, calls the cingulate gyrus 'fornicate gyrus'. MacLean (1949, 1955) upholds the choice of the term *visceral* on psychiatric parlance grounds and "its original 16th century meaning, because of the inferred primacy of the role of the limbic system in emotional behavior". MacLean (1954, 1955) subsequently reintroduced Broca's term 'limbic' to describe a neuroanatomical system mediating emotional behavior (Lambert, 2003). MacLean (1955) noted that the limbic system (including the superior limbic lobe and the hippocampal formation), a common denominator in the brains of all mammals, is also a common denominator – in physiological terms – of a variety of viscerosomatic and emotional reactions. Based on the exposed evidence in the present article, it becomes obvious that Jakob chronologically predated MacLean in using the term 'visceral brain' in his Spanish and German publications from the period 1907–1913 to denote the functional attributes of the cingulate cortex or superior limbic gyrus (cf. also Orlando, 1964).

---

[1] A philosophy savant, Jakob borrows the Kantian key concept of the 'categorical imperative' – the simplicity and profundity of which he exalted in a subsequent discourse (Jakob, 1925) – to convey the emotive drives of the individual (Jakob and Onelli, 1913) or, as Papez (1958) would describe them decades later as "the basic biologic needs of the individual". In the metaphysical system of Kant (1948), imperatives express the ethical laws that actions should follow. They are of two kinds: the technical *hypothetical imperative* presupposes an end and compels actions in a given circumstance; the absolute *categorical imperative* is a formal statement that lays claim to universal validity (Jaspers, 1962). The categorical imperative is the ultimate commandment of reason, from which all duties and obligations derive. It is of interest to note that, while Kant reserves the term for humans (whom he considers as the only rational beings), Jakob extends its use throughout the phylogenetic spectrum.

On p. 344 of his 1949 paper, MacLean (1949) writes:

"In primitive forms the visceral brain provides the highest correlation center for ordering the affective behaviour of the animal in such basic drives as obtaining and assimilating food, fleeing from or orally disposing of an enemy, reproducing, and so forth ... The visceral brain continues to subserve such functions in higher forms, including man. Some of the neuroanatomy of the visceral brain that may have to do with the correlation of feeding and sexual activities and their bearing on affective states will be dealt with presently in more detail ...".

In a remarkably convergent – but temporally disjoined – reasoning, Jakob and Onelli (1913) propose:

"The visceral splenial center, according to its importance for the conservation of the individual, is already well developed in lower mammals and becomes differentiated in proportion to the general advancement of an organism in the [phylogenetic] series. Visceral sensations, in their different components, contribute with a most important contingent to the affective atmosphere and the temperament of the individual. Here reside 'hunger and love', and from the splenial zone they emit their categorical imperatives from the opossum to human."

Jakob suggested that the supracallosal gyrus is associated with the 'visceral cortical center' (Jakob, 1911; Jakob and Onelli, 1911): here exist and have their highest central location the feelings from the visceral organs, especially in association with food intake, digestion, defecation and the sexual organs, that is, functions directly connected with the preservation of the individual and mating. By concluding that "from the opossum to humans, hunger and love reside in the limbic cortex, and from there they emit categorical imperatives that form individual temperament and affection", Jakob and Onelli (1913) conclusively couple the temporal rhinencephalon with emotional and affective behavior. Hunger and love, or feeding and mating, are regarded by modern neurobiology as the two most important animal expressions accommodated by visceral brains, standing out from the entire behavioral repertoire, other behaviors having their significance as preludes or consequences of these two fundamental activities (Shepherd, 1983). Thus, 'love' is no longer ignored by neuroscience, and there is evidence that the brain's opioids facilitate social bonding in several species (Panksepp, 2007).

In modern neuroanatomy, the fornicate gyrus is viewed as consisting of the subcallosal gyrus, the cingulate gyrus, the retrosplenial area, the hippocampal gyrus and the uncus (Nauta and Feirtag, 1986). The cingulate region forms one of the nine defined cortical regions by Brodmann; based on cytoarchitectonic criteria, Brodmann distinguished six areas, numbers 23–25 and 13–33 (Brodmann, 1909). According to von Economo and Koskinas (1925, 2008), the superior limbic lobe is one of seven defined cortical lobes; these investigators distinguished in it 13 cortical modification areas, namely, LA_1 [EK 36] through LF_2 [EK 48] (Triarhou, 2007). Further, von Economo (1927) provided the first subregional map of the posterior cingulate gyrus.

At the dome of area LA_2 and especially at the inner lip of the dome and the transition to area LA_3, some of the cells of layer Vb are strikingly elongated, being 'corkscrew-like' and stretched like 'spindles'. These are the rod (*Stäbchen*) and corkscrew cells (*Korkzieherzellen*) that von Economo and Koskinas (1925) and von Economo (1926, 1927) described in the anterior cingulate and transverse insular gyri, as well as in area FJ of the frontal lobe. Ngowyang (1932) appears to be the first author to call these spindle-shaped, fusiform neurons in the inferior ganglionic layer (Vb) of the dome of the transverse insular gyrus 'von Economo cells'. The name 'von Economo neurons' is being currently used for these large bipolar cells in the anterior cingulate and frontoinsular cortices, distinct from pyramidal cells, that are possibly involved in the fast intuitive assessment of complex situations (Nimchinsky et al., 1999; Allman et al., 2005; Watson et al., 2006). These neurons are a recent phylogenetic acquisition; they have only been found in the brain of humans, apes, and whales (Hof and Van Der Gucht, 2007) and appear to be vulnerable to degeneration in frontotemporal dementia, a neurodegenerative disorder that features an early injury to anterior cingulate and frontoinsular (i.e. paralimbic) cortices (Seeley et al., 2006).

3.2. Anatomical structures in the mechanism of emotion

The four components of the 'circuit of Papez' and their interconnections are all described by Jakob in his atlases (Jakob, 1911; Jakob and Onelli, 1911, 1913). The cingulate cortex as a visceral zone had been described even earlier (Jakob, 1907/1908, 1909).

The individual elements of the 'circuit of Papez' had been characterized before Papez (Pribram and Kruger, 1954; Meyer, 1971; Lautin, 2001). In examining "the rather sudden appearance of Papez's paper", Lautin (2001) goes over a discussion of four areas or prior developments in brain research that provided the underpinnings of Papez's circuit proposal and virtually set the

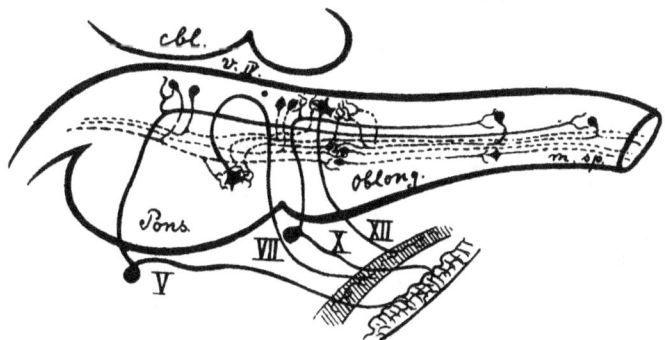

Fig. 10. Schematic drawing of the general structure of the important bulbar reflex pathways, whose mechanism appears to be very complicated. Various nuclear formations are interposed in their course (bulbar intercalated nuclei), which operate specifically as a multiplicator system and enable the conduction of a single, localized impulse and one reflex motion, derived from the combination of the multiple actions of spinal, bulbar, and other nuclei. *Abbreviations: cbl*, cerebellum; *oblong.*, medulla oblongata; *m. sp.*, spinal cord; *v.IV.*, fourth ventricle; *V, VII, X, XII*, corresponding cranial nerves. Originally Fig. 9 on page 11 in Jakob (1911).

stage for its birth. These areas include neurophysiological stimulation and transection studies; developments in comparative neuroanatomy; clinical observations; and an anatomical framework implicating cortical-subcortical connections as the neural substrate of conscious awareness. Lautin (2001) writes that Papez (1937a) does not reference the earlier electrical stimulation experiments. On the other hand, the concept of Broca's great limbic lobe – 'le grand lobe limbique' described in 1878 as encompassing the cingulate and the parahippocampal gyri (Broca, 1878; Nieuwenhuys, 1996) – does receive a mention by Papez (Lautin, 2001).

One may notice certain parallels in the thought of Jakob and Papez:

"The hypothalamus is the recipient of those widespread and ill-defined kinesthetic excitations which arise in association with all sorts of visceral, cerebral (e.g. hippocampal), and other somatic activities ..." and "The mamillary bodies form an important part of the hypothalamus. Each mamillary body is connected forward by a conspicuous bundle, the mamillothalamic tract or fasciculus of Vicq d'Azyr, to the anterior thalamic nuclei. These nuclei are in turn connected forward with the cortex of the gyrus cinguli, which forms a conspicuous feature on the medial surface of each cerebral hemisphere" (Papez, 1937b, p. 226).

MacLean (1949, p. 339) writes:

"The cortex of the cingular gyrus may be looked on as the receptive region for the experiencing of emotion ... Papez's delimitation of this region in the experiencing of emotion strikes one today as a considerable *tour de force*. For, other than the known comparative and neuroanatomy of this region, there was little experimental data to support his thesis, and the clinical evidence was more suggestive than definitive."

It may be of interest to reconsider such reasoning in the light of Jakob's observations (Jakob, 1907/1908, 1909, 1911; Jakob and Onelli, 1911, 1913), developed before the time of Papez (1937a, b, 1958) and MacLean (1949, 1954, 1955).

The basic concept of hemispheric wall function – that is, associating the lateral aspect with somatic function and the medial aspect with visceral function – a concept well understood by Papez and a principal aspect of his theory (Lautin, 2001), is explicitly related by Jakob and Onelli (1911) in the opening statement under the subheading *Organ projections*. Later, Kleist (1934), elaborating and extending the observations on cortical morphology and providing his own clues on cortical maps, assigned the 'somatic ego' to the median facies of the cerebral hemisphere.

In his book on the comparative anatomy of the nervous system of vertebrates, Papez (1929) describes the neopallium in terms of the [then recent] work of von Economo and Koskinas (1925) on cortical lamination. In particular, Papez was aware of the cytoarchitectonic work of von Economo and Koskinas (1925) that had established the cingulate cortex as more laminated than allocortical areas (Lautin, 2001). In his other 1937 paper, Papez (1937b) also cites specifically the 'general part' (pages 1–258) of the text volume of von Economo and Koskinas (1925) that actually makes plentiful references to Jakob's 'ingenious' concepts on brain development and evolution, referring to both of the 1911 German atlases. Whether Papez actually became aware of those original Jakob sources remains in the unwarranted domain of conjecture.

According to Jakob (1911; Orlando, 1964), the discharge of the superior limbic gyrus is achieved by means of the limbicotuberal bundle, which, being formed in the inner cortical layer, descends by the frontomedial corona radiata. At the height of the ansa

peduncularis it comes into contact with the internal peripeduncular nuclei of the nuclear mass laterally to the tuber cinereum (Figs. 10 and 11). This fascicle would regulate internal functions, acquired, like all those of a cortical nature, through the maturation of the individual and as a product of memory elaborations.

Jakob (1911) amply exposes the connections between the anterior nucleus of the thalamus and the limbic gyrus, a relationship effected by means of the frontomedial radiation and inclusively delimiting the frontocaudate fasciculus as foreign to the otherwise false frontooccipital system of Déjerine, being a part of the pathway projecting from the anterior thalamic nucleus to the limbic cortex (Orlando, 1964). Its myelination can be seen in Fig. 8.

Jakob later elaborated on the efferent fibers of the hippocampus and the composition of the fornix, including those structures in the anatomical framework of the functional expression of affect and temperament that he attributed to the visceral zone (Jakob, 1946b; Orlando and Goldar, 1970). In atypical Pick disease with degenerative changes of the fornix and limbic cortex (cingulate and hippocampal) bilaterally, Jakob described a cognitive-affective disorientation, and stressed the importance of the fornix as a central efferent pathway for the cognitive-emotive sphere (Jakob and Esteves Balado, 1944; Jakob, 1946b; Barraquer-Bordas, 1954).

Jakob (1945, 1946a, 1948) revisited the theme of the visceral brain (Figs. 4b and 12) and provided a comprehensive view of consciousness as a product of the synthesis of two sectors: the endogenous (vegetative-autonomic, which he termed 'introyental', represented on the medial facies of the mammalian cerebral hemispheres) and the exogenous (somatic or sensory–motor, which he termed 'ambiental', represented on the convexity of the cerebral hemispheres). In the philosophical tradition of Kant and Schopenhauer, he characterized the former as *a priori* and the latter as *a posteriori* (Orlando and Goldar, 1970; Szirko, 1995; Capizzano, 2006). In his 1949 psychophysiological review on the visceral brain, MacLean (1949) illustrated the possible convergence of interoceptive and exteroceptive systems in the hippocampus. In modern neuroscience, the distinction is made between receptors in the viscera and other internal organs, called *interoceptors* (or *visceroceptors*), and the olfactory, auditory and visual *exteroceptors* (or *teleceptors*), receiving signals from the external world (Shepherd, 1983).

Jakob's final paper (Jakob and Copello, 1949), published in the same year as MacLean's paper in *Psychosomatic Medicine* (MacLean, 1949), dealt with quantitive aspects of the limbic region, the afferent and efferent connections of the superior limbic and hippocampal regions, and their relation to the affective sphere (cf. Section 3.6, below).

3.3. The thalamocingulate projection

The thalamocingulate pathway received special attention by MacLean because of its recent evolution and unique appearance in mammals, and its involvement in advanced social behaviors and communication processes (Lambert, 2003). The first description of the thalamocingulate ('thalamosplenial') projection, generally attributed (e.g. Lautin, 2001) to Le Gros Clark and Boggon (1933), who studied this projection in the cat and the rat by means of degeneration methods, is mentioned by Jakob in the passage 'Splenial formation', published 22 years earlier (Jakob, 1911). It is worth mentioning that two of the earliest publications of Jakob during his Germany years were contributions to the understanding of cortical and thalamic pathology (Jakob, 1894, 1895), in which he supported the view held by Mahaim, von Monakow, Bekhterew and Bielschowsky – against Flechsig and

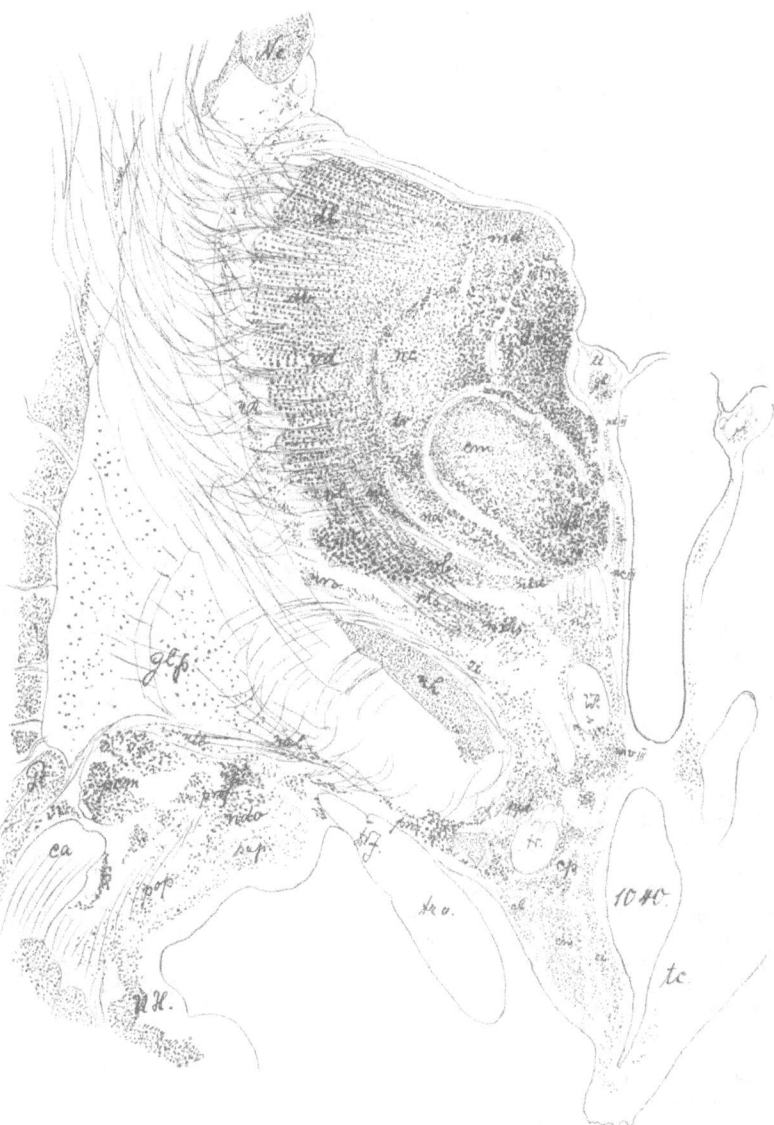

Fig. 11. Middle segment of the thalamus (caudal third). Diminution of the ventral nuclear groups (*vd., vl., vb.*) and hypothalamus (*hth.*), and increasing presence of the dorsal thalamus (*dl., dv., dm.*), tuber cinereum (*tc.*) and parolfactory area (*pop.*). Originally plate 61 in Jakob (1911).

Hösel – that the medial lemniscus terminates at the ventral tier of thalamic nuclei (subthalamus).

Although Jakob's (1911) is probably the earliest mention of a thalamocingulate projection, he did not detail the evidence to support his find. An important point for consideration is that the necessary methods were not available in the early 1900s, an era when conclusions were more often based on the acuity and speculative ability of the observer than the sophistication of the technique at hand. Even the anatomical evidence of Le Gros Clark and Boggon (1933) may appear tentative, were it to be judged with today's standards. Direct projections of the anterior, laterodorsal and ventrobasal thalamic nuclei to cingulate areas and their topographic distribution were unequivocally demonstrated by

means of tract-tracing techniques decades later (Horikawa et al., 1988; Yasui et al., 1988).

3.4. Some further developments on the limbic system

In the latter part of the twentieth century, the interconnected neural system network that subserves emotions and affective behavior, and forms the substrate of affective neuroscience, has been modified and expanded to include the amygdala, anterior insula, putamen, and orbitofrontal cortex (Amaral et al., 1992; Berthoz et al., 2002; LeDoux, 2003). In particular, the amygdala plays a critical role in fear conditioning, by linking external stimuli to defense responses (LeDoux, 2003). In that sense, it has replaced

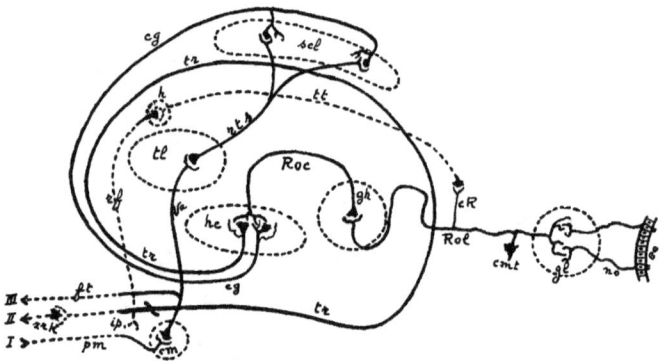

Fig. 12. A later drawing by Jakob (1945) of the circuit organization of the 'visceral olfactory-limbic-fornical system'. *Abbreviations:* cg, cingulum; cm, mamillary body; cmt, mitral cell; cR, reflex collateral; co, olfactory epithelium; ft, tegmental fasciculus; gh, uncus; gl, olfactory glomerulus; h, habenula; hc, hippocampus; ip, interpeduncular nucleus; no, olfactory nerves; pm, mamillary peduncle; rf, fasciculus retroflexus; Roc, central olfactory radiation; Rol, lateral olfactory radiation; rts, superior thalamic radiation; scl, supracallosal gyrus; tl, thalamus; tr, trigonum [fornix]; tt, taenia thalamica; Va, mamillothalamic bundle of Vicq d'Azyr; zrK, bulbar reticular zone; I, sympathetic afferent pathway; II, reticulospinal efferent pathway; III, sympathetic efferent pathway.

the hypothalamus (Karplus and Kreidl, 1927; Bard, 1928; Cannon, 1929; Eccles, 1980) as the centerpiece of subcortical structures involved in the detection and response to threats, especially since Weiskrantz (1956) documented in monkeys that the emotional components of the Klüver–Bucy syndrome (Klüver and Bucy, 1937) were associated with amygdala lesions in the temporal region of the forebrain (LeDoux, 2003). For an exposition of the added midbrain and ventral striatal components of the limbic system discovered by Walle Nauta and Lennart Heimer, respectively, the reader is referred to the reviews of Nieuwenhuys (1996) and Lautin (2001).

The further advances in the understanding of the physiology of emotion that were marked in the 1990s placed special emphasis on the nucleus of the amygdala and its interconnections with prefrontal cortical and hippocampal areas (Davidson and Sutton, 1995). These findings supplemented the already deciphered projections from the amygdala to the mediodorsal thalamic nucleus, the hypothalamus, and the septal nuclei, and their projections in turn to the habenula, tegmentum, and interpeduncular nucleus (Eccles, 1980). The reciprocal pathways between the limbic system and the neocortex provide an essential link in bringing about conscious experiences of affects and emotions (Eccles, 1980).

The anterior insula and the putamen seem to be involved in the recognition of disgust expression; the orbitofrontal and anterior cingulate cortices in the facial expression of anger (Berthoz et al., 2002). The brain areas that may influence the function of the circuit have been expanded to include the cerebellum, ventral tegmental area, interpeduncular and periaqueductal gray areas, and locus coeruleus (Snider and Maiti, 1976).

One related field of research concerns the central sensory pathways that process visceral and somatic pain and terminate, via the thalamus, to insular, cingulate, and parietal cortical areas (Derbyshire, 2003). The reverse, i.e. visceral phenomena elicited secondarily to focal brain lesions – be they as it may pathological, such as gliosis and tumors, or surgical from cortical ablation operations, as well as electrical stimulation procedures – has been a topic of importance to clinical medicine (Pool, 1954). As a matter of fact, Pool (1954) addressed only within a few years one of the actual problems posed by MacLean (1949, p. 351), who had concluded that "a notable deficiency attendant on psychosomatic theory at the present time is the inability to point to a mechanism of emotion that would account for the variety of

ways the effective qualities of experience may act on autonomic centers."

Much of the progress in understanding the neurobiological basis of affect has been made through experimental manipulations in animals and neuroimaging studies in healthy human subjects (Berthoz et al., 2002; LeDoux, 2003; Panksepp, 2003). In the modern view, the behavioral components encompassed under the general term 'emotional processes' include motor-expressive, sensory-perceptual, affective-feeling, cognitive-attentional, and autonomic-hormonal attributes (Panksepp, 2003). All these attributes were at some point considered by Jakob under terms like sensorimotor, musculoskeletal and 'gnosioaffective' functions (Jakob, 1910a, 1910b, 1911, 1945, 1946a, b; Jakob and Onelli, 1910, 1911, 1913; Jakob and Esteves Balado, 1944; Barraquer-Bordas, 1954).

3.5. Emendations to the 'circuit'

It is a fact that, after decades of vigorous research, the views embodied in the so-called 'circuit of Papez' have been substantially modified.

(a) The 'circuit' as proposed by Papez (1937a) does not hold today. The projection between the cingulate cortex and the hippocampus is weak and is mainly with the parahippocampal cortex (Room and Groenewegen, 1986; Smith et al., 2004; Jones and Witter, 2007). Moreover, the 'circuit' has been redrawn to include a projection from the hippocampus to the amygdala, and from there to serotoninergic pacemaker neurons of the dorsal raphé nuclei; in turn, the raphé nuclei project back to the dentate gyrus both directly and via an indirect route with a relay in the entorhinal cortex (Köhler and Steinbusch, 1982; Eggers, 2007).

(b) There are no outlets in the original 'circuit' to drive autonomic functions. The synaptic connections between central autonomic neurons and the forebrain have been traced to areas such as the prefrontal cortex, insula, hypothalamus, amygdala, and bed nucleus of the stria terminalis (Saper, 1982; Groenewegen and Uylings, 2000; Rinaman et al., 2000). The hippocampus, amygdala, and dorsal raphé nuclei all project to the hypothalamus, which are branches off the basic loop subserving the autonomic expression of emotion (Eggers, 2007).

(c) Elements of the 'circuit' such as the anterior thalamus and hippocampus are involved in memory and visuospatial functions rather than emotion (Lacquaniti et al., 1997; Kobayashi and Amaral, 2003; Carrozzo et al., 2005; Línek et al., 2005; Malin and McGaugh, 2006). Areas involved in emotion are the anterior insular and cingulate cortices, the hypothalamus, and the amygdala. The amygdala receives afferent signals related to emotion from four pathways: (i) *olfactory information* directly from the olfactory cortex, without a thalamic relay; (ii) *visceral information* from the hypothalamus and septal area through the stria terminalis; (iii) *internal affect-relevant information* from the hypothalamus, thalamus, brainstem, and from the orbital and anterior cingulate cortices via the ventral pathway; and (iv) *sensory information* from the temporal cortex and the hippocampus (McGovern, 2007). The efferent connections of the amygdala that are of interest for cognition-emotion considerations include pathways to the entorhinal, inferior temporal, and visual cortices, including the fusiform face area (McGovern, 2007).

(d) One other problem with the 'Papez circuit' is that the connections of the mamillary body to the rest of the hypothalamus are meager (Eggers, 2007): there is a small downward projection from the mamillary body to the midbrain via the mamillotegmental tract, but the mamillary body does not project to or control autonomic or endocrine hypothalamic regions; the hippocampus projects to those regions via the fornix (this is the case in the reptilian brain, which lacks a mamillary body), while non-mamillary hypothalamic regions do not project to the anterior thalamic nucleus.

(e) A further problem with Papez's theory is that the purpose of the loop per se is unclear, as is the necessity of the cingulate cortex to project back to the hippocampus. Eggers (2007) thinks that locating emotion in the hippocampus makes sense because of the commonly recognized relationship between emotion and memory; instead of hippocampus → mamillary body → anterior anterior thalamic nucleus → cingulate gyrus → hippocampus, Eggers (2007) suggests a redrawing of the circuit as: hippocampus → amygdala → dorsal raphé → hippocampal formation and entorhinal cortex. Moreover, the subiculum has bidirectional connections with the mamillary body, hippocampus, and entorhinal cortex (Oikawa et al., 2001; McNaughton, 2006).

(f) In clinical studies on the emotional outcome of stroke cases, the combined symptoms of different, overlapping behavior categories, such as depression, modified judgment, attention or memory dysfunction, athymormia, and poststroke fatigue, may confound the functional role of any part of the limbic system in emotion (Bogousslavsky, 2003).

3.6. Closing comments

This manuscript aims at rectifying a historical neglect suffered by the German-Argentinian anatomist Christfried Jakob, who preceded by around 30 years the concept of 'visceral brain', the 'Papez circuit' and parts of the 'limbic system' with a series of articles written in German and Spanish between 1894 and 1913. His work was apparently not cited by later workers, including Papez (1937a), Le Gros Clark and Boggon (1933), MacLean (1949, 1955), and Isaacson (1982), who in his book 'The Limbic System' provides a historical overview of this concept but makes no mention of Jakob.

The following points of discussion attempt to address certain aspects on the importance of Jakob's ideas and their bearing on current neurobiological research.

(a) The scientific interest in the 'Papez circuit' has not waned, as evidenced from the fact that it continues to figure among the milestones in the psychobiology of emotions and the anatomical systems underlying the 'emotional brain' (King, 2001), as well as in epistemological analyses on the convergence of brain research with the physiology of emotions in deciphering the relationships between physiological and psychological forms of knowledge (Dror, 2001). Despite its substantial amendments over the past 40 years, the 'circuit of Papez' continues to be taken into account by scientists as a model for laying down basic and clinical premises in studying pathogenetic mechanisms in conditions such as anxiety (McNaughton, 2006), temporal lobe epilepsy and its commonest neuropathological substrate of mesial temporal sclerosis (Oikawa et al., 2001), and viral encephalitides (Kapur et al., 1994), and in devising novel therapeutic interventions and drug-targeting for diseases such as the spectrum of chronic stress that includes migraine, essential hypertension, depression, and metabolic syndrome (Eggers, 2007).

(b) Jakob uses the terms 'visceral' and 'affective' virtually interchangeably. This was likely a common trend around the fin-du-siècle, reflecting on the bodily theory of emotion formulated by William James (1884). That theory is best remembered today for posing the dilemma whether bodily reactions create emotions or vice versa (King, 2001). Panksepp (2007) argues that James (1884) was correct in that emotions are linked to visceral arousals, but could not have had knowledge of the host of the chemical systems robustly represented within the visceral regions of the brain; this led to the wrongful view – which prevailed for a century – that feelings are simply sensory readouts of peripheral bodily states, whereas in reality, they are created by CNS circuits that sustain a rich interchange with visceral bodily changes. The basic emotional systems of the brain appear to be built around neuropeptidergic and serotoninergic systems, enriched in the visceral organs and the gastrointestinal tract, denoting the evolutionary roots of many emotional feelings, traceable to the large-scale dynamic visceral changes that mark emotional states (Panksepp, 2007). Thus, Panksepp (2007) concludes that "the energetic regulators of visceral organ activities are still represented in the emotional circuits of the brain, helping to explain why all basic emotions have such a visceral feel to them and why psychosomatic disorders can lead to imbalances in our vital organs."

(c) Throughout his writings, from the early 1900s until his last publication of 1949 (Jakob and Copello, 1949), Jakob made the dichotomy between an interoceptive ('introyental') affective sphere, represented in the medial/limbic side of the hemispheric cortex, and the lateral cognitive ('intelectualizante') sphere interacting with the outer environment. Today, psychobiological differences between emotions and cognitions instantiated in brain mechanisms are emphasized (Panksepp, 2003), and a dichotomy is made between distinct *energetic state processes* and *informational channel functions* of the brain, considered essential for understanding the neurology of the affective mind (Mesulam, 2000; Panksepp, 2007). Emotional *state processes* globally control vast brain regions by means of particular (aminergic and peptidergic) neurochemical systems, whereas *channel functions* refer to the discrete cognitive processing of exteroceptive information; the former reflect non-linear, large-scale analogue neurodynamics that are the bedrock of emotional experiences, whereas the later reflect fast and precise excitatory (glutamate) neurotransmission along the line of classical computational metaphors (Panksepp, 2007).

(d) In today's behavioral, cognitive and affective neuroscience, emotional brain functions are considered from two complementary perspectives, a neurodynamic and a psychodynamic, in what is termed a 'dual-aspect monism strategy' (Panksepp, 2007). Another of Jakob's unheeded treasures, his 1923 masterful *Elements of Neurobiology* (Jakob, 1923), abounds with terms such as cortical dynamisms, neural and psychic dynamics, and affective states. According to Jakob, in the phylogenetic scale, 'psychodynamic' nervous functions are preceded by 'plasmodynamic' activities, which encompass tropism and pulsatility; 'neurodynamic' processes are subserved by three underlying hierarchical levels of the vertebrate CNS, designated as archineuronal, paleoneuronal and neoneuronal, and encompass *archikinesias* or reflex actions similar to invertebrates, *paleokinesias* or instinctual-automated reactions, and *neokinesias* or conscious motor reactions (Jakob, 1923, 1935). In the framework of the dynamic workings of the human cerebral cortex, neokinesias include *gnosias* (cognitive processes related to conscious orientation in one's environment), *praxias* (active individual intervention) and *symbolias* (ideative abstraction to facilitate interindividual communication, such as the sociogenetic processes on which human culture is based) (Jakob, 1921, 1923). Jakob (1945) further developed his views on neocortical histotopography, with its macrodynamic and microdynamic events, in a scheme depicting the most probable trajectories of nervous current in cortical layers, the termination of thalamocortical fibers and large motor neuron projections to the pes pedunculi, incoming afferent fibers in association with small interneurons (equivalent to the cortical 'microdynamic apparatus'), and probable circuits intercalated among afferent and efferent fibers in the various cortical layers (cf. Triarhou and del Cerro, 2006b).

(e) Finally, many of the above elements of Jakob's thought resonate in his final paper, entitled 'The neuronal quantification of the limbic region in its relation to the inner affective sphere' (Jakob and Copello, 1949). An excerpt from that paper follows:

"A fundamental concept emerges on the existence of distinct functions in two cortical spheres: the lateral cortex is linked to the outer environment, whereas the inner medial limbic sphere is related to sensitivity and viscerovascular motility … Since the latter is intimately associated with our emotive sphere (circulation, nutrition, inner secretions, etc.), it would represent our inner affective cortex. Both together generate, by being intimately combined, the individual consciousness of the self (neopsychisms).

Thus closes the *limbic internal* functional circuit, through which visceral affective states, from euphoric to anxious, are mediated, accumulated and discharging in the personal experience during the normal life of the individual, biologically relating the limbic cortex to visceroendocrine functions. This is evidently a complex psychophysiological problem still not solved, and common to the entire animal series, to which belong, among other, the care for personal hygiene and the young; cave and nest in animals; elimination of organic wastes via sphincters and regulation of nutritive and sexual functions. The training for such functions begins in early childhood and never ceases; it is the evoked inner affect, which forms the biological basis of being, and which we connect foremost with cortical functions of the superior limbic gyrus. In these studies, human pathology has a most important part with its clinico-anatomical correlations.

The limbic gyrus maintains, as far as its afferent–efferent projections are concerned, a functional independence from the remaining gyri of the lateral neocortex, which are related to the outer environment. This important circuit is not fully understood yet; being phylogenetically older cortex, it provides us with viscerovasomotor signals and the feeling of well-being or of vegetative malaise, oscillating between euphoria and anxiety. Thus, we can deduce that here reside the bases for the affectively colored feelings of inner mental life, as opposed to those subserved by the lateral cortex, thus providing the association for integrated conscious mentality. In spite of these centers existing in lower vertebrates, it is in mammals that they successively become associated with the lateral outer effects, intensifying the intellectualization of states in higher consciousness in relation to the elementary nature of the others. Only in primates, and especially in humans with their verbal symbolization, one rises to a supreme mental degree, where the sphere of inner feelings of visceral origin becomes amalgamated with the environmental gnosio-praxic experience; in that superior synthesis arises the neopsychic sphere, from the concrete to the most abstract.

In this study, we attempted to achieve a constructive synthesis for the creation of those inner affective mental phenomena in their relation to external cognitive phenomena. The associative process between both systems creates the conscious and total personality: affect and cognition, respectively located at the medial fornical-limbic and at the lateral pyramidal systems."

Acknowledgments

The author gratefully acknowledges Michael C. Triarhou, LL.M. for invaluable help with the German language, and the anonymous reviewers for generously offering their constructive criticism and expertise, which led to an improved manuscript. Supported in part by the Intramural Research Funding Operations of the University of Macedonia, Greece.

References

Allen, G.V., Saper, C.B., Hurley, K.M., Cechetto, D.F., 1991. Organization of visceral and limbic connections in the insular cortex of the rat. Journal of Comparative Neurology 311, 1–16.

Allman, J.M., Watson, K.K., Tetreault, N.A., Hakeem, A.Y., 2005. Intuition and autism: a possible role for von Economo neurons. Trends in Cognitive Sciences 9, 367–373.

Amaral, D.G., Price, J.L., Pitkänen, A., Carmichael, T., 1992. Anatomical organization of the primate amygdaloid complex. In: Aggleton, J.P. (Ed.), The Amygdala: Neurobiological Aspects of Emotion, Memory, and Mental Dysfunction. Wiley-Liss, New York, pp. 339–352.

Bagaev, V., Aleksandrov, V., 2006. Visceral-related area in the rat insular cortex. Autonomic Neuroscience 125, 16–21.

Bard, P., 1928. A diencephalic mechanism for the expression of rage with special reference to the sympathetic nervous system. American Journal of Physiology 84, 490–515.

Barraquer-Bordas, L., 1954. Fisiología y Clínica del Sistema Límbico: Aportaciones Recientes al Conocimiento de las Bases Neurofisiológicas de la Personalidad. Editorial Paz Montalvo, Madrid, pp. 136–137.

Barraquer-Bordas, L., 1976. Neurología Fundamental: Fisiopatología, Semiología, Síndromes, Exploración, third ed. Ediciones Toray, Barcelona, p. 596.

Berthoz, S., Blair, R.J.R., Le Clec'h, G., Martinot, J.-L., 2002. Emotions: from neuropsychology to functional imaging. International Journal of Psychology 37, 193–203.

Bogousslavsky, J., 2003. Emotions, mood, and behavior after stroke. Stroke 34, 1046–1050.

Broca, P., 1878. Anatomie comparée des circonvolutions cérébrales: le grand lobe limbique et la scissure limbique dans la série des mammifères. Revue d'Anthropologie – Deuxième Série (Paris) 1, 385–498.

Brodmann, K., 1909. Vergleichende Lokalisationslehre der Grosshirnrinde in ihren Prinzipien Dargestellt auf Grund des Zellenbaues. Johann Ambrosius Barth, Leipzig.

Cannon, W.B., 1929. Bodily Changes in Pain, Hunger, Fear, and Rage: An Account of Recent Researches into the Function of Emotional Excitement, second ed. D. Appleton and Company, New York, London.

Capizzano, A.A., 2006. Actualidad del pensamiento de Cristofredo Jakob. Revista del Hospital Italiano de Buenos Aires 26, 71–73.

Carrozzo, M., Koch, G., Turriziani, P., Caltagirone, C., Carlesimo, G.A., Lacquaniti, F., 2005. Integration of cognitive allocentric information in visuospatial short-term memory through the hippocampus. Hippocampus 15, 1072–1084.

Critchley, H.D., Elliott, R., Mathias, C.J., Dolan, R.J., 2000. Neural activity relating to generation and representation of galvanic skin conductance responses: A functional magnetic resonance imaging study. Journal of Neuroscience 20, 3033–3040.

Crocco, M.F., 1994. Alberto Alberti y el primer mapeo con electricidad ¡durante ocho meses! de un cerebro humano consciente: hazaña científica silenciada durante un siglo Electroneurobiología (Buenos Aires) 1, 73–82.

Dalgleish, T., 2004. The emotional brain, Nature Reviews Neuroscience 5, 582–589.

Dana, C.L., 1921. The anatomic seat of emotions: a discussion of the James–Lange theory. Archives of Neurology and Psychiatry 6, 634–639.

Davidson, R.J., Sutton, S.K., 1995. Affective neuroscience: the emergence of a discipline. Current Opinion in Neurobiology 5, 217–224.

Déjerine, J.J., 1895. Anatomie des Centres Nerveux. Rueff et Cie, Paris.

DeMyer, W., 1988. Neuroanatomy. Williams & Wilkins/Harwal Publishing Company, Baltimore–Malvern, pp. 287–297.

Derbyshire, S.W.G., 2003. Visceral afferent pathways and functional brain imaging. Scientific World Journal 3, 1065–1080.

Dror, O.E., 2001. Techniques of the brain and the paradox of emotions, 1880–1930. Science in Context 14, 643–660.

Eccles, J.C., 1980. The emotional brain. Bulletin et Mémoires de l'Académie Royale de Médecine de Belgique (Bruxelles) 135, 697–711.

von Economo, C., 1926. Eine neue Art Spezialzellen des Lobus cinguli und Lobus insulae. Zeitschrift für die Gesamte Neurologie und Psychiatrie (Berlin) 100, 706–712.

von Economo, C., 1927. Zellaufbau der Grosshirnrinde des Menschen. Zehn Vorlesungen. Verlag von Julius Springer, Berlin.

von Economo, C., Koskinas, G.N., 1925. Die Cytoarchitektonik der Hirnrinde des Erwachsenen Menschen. Verlag von Julius Springer, Wien–Berlin.

von Economo, C., Koskinas, G.N., 2008. Atlas of Cytoarchitectonics of the Adult Human Cerebral Cortex (Transl. by L.C. Triarhou). Karger, Basel.

Eggers, A.E., 2007. Redrawing Papez' circuit: a theory about how acute stress becomes chronic and causes disease. Medical Hypotheses 69, 852–857.

Faccio, E.J., 1991. Christofredo Jakob y el origen del psiquismo. Alcmeón – Revista Argentina de Clínica Neuropsiquiátrica 3, 331–348.

Fulton, J.F., 1953. The limbic system: a study of the visceral brain in primates and man. Yale Journal of Biology and Medicine 26, 107–118.

Goldar, J.C., 1975. Cerebro Límbico y Psiquiatría. Editorial Salerno, Buenos Aires, pp. 10, 84.

Goldar, J.C., 1997. A sesenta años del "circuito emocional" de Papez. Alcmeón – Revista Argentina de Clínica Neuropsiquiátrica 23, 1–8.

Groenewegen, H.J., Uylings, H.B.M., 2000. The prefrontal cortex and the integration of sensory, limbic and autonomic information. Progress in Brain Research 126, 3–28.

Hanamori, T., Kunitake, T., Kato, K., Kannan, H., 1998. Responses of neurons in the insular cortex to gustatory, visceral, and nociceptive stimuli in rats. Journal of Neurophysiology 79, 2535–2545.

Herrick, C.J., 1933. The functions of the olfactory parts of the cerebral cortex. Proceedings of the National Academy of Sciences of USA 19, 7–14.

Hess, W.R., 1932. Beiträge zur Physiologie des Hirnstammes. I. Die Methodik der Lokalisierten Reizung und Ausschaltung Subkortikaler Hirnabschnitte. Georg Thieme Verlag, Leipzig.

Hof, P.R., Van Der Gucht, E., 2007. Structure of the cerebral cortex of the humpback whale, Megaptera novaeangliae (Cetacea, Mysticeti, Balaenopteridae). The Anatomical Record–Advances in Integrative Anatomy and Evolutionary Biology 290, 1–31.

Horikawa, K., Kinjo, N., Stanley, L.C., Powell, E.W., 1988. Topographic organization and collateralization of the projections of the anterior and laterodorsal thalamic nuclei to cingulate areas 24 and 29 in the rat. Neuroscience Research 6, 31–44.

Isaacson, R.L., 1982. The Limbic System, second ed. Plenum Press, New York.

Ito, S.I., 1998. Possible representation of somatic pain in the rat insular visceral sensory cortex: a field potential study. Neuroscience Letters 241, 171–174

Jakob, C., 1894. Ueber einen Fall von Hemiplegie und Hemianästhesie mit gekreuzter Oculomotoriuslähmung bei einseitiger Zerstörung des Thalamus opticus, des hinteren Theiles der Capsula interna, der vorderen Vierhügel- und Haubengegend, mit besonderer Berücksichtigung der secundären Degenerationen. Deutsche Zeitschrift für Nervenheilkunde (Leipzig) 5, 188–224.

Jakob, C., 1895. Ein Beitrag zur Lehre vom Schleifenverlauf (obere, Rinden, Thalamusschleife). Neurologisches Centralblatt (Leipzig) 14, 308–310.

Jakob, C., 1899. Atlas des Gesunden und Kranken Nervensystems nebst Grundriss der Anatomie. Pathologie und Therapie Desselben, second ed. Verlag von J. Lehmann, München.

Jakob, C., 1902. Sífilis medular. Revista del Centro de Estudiantes de Medicina (Buenos Aires) 2, 351–353.

Jakob, C., 1906a. ¿Existe ó no un centro de Broca? La Semana Médica (Buenos Aires) 13, 677–678.

Jakob, C., 1906b. Consideraciones anátomo-biológicas sobre los centros del lenguaje. La Semana Médica (Buenos Aires) 13, 733–737.

Jakob, C., 1907/1908. Localización del alma y de la inteligencia. El Libro (Buenos Aires) 1, 151/281/433/553; 2, 3/171/293/537/695 (published in nine sequels).

Jakob, C., 1909. Curso sobre Enfermedades del Sistema Nervioso en Relación con su Anatomía Patológica Dictado en la Clínica del Dr. Ramos Mejía (Recog. por P.M. Barlaro). La Ciencia Médica, Buenos Aires.

Jakob, C., 1910a. La célula cortical en la locura. (Estudios histopatológicos sobre las células piramidales en las enfermedades mentales). Anales de la Administración Sanitaria y Asistencia Pública. Ediciones de La Semana Médica–Imprenta de E. Spinelli, Buenos Aires, pp. 263–267.

Jakob, C., 1910b. La afasia motriz y su localización: Estudios biológicos y biopatológicos sobre los centros del lenguaje. Revista de la Sociedad Médica Argentina (Buenos Aires) 18, 353–374.

Jakob, C., 1911. Das Menschenhirn: Eine Studie über den Aufbau und die Bedeutung seiner Grauen Kerne und Rinde. J. F. Lehmann's Verlag, München, pp. 48–49.

Jakob, C., 1918. La filogenia cortical: Sobre la corteza cerebral de gimnofiones y anfisbenas Argentinas. Actas y Trabajos del Premier Congreso Nacional de Medicina (Buenos Aires) 4, 82.

Jakob, C., 1921. La teoría actual de las gnosias y praxias como factores fundamentales en el dinamismo de la corteza cerebral. Crónica Médica (Lima) 38, 17–24.

Jakob, C., 1923. Elementos de Neurobiología, vol. I: Parte Teórica. Biblioteca Humanidades, Facultad de Humanidades y Ciencias de la Educación de la Universidad Nacional de La Plata, La Plata, Argentina.

Jakob, C., 1925. La filosofía de la naturaleza según Kant. Revista de Filosofía (Buenos Aires) 11, 45–57.

Jakob, C., 1935. La filogenia de las kinesias (Sobre su organización y dinamismo evolutivo). Anales del Instituto de Psicología de la Facultad de Filosofía y Letras de la Universidad de Buenos Aires 1, 109 127.

Jakob, C., 1939. El Cerebro Humano, su Anatomía Sistemática y Topográfica (Folia Neurobiológica Argentina, Atlas I). Aniceto López, Buenos Aires.

Jakob, C., 1945. El cerebro humano: su significación filosófica. Ensayo de un programa psico-bio-metafísico. Revista Neurológica de Buenos Aires 10, 89–110.

Jakob, C., 1946a. Documenta Biofilosófica (Folia Neurobiologica Argentina, Tomo V). López & Etchegoyen, Buenos Aires.

Jakob, C., 1946b. El trígono cerebral: su significación neurobiológica. (Vía central eferente para la musculatura lisa víscero-vascular de la esfera gnósica-emotiva). Revista Neurológica de Buenos Aires 11, 2–36.

Jakob, C., 1948. La psicointegración introyento-ambiental orgánica y sus problemas para la neuropsiquiatría y psicología, primera parte: su filogenia constructiva. Revista Neurológica de Buenos Aires 13, 115–141.

Jakob, C., Onelli, C., 1910. Anatomía comparada del encéfalo de los mamíferos de la República Argentina, atlas y planchas. In Barabino, S.E., Besio Moreno, N. (Eds.), Congreso Científico Internacional Americano. vol. 1. Relación General del Funcionamiento del Congreso. Imprenta y Casa Editora de Coni Hermanos, Buenos Aires, pp. 455–456.

Jakob, C., Onelli, C., 1911. Vom Tierhirn zum Menschenhirn: Vergleichend Morphologische, Histologische und Biologische Studien zur Entwicklung der Grosshirnhemisphären und ihrer Rinde. J. F. Lehmann's Verlag, München, pp. 16–17.

Jakob, C., Onelli, C., 1913. Atlas del Cerebro de los Mamíferos de la República Argentina: Estudios Anatómicos, Histológicos y Biológicos Comparados sobre la Evolución de los Hemisferios y de la Corteza Cerebral, 20. Guillermo Kraft, Buenos Aires, pp. 38–39.

Jakob, C., Esteves Balado, L., 1944. Encefalosis progresiva simétrica frontocaudal (Una forma no descripta de la enfermedad de "Pick" atípica combinada). Boletín de la Academia Nacional de Medicina (Buenos Aires) 25, 11–28.

Jakob, C., Copello, A.R., 1949. La cuantificación neuronal de la región limbica en su relación con la esfera introyental afectiva. Archivos de Neurocirugía 6, 475–481.

James, W., 1884. What is an emotion? Mind 19, 188–205.

Jaspers, K., 1962. The Great Philosophers, Volume 1: Kant (Transl. by R. Manheim). Harcourt Brace Jovanovich, San Diego, pp. 64–66.

Jones, B.F., Witter, M.P., 2007. Cingulate cortex projections to the parahippocampal region and hippocampal formation in the rat. Hippocampus 17, 957–976.

Kant, I., 1948. The Moral Law, or Groundwork of the Metaphysic of Morals [1785] (Transl. by H.J. Paton). Hutchinson, London.

Kapur, N., Barker, S., Burrows, E.H., Ellison, D., Brice, J., Illis, L.S., Scholey, K., Colbourn, C., Wilson, B., Loates, M., 1994. Herpes simplex encephalitis: long term magnetic resonance imaging and neuropsychological profile. Journal of Neurology Neurosurgery and Psychiatry 57, 1334–1342.

Karplus, J.P., Kreidl, A., 1927. Gehirn und Sympathicus. VII. Mitteilung. Über Beziehungen der Hypothalamuszentren zu Blutdruck und innerer Sekretion. Pflügers Archiv für die Gesamte Physiologie des Menschen und der Tiere (Berlin) 215, 667–670.

King, M.G., 2001. Emotions in the workplace: biological correlates. In: Payne, R.L., Cooper, C.L. (Eds.), Emotions at Work: Theory, Research and Applications for Management. John Wiley & Sons Ltd., Chichester, pp. 85–106.

Kleist, K., 1934. Gehirnpathologie. Johann Ambrosius Barth, Leipzig.

Klüver, H., Bucy, P.C., 1937. "Psychic blindness" and other symptoms following bilateral temporal lobectomy in Rhesus monkeys. American Journal of Physiology 119, 352–353.

Kobayashi, Y., Amaral, D.G., 2003. Macaque monkey retrosplenial cortex. II. Cortical afferents. Journal of Comparative Neurology 466, 48–79.

Köhler, C., Steinbusch, H.W.M., 1982. Identification of serotonin and non-serotonin-containing neurons of the midbrain raphé projecting to the entorhinal area and the hippocampal formation: a combined immunohistochemical and fluorescent retrograde tracing study in the rat brain. Neuroscience 7, 951–975.

von Kölliker, A., 1902. Handbuch der Gewebelehre des Menschen. Engelmann, Leipzig.

Krushel, L.A., van der Kooy, D., 1988. Visceral cortex: integration of the mucosal senses with limbic information in the rat agranular insular cortex. Journal of Comparative Neurology 270, 39–54 and 62–63.

Lacquaniti, F., Perani, D., Guigon, E., Bettinardi, V., Carrozzo, M., Grassi, F., Rossetti, Y., Fazio, F., 1997. Visuomotor transformations for reaching to memorized targets: a PET study. Neuroimage 5, 129–146.

Lambert, K.G., 2003. The life and career of Paul MacLean: A journey toward neurobiological and social harmony. Physiology and Behavior 79, 343–349.

Lane, R.D., Reiman, E.M., Axelrod, B., Yun, L.-S., Holmes, A., Schwartz, G.E., 1998. Neural correlates of levels of emotional awareness: Evidence of an interaction between emotion and attention in the anterior cingulate cortex. Journal of Cognitive Neuroscience 10, 525–535.

Lautin, A., 2001. The Limbic Brain. Kluwer Academic/Plenum Publishers, New York, pp. 55–68.

LeDoux, J., 2003. The emotional brain, fear, and the amygdala. Cellular and Molecular Neurobiology 23, 727–738.

Lee, G.P., Meador, K.J., Loring, D.W., Allison, J.D., Brown, W.S., Paul, L.K., Pillai, J.J., Lavin, T.B., 2004. Neural substrates of emotion as revealed by functional magnetic resonance imaging. Cognitive and Behavioral Neurology 17, 9–17.

Le Gros Clark, W.E., Boggon, R.H., 1933. On the connections of the anterior nucleus of the thalamus. Journal of Anatomy (London) 67, 215–226.

Línek, V., Sonka, K., Bauer, J., 2005. Dysexecutive syndrome following anterior thalamic ischemia in the dominant hemisphere. Journal of the Neurological Sciences 229/230, 117–120.

MacLean, P.D., 1949. Psychosomatic disease and the "visceral brain": Recent developments bearing on the Papez theory of emotion. Psychosomatic Medicine 11, 338–353.

MacLean, P.D., 1954. The limbic system and its hippocampal formation: studies in animals and their possible application to man. Journal of Neurosurgery 11, 29–44.

MacLean, P.D., 1955. The limbic system ("visceral brain") and emotional behavior. Archives of Neurology and Psychiatry 73, 130–134.

Malin, E.L., McGaugh, J.L., 2006. Differential involvement of the hippocampus, anterior cingulate cortex, and basolateral amygdala in memory for context and footshock. Proceedings of the National Academy of Sciences of USA 103, 1959–1963.

Marshall, L.H., Magoun, H.W., 1998. Discoveries in the Human Brain: Neuroscience Prehistory, Brain Structure, and Function. Humana Press, Totowa, NJ, pp. 199–223.

McGovern, K., 2007. Emotion. In: Baars, B.J., Gage, N.M. (Eds.), Cognition, Brain, and Consciousness: Introduction to Cognitive Neuroscience. Elsevier Academic Press, London, pp. 369–389.

McNaughton, N., 2006. The role of the subiculum within the behavioural inhibition system. Behavioural. Brain Research 174, 232–250.

Mesulam, M.-M., 2000. Principles of Behavioral and Cognitive Neurology, second ed. Oxford University Press, Oxford, New York.

Meyer, A., 1971. Historical Aspects of Cerebral Anatomy. Oxford University Press, Oxford.

Meyer, L., 1981. Cristofredo Jakob. A veinticinco años de su muerte. Acta Psiquiátrica y Psicológica de América Latina 27, 13–14.

Moyano, B.A., 1957. Christfried Jakob, 25/12/1866-6/5/1956. Acta Neuropsiquiátrica Argentina 3, 109–123.

Müller, L.R., Feirtag, M., 1926. Bilder zur Makroskopischen Anatomie des Gehirns und zum Bahnenverlauf. Erläuterungen zur 2. Auglage der Icones Neurologicae von Strümpell und Jakob. J. F. Lehmanns Verlag, München.

Nauta, W.J.H., Feirtag, M., 1986. Fundamental Neuroanatomy. W. H. Freeman and Company, New York, 120–131.

Neylan, T.C., 1995. Classic articles in neuropsychiatry: introduction to the series. Journal of Neuropsychiatry and Clinical Neurosciences 7, 102–103.

Ngowyang, G., 1932. Beschreibung einer Art von Spezialzellen in der Inselrinde zugleich Bemerkungen über die v. Economoschen Spezialzellen. Journal für Psychologie und Neurologie (Leipzig) 44, 671–674.

Nieuwenhuys, R., 1996. The greater limbic system, the emotional motor system and the brain. Progress in Brain Research 107, 551–580.

Nimchinsky, E.A., Gilissen, E., Allman, J.M., Perl, D.P., Erwin, J.M., Hof, P.R., 1999. A neuronal morphologic type unique to humans and great apes. Proceedings of the National Academy of Sciences of USA 96, 5268–5273.

Oikawa, H., Sasaki, M., Tamakawa, Y., Kamei, A., 2001. The circuit of Papez in mesial temporal sclerosis: MRI. Neuroradiology 43, 205–210.

Orlando, J.C., 1964. Sobre el cerebro visceral. Documentación histórica de una prioridad científica. Revista Argentina de Neurología y Psiquiatría 1, 197–201.

Orlando, J.C., 1966. Christofredo Jakob: su Vida y Obra. Editorial Mundi, Buenos Aires.

Orlando, J.C., Goldar, J.C., 1970. Rinencéfalo o sistema límbico. Archivos de la Fundación Roux-Ocefa (Buenos Aires) 4, 33–40.

Outes, D.L., 2006. A medio siglo de la muerte de Christofredo Jakob, 1956–2006: Fuentes de la concepción biológica de la doble corteza. Electroneurobiología (Buenos Aires) 14, 3–35.

Owen, R., 1857. On the characters, principles of division, and primary groups of the class Mammalia. Journal of the Linnean Society of London – Zoology 2, 1–37.

Panksepp, J., 2003. At the interface of the affective, behavioral, and cognitive neurosciences: Decoding the emotional feelings of the brain. Brain and Cognition 52, 4–14.

Panksepp, J., 2007. The neuroevolutionary and neuroaffective psychobiology of the prosocial brain. In: Dunbar, R.I.M., Barrett, L. (Eds.), The Oxford Handbook of Evolutionary Psychology. Oxford University Press, Oxford, pp. 145–162.

Papez, J.W., 1929. Comparative Neurology: A Manual and Text for the Study of the Nervous System of Vertebrates. Thomas Y. Crowell Company, New York, pp. 336-355.

Papez, J.W., 1937a. A proposed mechanism of emotion. Archives of Neurology and Psychiatry 38, 725–743.

Papez, J.W., 1937b. The brain considered as an organ: Neural systems and central levels of organization. American Journal of Psychology 49, 217–232.

Papez, J.W., 1958. Visceral brain, its component parts and their connections. Journal of Nervous and Mental Disease 126, 40–56.

Papez, J.W., 1995. Neuropsychiatry classics: A proposed mechanism of emotion [1937]. Journal of Neuropsychiatry and Clinical Neurosciences 7, 103–112.

Pelliza, S., 2006. Neurociencias—El Sistema Nervioso. Provincia de Santa Cruz: Honorable Cámara de Diputados, Departamento Apoyatura Académica del Instituto Saleciano de Estudios Superiores (www.hcdsc.gov.ar/biblioteca/ISES/neurociencias.asp).

Peper, M., Markowitsch, H.J., 2001. Pioneers of affective neuroscience and early concepts of the emotional brain. Journal of the History of the Neurosciences 10, 58–66.

Piva, J.R., Virasoro, C., 2004. Christofredo Jakob, neurobiólogo: científico en diálogo filosófico. Comunicaciones del Museo Provincial de Ciencias Naturales 'Florentino Ameghino' – Nueva Serie 9, 1–18.

Pool, J.L., 1954. The visceral brain of man. Journal of Neurosurgery 11, 45–63.

Pribram, K.H., Kruger, L., 1954. Functions of the "olfactory brain". Annals of the New York Academy of Sciences 58, 109–138.

Ramón y Cajal, S., 1903. Plan de estructura del tálamo óptico. Revista de Medicina y Cirugía Prácticas (Madrid) 59, 329–348.

Ranke, O., 1911. Bücheranzeigen und Referate: Chr. Jakob und Cl. Onelli: Vom Tierhirn zum Menschenhirn; Chr. Jakob: Das Menschenhirn. Münchener Medizinische Wochenschrift 58, 2510–2512.

Rinaman, L., Levitt, P., Card, J.P., 2000. Progressive postnatal assembly of limbic-autonomic circuits revealed by central transneuronal transport of pseudorabies virus. Journal of Neuroscience 20, 2731–2741.

Room, P., Groenewegen, H.J., 1986. Connections of the parahippocampal cortex. I. Cortical afferents. Journal of Comparative Neurology 251, 415–450.

Saper, C.B., 1982. Convergence of autonomic and limbic connections in the insular cortex of the rat. Journal of Comparative Neurology 210, 163–173.

Seeley, W.W., Carlin, D.A., Allman, J.M., Macedo, M.N., Bush, C., Miller, B.L., DeArmond, S.J., 2006. Early frontotemporal dementia targets neurons unique to apes and humans. Annals of Neurology 60, 660–667.

Shepherd, G.M., 1983. Neurobiology. Oxford University Press, New York-Oxford, pp. 187–202, 487–544.

Showers, M.J., Crosby, E.C., 1958. Somatic and visceral responses from the cingulate gyrus. Neurology 8, 561–565.

Siemerling, E., 1911. Referate—Kleinere Mitteilungen: Chr. Jakob und Cl. Onelli. Vom Tierhirn zum Menschenhirn. Archiv für Psychiatrie und Nervenkrankheiten (Berlin) 49, 353–355.

Sikes, R.W., Vogt, L.J., Vogt, B.A., 2008. Distribution and properties of visceral nociceptive neurons in rabbit cingulate cortex. Pain 135, 160–174.

Smith, A.P., Henson, R.N., Dolan, R.J., Rugg, M.D., 2004. fMRI correlates of the episodic retrieval of emotional contexts. Neuroimage 22, 868–878.

Snider, R.S., Maiti, A., 1976. Cerebellar contributions to the Papez circuit. Journal of Neuroscience Research 2, 133–146.

Striedter, G.F., 2005. Principles of Brain Evolution. Sinauer Associates, Inc., Sunderland, MA, pp. 23–24.

Szirko, M., 1995. A la antropología ganglionar desde la kinesiología: un fallido ensayo de extrapolar lo orgánico. Electroneurobiología (Buenos Aires) 2, 104–169.

Triarhou, L.C., 2007. A proposed number system for the 107 cortical areas of Economo and Koskinas, and Brodmann area correlations. Stereotactic and Functional Neurosurgery 85, 204–215.

Triarhou, L.C., del Cerro, M., 2006a. Semicentennial tribute to the ingenious neurobiologist Christfried Jakob (1866–1956). 1. Works from Germany and the first Argentina period, 1891–1913. European Neurology 56, 176–188.

Triarhou, L.C., del Cerro, M., 2006b. Semicentennial tribute to the ingenious neurobiologist Christfried Jakob (1866–1956). 2. Publications from the second Argentina period, 1913–1949. European Neurology 56, 189–198.

Tsapkini, K., Vivas, A.B., Triarhou, L.C., 2008. 'Does Broca's area exist?' Christofredo Jakob's 1906 response to Pierre Marie's holistic stance. Brain and Language 105, 211–219.

Ure, J., 2001. Jacinto C. Orlando. Revista Neurológica Argentina 26, 197.

Vicq d'Azyr, F., 1786. Traité d'Anatomie et de Physiologie, avec des Planches Coloriées Représentant au Naturel les Divers Organes de l'Homme et des Animaux. Didot, Paris.

Vivas, A.B., Tsapkini, K., Triarhou, L.C., 2007. 'Anatomo-biological considerations on the centers of language': an Argentinian contribution to the 1906 Paris debate on aphasia. Brain and Development 29, 455–461.

Watson, K.K., Jones, T.K., Allman, J.M., 2006. Dendritic architecture of the von Economo neurons. Neuroscience 141, 1107–1112.

Weiskrantz, L., 1956. Behavioral changes associated with ablation of the amygdaloid complex in monkeys. Journal of Comparative and Physiological Psychology 49, 381–391.

Woodsworth, R.S., Sherrington, C.S., 1904. A pseudaffective reflex and its spinal path. Journal of Physiology 31, 234–243.

Yasui, Y., Itoh, K., Kamiya, H., Ino, T., Mizuno, N., 1988. Cingulate gyrus of the cat receives projection fibers from the thalamic region ventral to the ventral border of the ventrobasal complex. Journal of Comparative Neurology 274, 91–100.

In: *Cognitive Psychology Research Developments*   ISBN 978–1–60692–197–5
Editors: Stella P. Weingarten and Helena O. Penat   © 2009 Nova Science Publishers, Inc.

Chapter 7

TRIPARTITE CONCEPTS OF MIND AND BRAIN, WITH SPECIAL EMPHASIS ON THE NEUROEVOLUTIONARY POSTULATES OF CHRISTFRIED JAKOB AND PAUL MACLEAN

Lazaros C. Triarhou[*]

Economo–Koskinas Wing for Integrative and Evolutionary Neuroscience,
University of Macedonia, Thessaloniki, Greece

ABSTRACT

The 'triune brain', conceived by Paul D. MacLean (1913–2007) in the late 1960s, has witnessed more attention and controversy than any other evolutionary model of brain and behavior in modern neuroscience. Decades earlier, in his book *Elements of Neurobiology* published in 1923 in La Plata, Argentina, neurobiologist Christfried (Christofredo) Jakob (1866–1956) had formulated a 'tripsychic' brain system, based on his deep understanding of biological and neural phylogeny. In a historical context, 1923 was also the year of publication of Sigmund Freud's *The Ego and the Id*, whereby the founder of psychoanalysis solidified his tripartite model of the mental apparatus. Tripartite systems of the human mind have been surmised since Plato and Aristotle; they continue to our era, an example being Robert J. Sternberg's triarchic theory of human intelligence. In view of the fact that both Jakob and MacLean invested a considerable part of their long and distinguished careers studying comparative, and particularly reptilian neurobiology, the present article revisits their neuroevolutionary models, underlining the convergence of their anatomical-functional propositions, in spite of a time distance of almost half a century.

[*] E-mail address: triarhou@uom.gr, phone +30 2310 891-387, fax +30 2310 891-388

INTRODUCTION

In the words of neurologist William E. DeMyer [1], "to understand the brain–thought–behavior triumvirate is the Holy Grail of neuroanatomy, as compelling to the researcher as a cyclonic vortex." Models to explain the mechanisms of operation of the human mental apparatus have been formulated since the era of classical Greece. The preponderance of such models appear to be 'tripartite,' without necessarily entailing that e.g. unipartite, bipartite or quintopartite models would be less meaningful.

Figure 1. Left: Portrait of Paul D. MacLean (1913–2007), taken in 2001 by Kelly G. Lambert [81] © 2003 Elsevier Inc., reproduced with permission; signature from the author's private archive. Right: Portrait of Christofredo Jakob (1866–1956), from Orlando's book cover [15]; signature from Jakob's 1924 notes of pathological anatomy and physiology (full reference in [17]), courtesy of Staatsbibliothek Berlin.

The 'triune brain', conceived by neurobiologist-psychiatrist Paul D. MacLean (1913–2007) (Fig. 1), has been popularized way beyond the neurosciences. This becomes evident in the treatments by Arthur Koestler in chapter 16 ('The three brains') of his *Ghosts in the Machine* [2] – the concluding sequel of the trilogy which includes the 1959 *Sleepwalkers: A History of Man's Changing Vision of the Universe* and the 1964 *Act of Creation* –, and by Carl Sagan in chapter 3 ('The brain and the chariot') of his evolutionary saga *The Dragons of Eden* [3]. Further, the triune brain featured in a documentary produced by the National Film Board of Canada [4]. In recent years, the triune modular brain concept has been accommodated by the social sciences: Cory [5] suggested an extended dynamic model of neural social architecture, the 'conflict systems neurobehavioral' (CSN) model, placing more emphasis on behavioral than on neurological terms.

Concerning neuroanatomical terminology, Elliot Smith [6] coined the term *neopallium* in 1901 to denote a cortical organ that had the ability, based on progressive evolution, to learn with experience through a mechanism of sensory perception, associative memory, consciousness and response (Fig. 2). Elliot Smith [7 (pp. 31–38)] regarded the neopallium as fulfilling all the conditions of the Aristotelian *sensorium commune*.

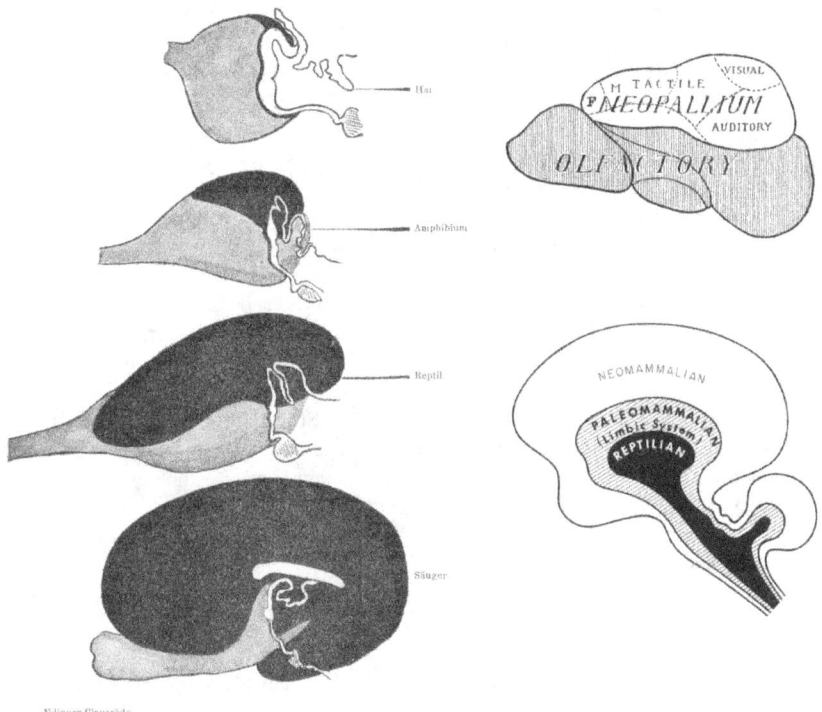

Figure 2. *Left column:* The first original idea of 1908, credited to Edinger, suggesting the evolutionary distinction of the brain into an 'old' *(gray)* and a 'new' part *(black)*. The drawing depicts, top to bottom, the brains of a shark, an amphibian, a reptile and a mammal. The *palaeëncephalon* exists from insects to man, and remains virtually unchanged from shark to elephant. The *neëncephalon* made its first appearance in fish and progressed all the way to humans, filling the entire skull. From figure 8 in Edinger [11 (p. 17)]. *Upper right:* Sir Grafton Elliot Smith introduced the term neopallium in 1901. The diagram shows the lateral aspect of the left cerebral hemisphere of a primitive mammal, the jumping shrew *(Macroscelides)*, and the relatively enormous extent of the primitive olfactory territories and small neopallium with its receptive areas for tactile, visual and auditory impressions. The anterior segment of the tactile area (from *M* back to the dotted line) represents the so-called 'motor area'. *F* represents the frontal area rudiment. From figure 7 in Elliot Smith [7 (p. 33)]. *Lower right:* Schematic representation of the theorized hierarchical organization in the triune brain concept of MacLean [73, 75]. The three basic components, in an ascending evolutionary order, are the *reptilian complex*, the *limbic system* (paleomammalian brain), and the *neocortex* (neomammalian brain), roughly subserving *instincts*, *emotions*, and *intellect*, respectively.

The designations *'paleo-'*, *'archi-'* and *'neo-'* for the cortex (pallium) on grounds of their sequential evolutionary appearance were introduced with a 1909 paper of Ariëns Kappers [8],[1] which subsequently led to similar designations for diverse brain regions, including the striatum, thalamus and cerebellum [10]. Kappers' mentor, Frankfurt anatomist Ludwig Edinger [11] had provided the fundamental intellectual stimulus the previous year, 1908, by systematizing the brain into *palaeëncephalon*, the highly conserved midbrain and hindbrain, and *neëncephalon*, the forebrain, which varies substantially among vertebrates, appearing for the first time in fish and increasing in size and complexity along the phylogenetic scale (Fig. 2). Edinger's core proposition, i.e. that the forebrain has evolved through a sequential addition of parts, is most likely the original idea from which MacLean's 'triune brain' concept derives [12 (p. 31)].

Tilney [13] used the term *archikinetic* to connote balance functions subserved by the medial group of the deep cerebellar nuclei (globose and fastigial), and *neokinetic* for the skilled movements of the face and limbs subserved by the lateral group (emboliform and dentate nucleus) in conjunction with the neopallium and the pyramidal system. According to Tilney [14], *neokinesis* is an extensive group of reactions comprising the externalized expression of the elaborate evolved coordination of eyes, head and hands.

Some of the most pertinent contributions to a neuroevolutionary approach of behavior have come from Christfried (Christofredo) Jakob (1866–1956) (Fig. 1), a German-born neuropathologist who spent most of his professional life in Argentina. Having studied under Friedrich Albert von Zenker (1825–1898) and Adolf von Strümpell (1853–1925) at Erlangen, Jakob subsequently became affiliated with the National Universities of La Plata and of Buenos Aires, where he established one of the most important neuropathological laboratories in South America. He is considered the father of Argentinian neurobiology and forensic histopathology and has left an invaluable legacy of 200 articles and 30 monographs in German and Spanish [15–17].

Jakob reached international renown through his early successful handbooks of clinical medicine [18] and neurology [19], as well as his atlases of human [20] and comparative neuroanatomy [21, 22] (Fig. 3 and 4). He later produced landmark works on evolutionary neurobiology, studying in detail dozens of species of the autochthonus Patagonian fauna, including the broad-snouted 'yacaré' (or *Caiman latirostris*), a reptile of the Alligatoridae family [23], and the 'pichiciego' (fairy armadillo or *Chlamyphorus truncatus*), a mammal of the Dasypodidae family [24].

---

1 In a footnote, Jakob [9 (p. 12)] argues that, to avoid confusion by various authors, the correct chronological order of the three terms on etymological grounds should be archi-, paleo- and neo-, since ἀρχή means origin or beginning, and should therefore precede παλαιός, which means old or ancient.

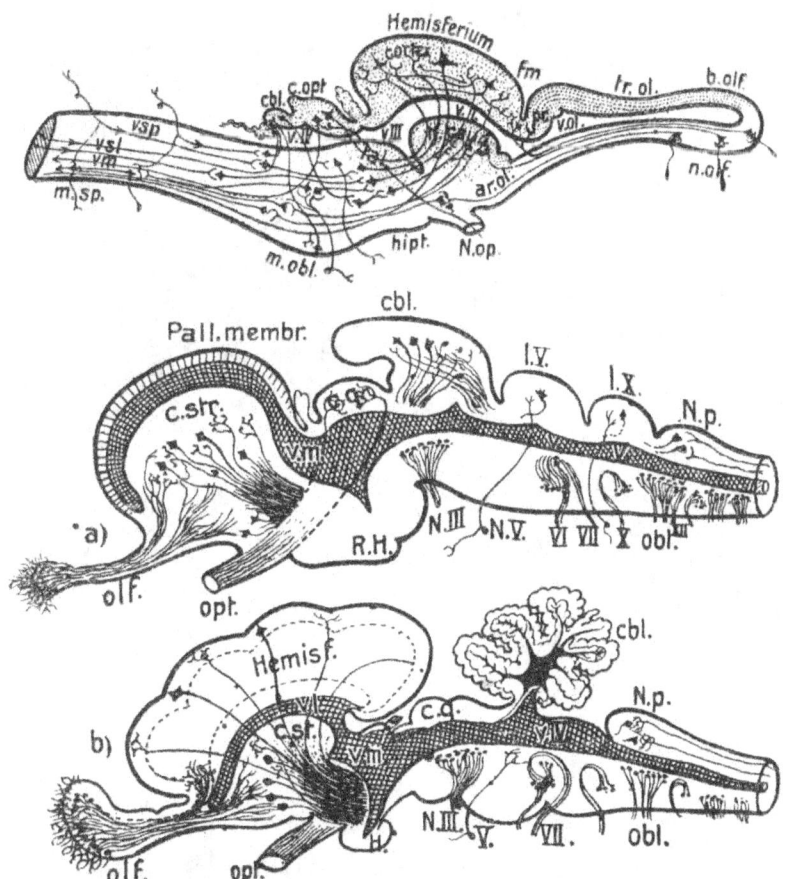

Figure 3. Sagittal schematic views of reflex and central nervous pathways in the brain of a reptile (top), a fish or 'lower vertebrate' (middle) and a mammal or 'higher vertebrate' (bottom). The atlases of comparative neuroanatomy by Jakob and Onelli [21, 22] are rare treasures of evolutionary information. In them, the anatomical properties and connections patterns of cortical (monostratified to polystratified), striatal, hypothalamic and mesencephalic structures, among others, are related from fish through reptiles to primates, as are the underlying functional correlations. Sources: top frame is from figure 7 on p. 8 in [21], also appearing on p. 13 in [22], middle and bottom frames are from figure 1 on p. 6 in [21], also appearing on p. 11 in [22] and in figure 13 on p. 32 in [66]. Note that in the top frame the olfactory bulb is oriented to the right side; in the middle and bottom frames to the left.

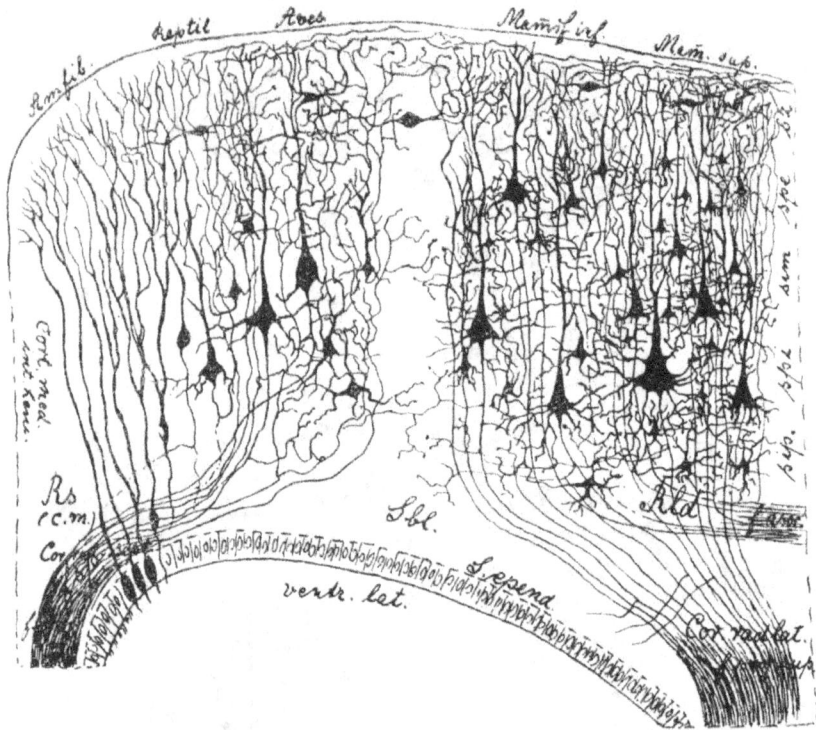

Figure 4. Phylogenetic evolution of the cerebral cortex from monostratified to polystratified. From left to right, amphibian, reptilian, avian, mammalian, and higher mammalian cortical layer structure. Source: from figure 32 on p. 23 in [21], also appearing on p. 16 in [22].

The aim of the present article is first, to revitalize an interest in the evolutionary concepts of Christofredo Jakob, a foremost neuroanatomist of the 20th century and perhaps one of the earliest pioneers in neurophilosophy; and second, to pinpoint the confluences between the anatomical-functional conceptions of Jakob and MacLean regarding their tripartite models of the human mind based on evolutionary criteria, despite a time distance of almost half a century. For a broader perspective, certain other tripartite models of the human mind from the classical antiquity through present are mentioned.

Both Jakob and MacLean invested a considerable part of their long and distinguished careers studying comparative neurobiology from reptiles (Fig. 5) to primates (Fig. 6), and both left their mark on scientific advances regarding the visceral and olfactory brain and the limbic system (Fig. 7). As a matter of fact, there is a whole line of evidence indicating that Jakob described the anatomical elements of the visceral brain some 30 years before the widely known 'circuit of Papez' came into being [25–31]. On the other hand, MacLean's contributions have been instrumental in both defining the limbic system and in sorting out some of its key emotional functions in normality and psychopathology [32–38].

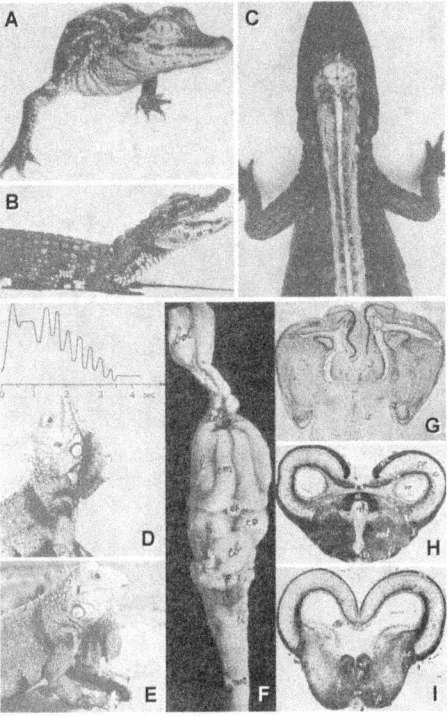

Figure 5. (**A**) A young alligator used for studies on temperature regulation and reproductive behavior at MacLean's Laboratory of Brain Evolution and Behavior at the National Institute of Mental Health [97]. (**B**) The yacaré *(Caiman latirostris)* displaying the innate 'terrorizing' reflex posture, an 'archineuronal brainstem reaction' [53 (p. 143)]. (**C**) Brain and spinal cord of a young yacaré ('alligator of Chaco') dissected in situ. One can discern the cerebral hemispheres, optic tectum, small cerebellum, and the dorsal medulla and spinal cord. From plate 37a in [9 (p. 97)]. (**D, E**) Onset and termination of a stereotypic 'head-nodding' behavior displayed by a male *Iguana iguana* lizard from South America; this type of fixed social signal may have evolved 150 million years ago. From work by MacLean's alumnus Detlev Ploog [100]. (**F**) Dorsal view of the iguana brain. The Ammonic zone or medial cortical *(cm)* and lateral cortical *(cl)* portion of the cerebral hemispheres can be seen, separated by the dorsal longitudinal lateral sulcus, which is the precursor of the hippocampal sulcus. *Abbreviations: bol,* olfactory bulb; *tol,* olfactory tubercle; *ep,* epiphysis (pineal); *co,* optic tectum; *cb,* cerebellum; X, vagus; *bl,* medulla oblongata; *mc,* spinal cord. From figure 86 in [9 (p. 136)]. (**G**) Coronal section of the anterior cerebrum of *Iguana iguana,* at the level of the basal commissure *(cb)* and its continuation towards the hypothalamus. Between the fascia dentata and Ammon's zone, the lateral sulcal depression is found. From figure 87 in [9 (p. 138)]. (**H**) The midbrain of *Iguana iguana.* Entry of the optic nerve *(II)* in the zonal layer *(stz)*; superior *(cs)* and inferior *(ci)* mesencephalic commissure in the tectum; at the bottom is the hypothalamic commissure *(cht).* Weigert stain, ×50. From figure 94 in [9 (p. 150)]. (**I**) Coronal section through the midbrain of *Iguana iguana,* showing the three layers of the optic body, the zonal *(sz),* the intermediate, and the deep *(stp),* as well as its commissure *(cc).* Laterally the optic bundle *(II)* enters into the zonal layer. From the deep layer derives the bulbar quadrigeminal descending fasciculus *(fcb)* also called by various authors tectospinal. In the ventral half one can distinguish the posterior longitudinal fasciculus *(fl),* the reticular formation and the basal (striodiencephalobulbar) fasciculus. Weigert stain, ×90. From plate XXXVIIa in [9 (p. 95)].

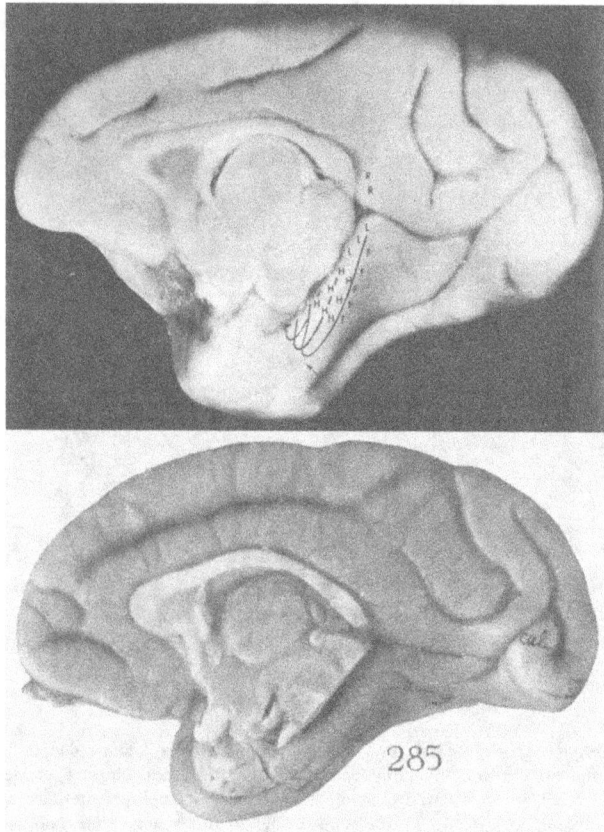

Figure 6. *Upper:* Median facies of the brain of a squirrel monkey to show the approximate location of light-responsive areas. *Abbreviations: H*, posterior hippocampal gyrus; *L*, parahippoampal portion of lingual gyrus; *R*, retrosplenial cortex; *F*, fusiform gyrus. Arrow shows the caudal extreme of rhinal sulcus at the caudal limit of the entorhinal area. From figure 5 in [75 (p. 343)]. *Lower:* Macroscopic brain specimen of a cebus monkey from Paraguay. From plate LXII in [22].

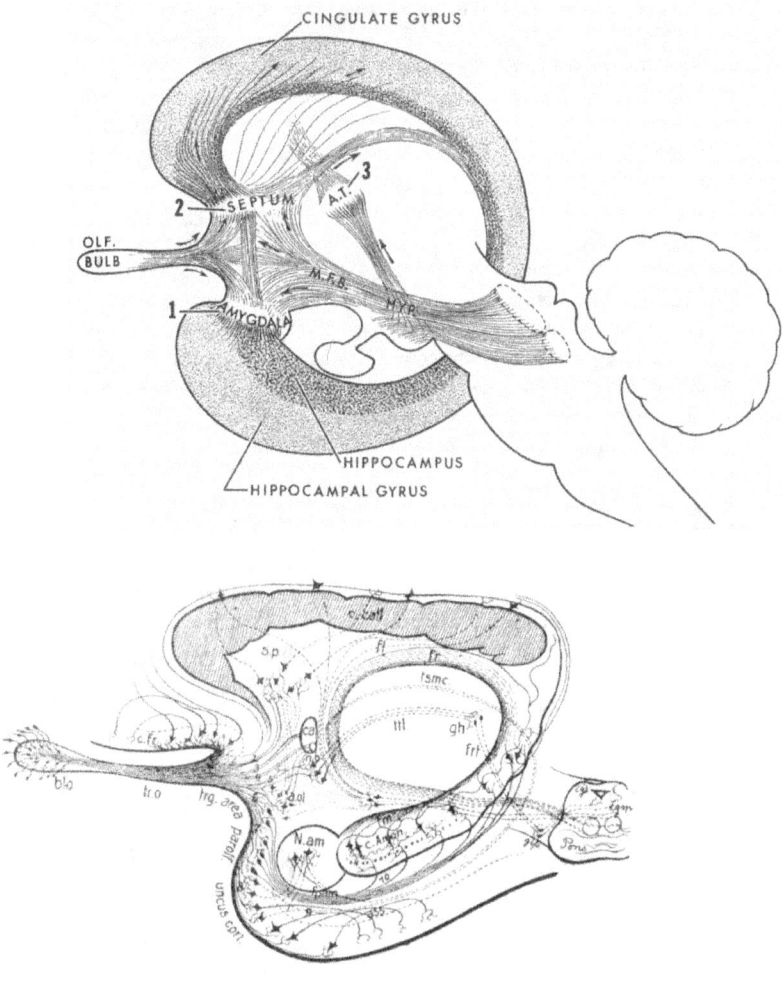

Figure. 7. Drawings of the limbic system and the rhinencephalon (olfactory brain) by MacLean [75] and Jakob [20]. *Upper:* Limbic cortex and its connections with brainstem structures. *Abbreviations:* *A.T.*, anterior thalamic nuclei; *HYP*, hypothalamus; *M.F.B.*, median forebrain bundle; *OLF*, olfactory [75 (figure 4 on p. 341)]. *Lower:* The rhinencephalon and the olfactory radiation, with the anatomical connections of the olfactory bulb, septum, anterior commissure and uncus hippocampi [20 (figure 15 on p. 19)].

PHILOSOPHICAL TRIPARTITE MODELS OF THE PSYCHE

Plato of Athens developed a tripartite concept of the human psyche in *Phaedrus* (370 B.C.) and in the *Republic* (360 B.C.). For him, the cerebrospinal marrow is the organic seat of the 'rational' or 'intelligent' ($\lambda o \gamma \iota \sigma \tau \iota \kappa \acute{o} \nu$), the 'temperamental' or 'courageous' ($\theta \nu \mu o \epsilon \iota \delta \acute{\epsilon} \varsigma$), and the passionate or appetitive ($\acute{\epsilon} \pi \iota \theta \nu \mu \eta \tau \iota \kappa \acute{o} \nu$), respectively occupying the cerebral, thoracic, and abdominal portion [39 (pp. 270–271), 40, 41]. Atomic philosopher Democritus of Abdera (c. 460–370 B.C.) may have anticipated Plato's tripartite division of the psyche [39 (pp. 254–255)], naming the brain the 'guard of the mind' ($\phi \acute{v} \lambda \alpha \xi \ \delta \iota \alpha \nu o \acute{\iota} \eta \varsigma$), the heart the 'control of passion' ($\acute{o} \rho \gamma \tilde{\eta} \varsigma \ \tau \iota \theta \eta \nu \acute{o} \varsigma$), and the liver the 'cause of desire' ($\acute{\epsilon} \pi \iota \theta \nu \mu \acute{\iota} \eta \varsigma \ \alpha \check{\iota} \tau \iota o \nu$).

Burnet [42 (p. 149)] argues that the doctrine of the tripartite psyche was in reality Pythagorean. Among existing suggestions at the time, Plato adopted in *Timaeus* (360 B.C.) the landmark discovery made by Alcmaeon of Croton (500 B.C.) that the brain is the faculty of the mind [43]. Based on that thesis, Socrates speculates in *Phaedo* that the brain provides sensations ($\alpha \check{\iota} \sigma \theta \acute{\eta} \sigma \epsilon \iota \varsigma$); from these arise memory ($\mu \nu \acute{\eta} \mu \eta$) and opinion ($\delta \acute{o} \xi \alpha$); and when stabilized, these become knowledge ($\acute{\epsilon} \pi \iota \sigma \tau \acute{\eta} \mu \eta$) [39 (p. 269)].

Cohen [44] surmises that, because the Bible makes God tripartite with the three persons of the Trinity and teaches that man is made in God's image, theologians have looked for a Biblical tripartite nomenclature for man: a unified totality, comprising the components of 'spirit' ($\pi \nu \epsilon \tilde{v} \mu \alpha$), 'psyche' ($\psi \nu \chi \acute{\eta}$), and 'body' ($\sigma \tilde{\omega} \mu \alpha$), the only Biblical support for a doctrine of human tripartiteness found in the First Epistle to the Thessalonians ($\epsilon'23$).

A noteworthy association from the domain of classical drama – not far from neuroscience in an essential way[2] – is a metaphorical tripartite model of the human psyche discerned in the Karamazov brothers of Fyodor Dostoyevsky's 1880 crowning success, Dmitri representing 'passion', Ivan 'intellect', and Alyosha 'contemplation' [46].

THE TRIPSYCHIC BRAIN SYSTEM OF JAKOB

In his 1923 *Elements of Neurobiology*, Jakob establishes three categories of 'neurodynamic' functions [47 (pp. 197–209)]:

(1) *plasmopsychisms* or *plasmodynamisms* (tropism, taxism, and pulsatile rhythms);
(2) *phylopsychisms* or *neurodynamisms* (simple serial or organized reflexes, otherwise instincts); and
(3) *ontopsychisms* or *psychodynamisms* ('gnosias', 'praxias' and 'symbolias').

Jakob [48] viewed motor and sensory elements and their connections as constituting a unit and a fundamental functional cortical arc that forms the basis of psychological phenomena. In the phylogenetic scale, *plasmopsychic* or *plasmodynamic* activities precede all other nervous functions [9].

---

[2] As highlighted by Steven Pinker [45], Dostoyevsky was not foreign to the nervous underpinnings of behavior. In his prison cell, prompted by a visit by the academician Rakitin, Dmitri Karamazov mulls over the fact that thinking results from quivering nerve tails and the chemistry inside the brain, rather than an immaterial soul.

The above three neurodynamic graduations are all primarily 'biophylactic', i.e. they serve to preserve life, respectively, at its fundamental, the species, and the organism levels [47]. They are found in varying distribution and combination among actual living beings: in the world of plants, as well as in protozoans and sponges one finds only *plasmopsychisms*; in metazoans with ganglionic nervous systems (e.g. hydropolyps, worms, molluscs and arthropods) through the inferior vertebrates (cyclostomes and fish) exist both *plasmopsychisms* and *phylopsychisms*; insects, having a cerebral ganglion, and fish, with their mesencephalon, might potentially exercise elemental *ontopsychic* functions; in amphibians through the higher vertebrates, including humans, one finds all three categories amalgamated and combined, with a predominance of the *ontopsychic dynamism*, which in humans has culminated into the symbolic psychisms (the elements of the intellectual world, of aesthetics and ethics), without though diminishing the concomitant existence of the more elemental levels.

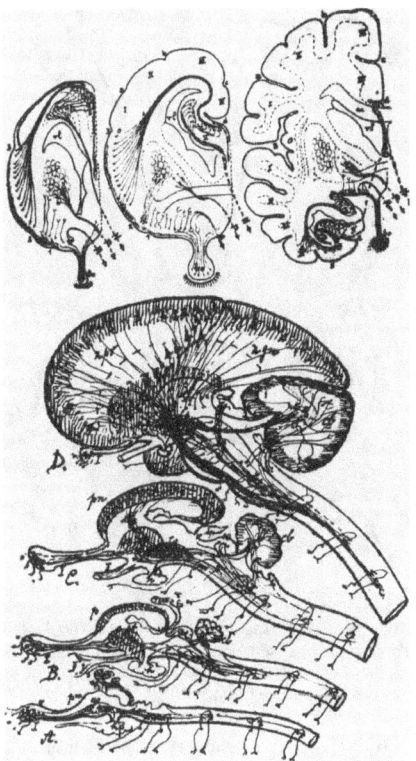

Figure 8. *Upper row:* Schematic drawing of the phylogeny of *archineuronal, paleoneuronal* and *neoneuronal* olfactory systems in a reptile *(left)*, a marsupial *(center)* and a primate *(right)*. From figure 110 in *El Yacaré* [23 (p. 100)]. *Lower group:* Schematic drawing of *primordial neural (A), archineural (B), paleoneural (C)* and *neoneural (D)* systems. From *The Neoencephalon, its Organization and Dynamism* [96 (plate 2)].

There are three underlying structural-functional hierarchical levels of the nervous system, designated as (*a*) *archineuronal*, (*b*) *paleoneuronal* and (*c*) *neoneuronal* (Fig. 8). *Archineuronal* and *paleoneuronal* levels are inherited[3] and constitute the substrate of *phylopsychic* or *neurodynamic* processes; the *neoneuronal* level subserves *ontopsychic* or *psychodynamic* processes, where individual experience becomes possible, and where the will resides [47, 50, 51 (pp. 18–30)].

The *archineuronal* system has a reflex function similar to the invertebrates, comprising simple visceral and somatic reflex arcs ('archikinesias') (Fig. 9, 10A). The *paleoneuronal* system hosts instinctive-automated reactions ('paleokinesias') and it becomes able to prolong the effects of stimuli over time, thus instigating a *chronotropic* ability (Fig. 9). Examples of paleokinetic systems [50, 52] can be found in the cerebellum and the corpus striatum (Fig. 10B, C).

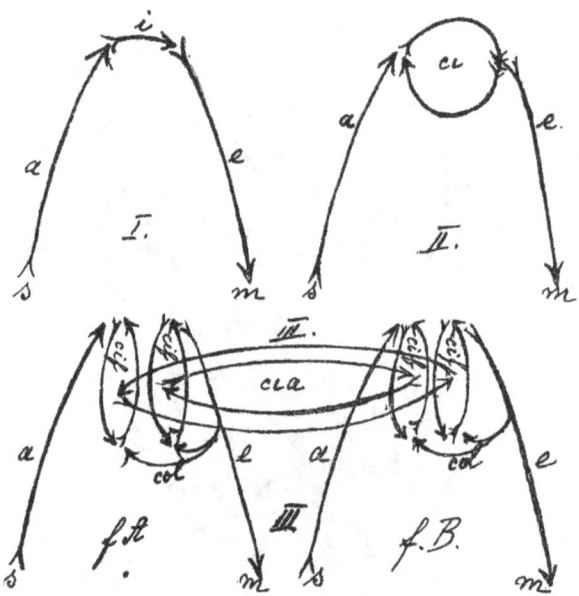

Figure 9. Diagrammatic scheme of *archikinesias (I)*, *paleokinesias (II)* and *neokinesias (III)* according to Jakob [50 (p. 116)]. *Abbreviations: a, s*, sensory afferent system; *e, m*, motor efferent system; *i*, intercalated system; *ci*, intercalated microdynamism; *fA, fB*, cortical foci A and B; *cif*, focal intercalated microdynamism; *cia*, associative intercalated circuit; *col*, motor collateral

The higher *neoneuronal* system subserves conscious acts, individuality and consciousness ('neokinesias') (Fig. 9, 11); it comprises two sectors, the limbic cortex ('introyente' or endogenous sphere) and the lateral cortex ('ambiente' or external environment). Time responses vary approximately from 20–30 msec for the 'archikinesias'

---

[3] In his 1969 Hincks Memorial Lecture at Queen's University in Ontario, MacLean [49] begins his tripartite diatribe with 'Man's reptilian and limbic *inheritance*' (my italics).

and from 200–300 msec for the 'neokinesias' [53 (pp. 35–37)];[4] in the latter instance, the major distance can be established through a central or volitional neokinetic transformation in a reaction time of 110–120 msec (the "fourth dimension of thought").

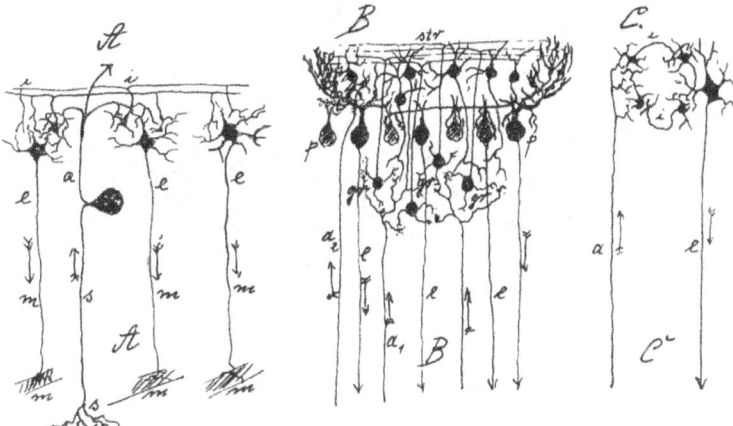

Figure 10. Histological organization of intercalated systems in (A) archikinesias, (B) cerebellar paleokinesias, and (C) striatal paleokinesias according to Jakob [50 (p. 119)]. *Abbreviations: a*, afferent; *e*, efferent; *m*, motor; *s*, sensory; a_1, mossy fiber; a_2, climbing fiber; *P*, Purkinje cells; *gr*, granule cells; *str*, stellate cells; *i*, intercalated element (interneuron).

Within the framework of the dynamic workings of the human cerebral cortex, the 'neokinesias' include (*i*) 'gnosias' (cognitive processes related to conscious orientation in one's environment), (*ii*) 'praxias' (individual active intervention) and (*iii*) 'symbolias' (ideative abstraction to facilitate interindividual communication, such as the sociogenetic processes on which human culture is based) [47, 50, 53 (pp. 41–66), 60; cf. also 29, 61, 62 (pp. 297–303), 63–65]. Jakob [66 (p. 16)] later elaborated on these three concepts in his *Documenta Biofilosófica*, commenting that 'gnosia' (attentive orientation), 'praxia' (active or passive intervention) and 'symbolia' (communicative verbal formulation) represent three intimately linked sensory-motor phases that together form the true psychogenetic trinity ('la verdadera trinidad psicocreadora'), from which "experience and thought are revealed as amalgamated neurodynamic realities."

Jakob treated the theme of the phylogeny of the *archicortex*, *paleocortex* and *neocortex* in greater detail in the second part of his 1945 monograph on the yacaré [23 (pp. 99–109)].

---

4 These are remarkable calculations for having been written in the 1940s, if one considers that current views hold that consciousness arises from neurons in about 500 msec – the 'neural time factor' [54] – or less [55]; in other words, it takes a fraction of a second between the occurrence of a physiological stimulus in the parietal cortex and a subject becoming conscious of it [56], with the mind compensating for real time through a 'backward referral' experience [57]. On the efferent side, brain potentials fire a little over 300 msec before one has the conscious intention to act [58] and cerebral potentials may precede finger movement by up to 1 sec [59].

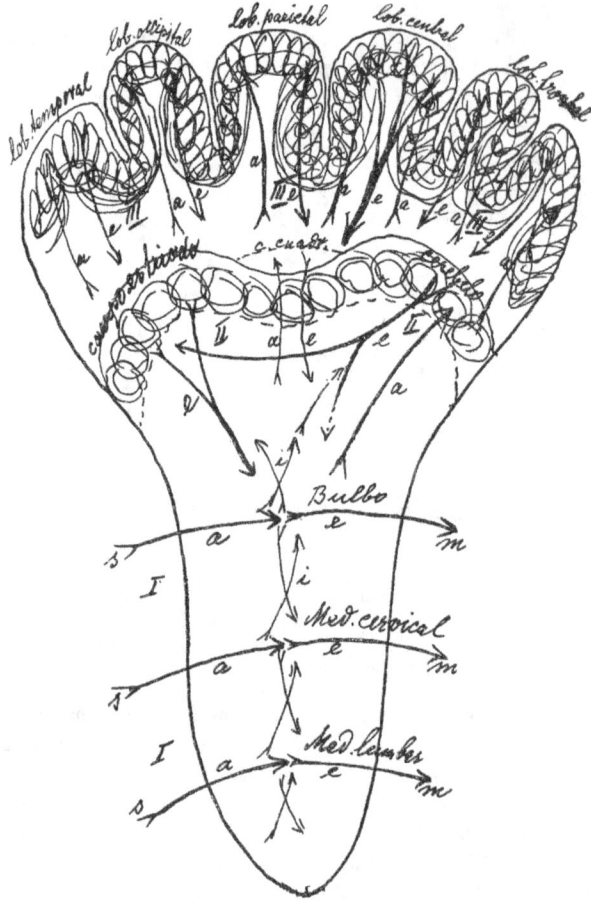

Figure. 11. Topographic depiction of the systems of *archipsychism (I)*, *paleopsychism (II)* and *neopsychism (III)* according to Jakob [50 (p. 127)].

THE TRIPARTITE MENTAL APPARATUS OF PSYCHOANALYSIS

Freud formally introduced the full scope of his infamous three-fold model of the mental apparatus – comprising the id ('das Es'), the ego ('das Ich'), and the super-ego or ego ideal ('das Über-Ich' oder 'Ich-Ideal') – in his 1923 book *Das Ich und das Es* [67], which was translated into English as *The Ego and the Id* [68]. Freud revisited and supplemented the theme a decade later (1933) in his *New Introductory Lectures* [69].

Jacobson [70] argues that Freud's contributions to the neuron doctrine may have been overestimated as part of "the blatant hero worship," which led to a reverberation of psychoanalytic thought in the new field of neuroscience. The theory of MacLean [71] on the processing of emotions by a network of deep brain structures seems to have found a general acceptance in the 1950s because of its conceptual correspondence to psychoanalytic theory [72]. As a matter of fact, MacLean [71] makes the point that "considered in the light of Freudian psychology, the visceral brain would have many of the attributes of the unconscious id." Regarding the assumed phylogenetic components of the 'triune brain,' MacLean [49, 73] does mention a correspondence to Freud's id, ego, and super-ego [74].

Sagan [3 (p. 79)] describes MacLean's triune brain model as being only in weak accord with the tripartite mind of psychoanalytic theory and stresses that, owing to the neuroanatomical connections between its three component parts, the triune brain must be useful, just like the metaphor for the human psyche found in Plato's *Phaedrus*. In that dialogue, Socrates likens the human soul to a chariot drawn by two horses, a black and a white, pulling in different directions and weakly controlled by the charioteer. Freud also described the *ego* as the rider of an unruly horse. Sagan [3] argues that an even better metaphor could be Freud's division of the mind into *conscious, preconscious* and *unconscious*.[5] A remarkable similarity can be seen between the metaphor of Plato's chariot and MacLean's neural substrates, the reptilian and paleomammalian brains corresponding to the two horses, and the neomammalian brain to the charioteer.

Freud's and Jakob's books, describing their respective landmark models of the tripartite mind and the tripsychic brain, were both published in the year 1923. Both investigators attribute a special emphasis to the body's projection pattern on the cerebral cortex: having a sound background and an early successful career in neuroanatomy, Freud [68] describes the ego as being "first and foremost a bodily ego, not merely a surface entity, but itself the projection of a surface." According to Freud, a neurological analogy for the ego is its identification with the "cortical homunculus of the anatomists." In a footnote first appearing in the authorized English translation of 1927 (absent from the German edition), this is further explained as the ego being "ultimately derived from bodily sensations, chiefly those springing from the surface of the body;" thus, it can be regarded as "a mental projection of the surface of the body, besides representing the superficies of the mental apparatus."

In his atlases of comparative and human neuroanatomy, Jakob [20–22] stresses in more detail the topographic dissociation of the body's projection pathways onto the lateral and medial surfaces of the cerebral hemispheres, depending on the perception of exogenous or endogenous signals, and introduces his pioneering concepts of the formation of sectors and pre-gyri, as well as the concept of a visceral brain with an anatomical correlate in the cingulate gyrus (reviewed in [31]).

---

5 In his paper on the phylogeny of the kinesias, Jakob [50] mentions a correspondence of plasmopsychic through archipsychic functions with the unconscious, of paleopsychic functions with the preconscious (which genetically precedes conscious phenomena, and thus is differentiated from the subconscious, which he considers part of the conscious), and of neopsychic functions with the conscious (Fig. 10).

THE TRIUNE BRAIN CONCEPT OF MACLEAN

Paul MacLean's conception of a 'triune' pattern and organization of the human brain includes three fundamental brain types or subsystems – each with its own subjectivity, intelligence, spacetime sense, memory, motor and other functions – intermeshing as one. Going phylogenetically and hierarchically from older to newer, over the past 200 million years, these are [49, 73, 75–78]:

(1) The *reptilian* brain at the base, the most ancient heritage, first evolved in primeval reptiles and the great lizards, forms the matrix of the upper brainstem and comprises most of the reticular formation, the midbrain, the hypothalamus, the basal ganglia (archipallium), and the cerebellum; it mediates biological and endocrine equilibrium, via stereotypic behaviors and vital functions, such as survival and self-preservation, sleep-wakefulness regulation, drinking and feeding, mating, territorial possession, imitation, aggression, flight and ritualized combat, and the establishment of social hierarchies. It assures an immediate response in the present and it is privileged concerning olfaction.

(2) The *paleomammalian* brain, the intermediate type that is distinguished by the presence of a primitive cortex, the limbic cortex; the limbic system is central in mediating feelings, emotions and affect – which necessitates a long-term memory and the motivation associated with it –, play, the sense of reality of oneself and the environment, the conviction of what is true or important, the recognition of offspring and parental care. It makes its appearance in the early mammals and it is based on the importance of vocalization and audition.

(3) The *neomammalian* brain, which 'mushroomed' late in evolution and is characterized by the more highly differentiated form of the neocortex (neopallium); it forms the basis of interpersonal communication via spoken and written language, arithmetic, rationality, creative abilities, and the intellect. Typical of humans, the neomammalian brain with its evolved frontal lobes, subserves (rather than 'understands', cf. [79]) reason and symbolic language; it is 'privileged' with regards to vision, abstraction, association, imagination and future anticipation.

According to the triune brain theory, the integrated human brain is a synthesis of the three successively evolved 'component' brains, which, while operating as a whole, retain their original attributes and functions. Limbic regions (e.g. anterior cingulate, medial frontal and insular), phylogenetically conserved to a higher degree than the neocortical regions which mediate cognitive capacities, are associated with emotional responses and their intrinsic affective attributes [73, 80]. The types of mental function for which the reptilian, limbic and neomammalian brain are responsible, are respectively assigned by MacLean [73] as 'protomentation', 'emotional mentation' and 'rational mentation'. In the relatively rapid human evolution, the three brain subsystems are imperfectly integrated, influencing individual and social behaviors that are under the commands of each individual or collective act.

MacLean chose the term 'triune' for his evolutionary brain theory because of the literal meaning of the word (three-in-one); the fact that his father was a Presbyterian minister may have something to do with such a choice, although MacLean reportedly regretted it [81], due to the potential confounding with the Christian doctrine.

FURTHER TRIPARTITE MODELS OF THE HUMAN MIND

In a temporal confluence, Knopp [82], studying patients with Gilles de la Tourette disease and acute schizophrenia, proposed – in the same year as MacLean [75] formally introduced the triune brain concept – a tripartite brain concept as well, whereby a balanced and continuous reconciliation among three components or at least two levels of central integration constitutes the 'rational brain'. According to Knopp's proposition, the limbic lobe, the hypothalamic network and the autonomic nervous system neurophysiologically subserve *visceral experience*, the nigrostriatal system and the limbic lobe *(conscious) emotional experience*, and the neocortex *social experience (effectuation)*.

In considering tripartite models of the mind, the synthetic neurophilosophical construct of Popper and Eccles [83] comes to mind, accounting for brain-mind interaction by means of their World 1 (liaison brain), World 2 (outer and inner sense and the self), and World 3 (cultural heritage).

In the cybernetic view, the nervous system operates as a tripartite system that involves *input* (sensory receptor), *integrating*, and *output* (muscle contraction) components [84]; this is based on a 19th century notion of anatomists and physiologists, and an ubiquitous principle of physiological psychology in the 20th century, of grouping brain structure and function into *sensory*, *association*, and *motor* systems [85]. Such properties accrete from the most fundamental properties of the nerve cell as a *receiving*, *conducting*, and *transmitting* functional unit, and the principle of dynamic polarization of neurons [86], which is at the core of neurobiology.

The triarchic model of the human mind propounded by Sternberg [87, 88] consists in *analytical*, *practical*, and *creative* sides of intelligence. Tigner and Tigner [89] argue that such a model may reflect the triptych developed by Aristotle of Stageira in his *Nicomachean Ethics* (350 B.C.). The Greek philosopher proposed a system of human intelligence comprising three virtues: 'theoretical' (διανόησις), 'practical' (φρόνησις), and 'productive' (ποίησις). The parallel conclusions of temporally disjoined inquiries by investigators working under disparate historical circumstances does not seem to be unusual in the historical evolution of psychological thought [89].

At the level of mind organization, Fodor [90] proposed a tripartite division of cognitive mechanisms, involving physical energy→symbol *transducers*, which record stimuli from the outside world and convert them into neural code, specialized modular *input systems*, which process raw data derived from the transducers, and a central *processor*, the site of higher cognitive processes that receives the outcome of input systems [91].

An established philosophical taxonomy, providing a tripartite structure of consciousness, divides it into *phenomenal*, *reflective*, and *access* consciousness [92, 93]. Phenomenal consciousness subserves the subjective feeling of mental content [94]; reflective consciousness is defined as the direct availability of the process of mental activities for access that allows one to access the steps of the reasoning process, whereas access consiousness is the direct availability of mental content that allows one to access the outcome of a reasoning process [93].

DISCUSSION

Based on the ideas exposed above, it becomes apparent that tripartite models abound in the attempt to shed light into the workings of the human brain and mind; they span over a spectrum from the philosophical to the biological. To attempt to draw parallels or pinpoint differences among various components in such a pleiad of systems forms a daunting task. Nonetheless, in considering the particular models in the evolutionary domain, Jakob's proposition of three hierarchically appearing 'psychisms' in the nervous system [47] seems to anticipate, by several decades, the 'triune brain' concept of MacLean [49, 75].

During his upbringing in Germany, Jakob acquired a strong background in philosophy [15] and throughout his career, Jakob [66, 95] maintained a philosophical perspective on biology in general and on neurobiology in particular (Fig. 12); he realized some of the earliest interpretations of the limbic or 'internal' brain as a viscero-emotional mechanism and the bidirectional communication between the internal organs and the splenial cortex, and the external environment and the lateral cortex [21, 22]. Although Jakob kept a distance from peripatetic metaphysics, he nevertheless adopted, in strictly biological terms, the Aristotelian notion of the psychic character (or 'psychism'): "That neurobiophylactic [neural life-protecting] complex of neuroenergetic reception, assimilation and reaction, which regulates the organism's vital necessities against variable factors in the external and inner environment, I call *psychism*" [96 (p. 8)]. In Jakob's synthesis, one can trace the influence of the precepts of his mentor von Strümpell, as well as of Italian positivism, particularly the views of philosopher Giovanni Marchesini (1868–1931) [30 (pp. 115–116)].

MacLean, on the other hand, planned to study philosophy before ending up studying medicine; he eventually proposed the creation of a new branch of knowledge ("epistemics") to look at things "from the inside out", combining neuroscience and psychology in an attempt to explain the subjective self and its relationship to the external and internal environment [97].

Jakob [47] attributed a life-preserving role ('biophylaxis') to phylopsychism – at the species – and ontopsychism – at the individual – level. MacLean highlighted the importance of the reptilian complex for integrating behaviors involved in self-preservation and in the preservation of the species [97].

Jakob [60] named 'praxias' the active individual intervention processes within the framework of the dynamic workings of the cerebral cortex, and treated 'tropism' within the spectrum of fundamental neurodynamic functions [47]; he further presented his ideas on life and mental experience relative to time at the basal, phylogenetic and ontogenetic levels in a paper under the encompassing title 'From tropism to the general theory of relativity' [98]. MacLean coined terms such as 'isopraxic' – to denote behaviors in which multiple members of a group perform the same thing – and 'tropistic' – to connote a behavior responding to partial representations of things, such as the marking on a prey [97].

In the 1990s, MacLean [99] placed special emphasis on the three cortical types that evolved in the forebrain of mammals, from mammalian-like reptiles (therapsids) through humans, to describe the idea of how resonance may contribute to the dynamic excitability of neural circuits by representing possible algorithms in the nervous system at the macroscopic, microscopic, molecular, and atomic level that underscore mental states and solutions for immediate or eventual actions. He suggested that the forebrain is particularly important as a 'central processor' for mental experience, including sensation, perception, drive, affect, thought, and the precise facts of science, and concluded that a refined picture of the structure and chemistry of the brain's circuitry accounting is needed, especially because of the central question of whether one may ever rely on the brain with its viscoelastic properties to reliably measure time and space and the general nature of things [99].

Over six decades earlier, in his classical exposé on the phylogeny of the kinesias, Jakob [50] employed the time spread of 'corollary discharge' – a mechanism proposed by Helmholtz, allowing a receiving system to either take into account or ignore self-generated sensory input during the monitoring of self-generated movements – in attempting to substitute *semovience* (the mind's capacity to generate new causal actions through experienced inner life). Jakob explained such a substitution with the mediation of axon collaterals, which upload neural activity into focal intercalated microcircuits, and an acquired 'associative system' [30 (pp. 121, 144)].

The ideas and propositions independently formulated by Jakob and by MacLean, despite their time distance and different environments, reflect a substantial convergence in their reasoning. Thus, studies attempting to penetrate human behavior by means of evolutionary hints, might benefit by taking a closer look at Jakob's contributions, which have largely remained unheeded in the English scientific literature and which contain information potentially pertinent to current problems in psychobiology.

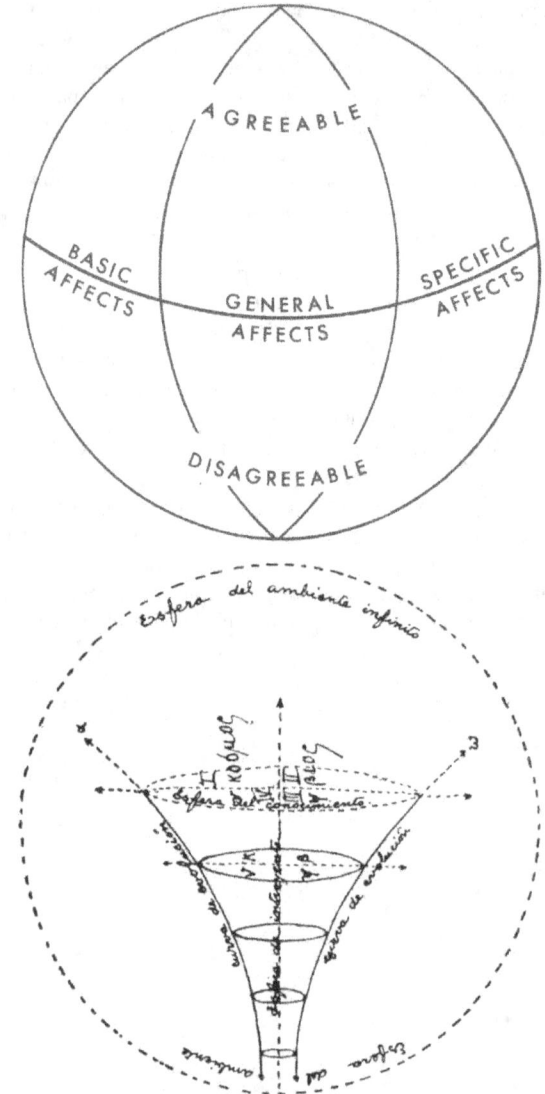

Figure 12. Upper: MacLean's [75 (p. 338)] scheme for viewing the world of affects, subjectively qualified into either agreeable or disagreeable. MacLean maintains that affects cannot be neutral, because "emotionally speaking, it is impossible to feel unemotional." He further subdivides affects into basic (informative about basic bodily needs), general (pertinent to situations, individuals or groups and the preservation of self or the species) and specific (occurring with activation of specific sensory systems). Lower: Jakob's [95] scheme of the empirical sphere with the four quadrants of the sciences: I. Cosmos (κόσμος), II. Life (βίος), III. Psyche (ψυχή) and IV. Order (νόμος). The sphere of the infinite environment is invaded by that of progressive knowledge, delimited by the curve of evolution.

CONCLUDING REMARKS

Born an ocean and five decades apart, Christofredo Jakob and Paul MacLean left their marks of productivity as interdisciplinary neuropsychiatrists, with special emphasis on comparative neurobiology and brain evolution. It is worth noting that the two investigators also died half a century apart, and published their tripartite models in 1923 and in 1970, respectively, when they both were 57 years of age. The fact that they never met, but formulated convergent propositions, can only lead one to recall the motto, "great minds think alike."

The extent to which the two neuroevolutionary constructs are empirically supported remains a controversial issue even today. Considering the recurring triune theories on the underpinnings of the mind, witnessed over twenty-five centuries of human inquiry, one may ponder over a putative tendency of the human brain itself to construct or understand tripartite models of reality. Perhaps the hypothesis of an evolutionary trend to develop tripartite models of existence could some day be put to rigorous experimental testing.

REFERENCES

[1]   DeMyer, W. (1988). *Neuroanatomy*. Baltimore–Malvern: Williams & Wilkins/Harwal Publishing Co., p. 315.

[2]   Koestler, A. (1967). *The Ghost in the Machine*. London: Hutchinson & Co. Ltd., pp. 267–296.

[3]   Sagan, C. (1977). *The Dragons of Eden: Speculations on the Evolution of Human Intelligence*. New York: Random House, pp. 49–79.

[4]   Thérien, G., Verrier, H., & Beaudet, M. (1984). *Les Trois Cerveaux*. Montréal: Office National du Film du Canada.

[5]   Cory, G. A. Jr. (2003). MacLean's evolutionary neuroscience and the conflict systems neurobehavioral model: some clinical and social policy implications. In A. Somit, & S. A. Peterson (Eds.), *Human Nature and Public Policy: An Evolutionary Approach*. New York–Hampshire: Palgrave Macmillan, pp. 161–180.

[6]   Elliot Smith, G. (1901). The natural subdivision of the cerebral hemisphere. *Journal of Anatomy and Physiology (London)*, 35, 431–454.

[7]   Elliot Smith, G. (1927). *The Evolution of Man: Essays*, 2nd edn. London: Humphrey Milford–Oxford University Press.

[8]   Ariëns Kappers, C. U. (1909). The phylogenesis of the paleocortex and archicortex compared with the evolution of the visual neocortex. *Archives of Neurology and Psychiatry (London)*, 4, 161–173.

[9]   Jakob, C. (1941). *El Cerebro Humano: Su Ontogenía y Filogenía* (Folia Neurobiológica Argentina, Atlas III). Buenos Aires: Aniceto López.

[10]   Swanson, L. W. (2000). What is the brain? *Trends in Neurosciences*, 23, 519–527.

[11]   Edinger, L. (1909). Die Beziehungen der vergleichenden Anatomie zur vergleichenden Psychologie, Neue Aufgaben. In L. Edinger, & E. Claparède (Eds.), *Über Tierpsychologie: Zwei Vorträge* (aus dem Bericht über den III. Kongreß für Experimentelle Psychologie, 1908). Leipzig: J. A. Barth, pp. 1–30.

[12] Striedter, G. F. (2005). *Principles of Brain Evolution*. Sunderland, MA: Sinauer Associates, Inc.

[13] Tilney, F. (1927). The chief intracerebellar and precerebellar nuclei, with a demonstration with models and charts. *Brain*, 50, 275–276.

[14] Tilney, F. (1927). The brain stem of tarsius. A critical comparison with other primates. *Journal of Comparative Neurology*, 43, 371–432.

[15] Orlando, J. C. (1966). *Christofredo Jakob: Su Vida y Obra*. Buenos Aires: Editorial Mundi. http://electroneubio.secyt.gov.ar/Vida_y_obra_de_Christofredo_Jakob.htm

[16] Triarhou, L. C., & del Cerro, M. (2006). Semicentennial tribute to the ingenious neurobiologist Christfried Jakob (1866–1956). 1. Works from Germany and the first Argentina period, 1891–1913. *European Neurology*, 56, 176–188.

[17] Triarhou, L. C., & del Cerro, M. (2006). Semicentennial tribute to the ingenious neurobiologist Christfried Jakob (1866–1956). 2. Publications from the second Argentina period, 1913–1949. *European Neurology*, 56, 189–198.

[18] Jacob, C. (1898). *A Klinikai Vizsgálati Módszerek Atlasza a Belgyógyászati Diagnostica és a Különös Kór- és Gyógytan Alapvonalaival* (fordította Ritoók Zsigmond). Budapest: Singer és Wolfner Kiadása.

[19] Jakob, C. (1899). *Atlante del Sistema Nervoso nello Stato Sano e nel Patologico con un Sunto di Anatomia Patologica e Terapia del Medesimo*. Milano: Società Editrice Libreria.

[20] Jakob, C. (1911). *Das Menschenhirn: Eine Studie über den Aufbau und die Bedeutung seiner Grauen Kerne und Rinde*. München: J. F. Lehmann.

[21] Jakob, C., & Onelli, C. (1911). *Vom Tierhirn zum Menschenhirn: Vergleichend Morphologische, Histologische und Biologische Studien zur Entwicklung der Grosshirnhemisphären und ihrer Rinde. I. Teil. Tafelwerk nebst Einführung in die Geschichte der Hirnrinde*. München: J. F. Lehmann.

[22] Jakob, C., & Onelli, C. (1913). *Atlas del Cerebro de los Mamíferos de la República Argentina. Estudios Anatómicos, Histológicos y Biológicos Comparados sobre la Evolución de los Hemisferios y de la Corteza Cerebral*. Buenos Aires: Imprenta de Guillermo Kraft.

[23] Jakob, C. (1945). *El Yacaré (Caiman latirostris) y el Orígén del Neocortex: Estudios Neurobiológicos y Folklóricos del Reptil más Grande de la Argentina* (Folia Neurobiológica Argentina, Tomo IV). Buenos Aires: Aniceto López.

[24] Jakob, C. (1943). *El Pichiciego (Chlamydophorus Truncatus): Estudios Neurobiológicos de un Mamífero Misterioso de la Argentina* (Folia Neurobiológica Argentina, Tomo II). Buenos Aires: Instituto de Biología de la Facultad de Filosofía y Letras.

[25] Barraquer-Bordas, L. (1954). *Fisiología y Clínica del Sistema Límbico: Aportaciones Recientes al Conocimiento de las Bases Neurofisiológicas de la Personalidad*. Madrid: Paz Montalvo, pp. 136–137.

[26] Barraquer-Bordas, L. (1976). *Neurología Fundamental: Fisiopatología, Semiología, Síndromes, Exploración*, 3rd edn. Barcelona: Toray, p. 596.

[27] Orlando, J. C. (1964). Sobre el cerebro visceral. Documentación histórica de una prioridad científica. *Revista Argentina de Neurología y Psiquiatría*, 1, 197–201.

[28] Goldar, J. C. (1975). *Cerebro Límbico y Psiquiatría*. Buenos Aires: Salerno.

[29] Faccio, E. J. (1991). Christofredo Jakob y el origen del psiquismo. *Alcmeón – Revista Argentina de Clínica Neuropsiquiátrica*, 3, 331–348.

[30] Szirko, M. (1995). A la antropología ganglionar desde la kinesiología: un fallido ensayo de extrapolar lo orgánico (Noticia preliminar). *Electroneurobiología (Buenos Aires)*, 2, 104–169.

[31] Triarhou, L. C. (2008). Centenary of Christfried Jakob's discovery of the visceral brain: An unheeded precedence in affective neuroscience. *Neuroscience and Biobehavioral Reviews*, 32, 984–1000.

[32] MacLean, P. D. (1969). The hypothalamus and emotional behaviour. In W. Haymaker, E. Anderson, & W. J. H. Nauta (Eds.), *The Hypothalamus*. Springfield, IL: Charles C Thomas, pp. 659–678.

[33] Marino, R. Jr. (1975). *Fisiologia das Emoçoes: Introduçao à Neurologia do Comportamento, Anatomia e Funçoes do Sistema Limbico*. São Paulo, Brasil: Sarvier S.A. Editora do Livros Médicos.

[34] MacLean, P. D. (1985). Evolutionary psychiatry and the triune brain. *Psychological Medicine*, 15, 219–221.

[35] Lautin, A. (2001). *The Limbic Brain*. New York: Kluwer Academic–Plenum Publishers, pp. 69–98.

[36] Ploog, D. (2003). The place of the triune brain in psychiatry. *Physiology and Behavior*, 79, 487–493.

[37] Perna, G. (2005). *Las Emociones de la Mente: Biología del Cerebro Emotivo*. Madrid: Tutor.

[38] Reep, R. L., Finlay, B. L., & Darlington, R. B. (2007). The limbic system in mammalian brain evolution. *Brain Behavior and Evolution*, 70, 57–70.

[39] Beare, J. I. (1906). *Greek Theories of Elementary Cognition from Alcmaeon to Aristotle*. Oxford: Clarendon Press.

[40] Stocks, J. L. (1915). Plato and the tripartite soul. *Mind – A Quarterly Review of Philosophy*, 24, 207–221.

[41] Georgulis, K. D. (1957). Plato. *Helios Encyclopaedical Lexikon (Athens)*, 16, 7–81.

[42] Burnet, J. (1920). *Early Greek Philosophy*, 3rd edn. London: A. & C. Black, Ltd..

[43] Doty, R. W. (2007). Alkmaion's discovery that brain creates mind: A revolution in human knowledge comparable to that of Copernicus and of Darwin. *Neuroscience*, 147, 561–568.

[44] Cohen, E. D. (1988). *The Mind of the Bible-Believer*. Amherst, NY: Prometheus Books, p. 121.

[45] Pinker, S. (2002). *The Blank Slate: The Modern Denial of Human Nature*. London: Allen Lane, p. 85.

[46] Edgeworth, R. J. (1994). The tripartite soul in Plato, Dostoyevsky, and Aldiss. *Neophilologus*, 78, 343–350.

[47] Jakob, C. (1923). *Elementos de Neurobiología, volumen I: Parte Teórica* (Biblioteca Humanidades, Tomo III). La Plata, Argentina: Facultad de Humanidades y Ciencias de la Educación de la Universidad Nacional de La Plata.

[48] Jakob, C. (1914). La psicología orgánica y su relación con la biología cortical (Referat von R. Allers). *Zeitschrift für die Gesamte Neurologie und Psychiatrie*, 9, 804–805.

[49] MacLean, P. D. (1973). *A Triune Concept of the Brain and Behaviour* (The Clarence Hincks Memorial Lectures, Volume 2, edited by T. J. Boag and D. Campbell).

Toronto–Buffalo: Ontario Mental Health Foundation–University of Toronto Press, pp. 1–66.

[50] Jakob, C. (1935). La filogenia de las kinesias: Sobre su organización y dinamismo evolutivo. *Anales del Instituto de Psicología de la Facultad de Filosofía y Letras de la Universidad de Buenos Aires*, 1, 109–127.

[51] Jakob, C. (1940). *Ontogenia del Sistema Nervioso Humano*. La Plata–Buenos Aires: Universidad Nacional de La Plata, Facultad de Humanidades y Ciencias de la Educación–Imprenta López.

[52] Jakob, C. (1936). *La Organización Subcortical del Sistema Nervioso Central de los Vertebrados Superiores: El Paleoencéfalo y sus Funciones Instintivas*. La Plata–Buenos Aires: Facultad de Humanidades y Ciencias de la Educación, Universidad Nacional de La Plata–Imprenta y Casa Editora «Coni».

[53] Jakob, C. (1941). *Neurobiología General* (Folia Neurobiológica Argentina, Tomo I). Buenos Aires: Aniceto López.

[54] Libet, B. (1999). How does conscious experience arise? The neural time factor. *Brain Research Bulletin*, 50, 339–340.

[55] Gazzaniga, M. (1998). *The Mind's Past*. Berkeley–London: University of California Press, pp. 69–82.

[56] Libet, B., Alberts, W. W., Wright, E. W. Jr., Delattre, L., Levin, G., & Feinstein, B. (1964). Production of threshold levels of conscious sensation by electrical stimulation of human somatosensory cortex. *Journal of Neurophysiology*, 27, 546–578.

[57] Libet, B., Wright, E. W. Jr., Feinstein, B., & Pearl, D. K. (1979). Subjective referral of the timing for a conscious sensory experience: a functional role for the somatosensory specific projection system in man. *Brain*, 102, 193–224.

[58] Libet, B., Gleason, C. A., Wright, E. W. Jr., & Pearl, D. K. (1983). Time of conscious intention to act in relation to onset of cerebral activities (readiness potential): the unconscious initiation of a freely voluntary act. *Brain*, 106, 623–642.

[59] Kristeva, R., Keller, E., Deecke, L., & Kornhuber, H. H. (1979). Cerebral potentials preceding unilateral and simultaneous bilateral finger movements. *Electroencephalography and Clinical Neurophysiology*, 47, 229–238.

[60] Jakob, C. (1921). La teoría actual de las gnosias y praxias como factores fundamentales en el dinamismo de la corteza cerebral. *Crónica Médica (Lima)*, 38, 17–24.

[61] Szirko, M. (1991). Definición de psiquismo y de conocimiento sensible, retención de las memorias, evolución del sistema nervioso, y relaciones mente-cuerpo o nexo psicofísico, en la Escuela Neurobiológica Argentino-Germana. *Electroneurobiología (Buenos Aires)*, 1 [Supl. 2], I–XIV.

[62] Kurowski, M. (2001). *La Obra Psicológica de Juan Cuatrecasas Arumí (1899–1990)*. Madrid: Universidad Complutense, Facultad de Psicología.

[63] Crocco, M. (2004). ¡Alma e' reptil! Los contenidos mentales de los reptiles y su procedencia filética. *Electroneurobiología (Buenos Aires)*, 12, 1–72.

[64] Capizzano, A. A. (2006). Actualidad del pensamiento de Cristofredo Jakob. *Revista del Hospital Italiano de Buenos Aires*, 26, 71–73.

[65] Crocco, M. (2007). Christofredo Jakob. In *Wikipedia, la Enciclopedia Libre*. http://es.wikipedia.org/wiki/Christofredo_Jakob

[66] Jakob, C. (1946). *Documenta Biofilosófica. Folleto I. Biología y Filosofía. A: Aspectos de sus divergencias y concomitancias. B: Ensayo de Psicogenia orgánica* (Folia Neurobiológica Argentina, Tomo V). Buenos Aires: López y Etchegoyen.

[67] Freud, S. (1978). Das Ich und das Es [1923]. In A. Freud, & I. Grubrich-Simitis (Eds.), *Sigmund Freud Werkausgabe in zwei Bänden, Band 1: Elemente der Psychoanalyse*, 2. Aufl. Frankfurt a.M.: S. Fischer Verlag GmbH, pp. 369–401.

[68] Freud, S. (1974). *The Ego and the Id* [1923] (transl. by J. Riviere, ed. by J. Strachey). London: The Hogarth Press and the Institute of Psycho-Analysis, pp. 9–17.

[69] Freud, S. (1978). Neue Folge der Vorlesungen zur Einführung in die Psychoanalyse [1933]. In A. Freud, & I. Grubrich-Simitis (Eds.), *Sigmund Freud Werkausgabe in zwei Bänden, Band 1: Elemente der Psychoanalyse*, 2. Aufl. Frankfurt a.M.: S. Fischer Verlag GmbH, pp. 402–418.

[70] Jacobson, M. (1993). *Foundations of Neuroscience*. New York–London: Plenum Press, pp. 204–205.

[71] MacLean, P. D. (1949). Psychosomatic disease and the "visceral brain": Recent developments bearing on the Papez theory of emotion. *Psychosomatic Medicine*, 11, 338–353.

[72] Kolb, B., & Whishaw, I. Q. (1996). *Fundamentals of Human Neuropsychology*, 4th edn. New York: W. H. Freeman and Co., pp. 417–420.

[73] MacLean, P. D. (1990). *The Triune Brain in Evolution: Role in Paleocerebral Functions*. New York–London: Plenum Press.

[74] Peper, M., & Markowitsch, H. J. (2001). Pioneers of affective neuroscience and early concepts of the emotional brain. *Journal of the History of the Neurosciences*, 10, 58–66.

[75] MacLean, P. D. (1970). The triune brain, emotion, and scientific bias. In F. O. Schmitt (Ed.), *The Neurosciences: Second Study Program*. New York: Rockefeller University Press, pp. 336–349.

[76] MacLean, P. D. (1977). The triune brain in conflict. *Psychotherapy and Psychosomatics*, 28, 207–220.

[77] MacLean, P. D. (1990). *Les Trois Cerveaux de l'Homme: Textes Traduits de l'Américain, Notes et Commentaires de Roland Guyot* (Collection «La Fontaine des Sciences» dirigé par Gérard Klein). Paris: Éditions Robert Laffont, S.A..

[78] Schoffeniels, E., Schmerling, P., & MacLean, P. D. (1994). Journée Commemorative du Bicentenaire de Pierre Schmerling: Colloque: Théorie des trois cerveaux de Paul Donald MacLean et incidences en Psychiatrie: vendredi 15 novembre 1991, Hôpital Psychiatrique du Petit Bourgogne, Liège (*Acta Psychiatrica Belgica*, 94, fasc. 4–6). Bruxelles, Société de Médecine Mentale.

[79] Bennett, M. R., & Hacker, P. M. S. (2001). Perception and memory in neuroscience: a conceptual analysis. *Progress in Neurobiology*, 65, 499 5⁄13.

[80] Panksepp, J. (2002). The MacLean legacy and some modern trends in emotion research. In G. A. Cory, Jr., & R. Gardner, Jr. (Eds.), *The Evolutionary Neuroethology of Paul MacLean*. Westport, CT: Praeger, pp. ix–xxvii.

[81] Lambert, K. G. (2003). The life and career of Paul MacLean: A journey toward neurobiological and social harmony. *Physiology and Behavior*, 79, 343–349.

[82] Knopp, W. (1970). Man's tripartite brain and psychosomatic medicine. *Psychotherapy and Psychosomatics*, 18, 130–136.

[83] Popper, K. R., & Eccles, J. C. (1977). *The Self and its Brain*. Berlin–Heidelberg: Springer International, pp. 358–365.

[84] Hubel, D. H. (1979). The brain. *Scientific American*, 241, 44–53.

[85] Uttal, W. R. (2001). *The New Phrenology: The Limits of Localizing Cognitive Processes in the Brain*. Cambridge, MA: MIT Press, p. 25.

[86] Ramón y Cajal, S. (1954). *Neuron Theory or Reticular Theory? Objective Evidence of the Anatomical Unity of Nerve Cells* (transl. by M. Ubeda Purkiss, C.A. Fox). Madrid: Consejo Superior de Investigaciones Científicas/S. Aguirre Torre.

[87] Sternberg, R. J. (1984). Toward a triarchic theory of human intelligence. *Behavioral and Brain Sciences*, 7, 269–315.

[88] Sternberg, R. J. (1988). *The Triarchic Mind: A New Theory of Human Intelligence*. New York–London: Viking Press.

[89] Tigner, R. B., & Tigner, S. S. (2000). Triarchic theories of intelligence: Aristotle and Sternberg. *History of Psychology*, 3, 168–176.

[90] Fodor, J. A. (1983). *The Modularity of Mind*. Cambridge, MA: MIT Press.

[91] Sterelny, K. (1990). *The Representational Theory of Mind*. Oxford: Basil Blackwell Ltd., p. 75.

[92] Block, N. (1995). On a confusion about a function of consciousness. *Behavioral and Brain Sciences*, 18, 227–287.

[93] Sun, R. (2002). *Duality of the Mind: A Bottom Up Approach Toward Cognition*. Mahwah, NJ: Lawrence Erlbaum Associates Inc., pp. 176–190.

[94] Nagel, T. (1974). What is it like to be a bat? *Philosophical Reviews*, 4, 435–450.

[95] Jakob, C. (1945). El cerebro humano: Su significación filosófica. Ensayo de un programa psico-bio-metafísico, después de 50 años de dedicación neurobiológica. *Revista Neurológica de Buenos Aires*, 10, 89–110.

[96] Jakob, C. (1939). *El Neoencéfalo: Su Organización y Dinamismo*. La Plata–Buenos Aires: Universidad Nacional de La Plata, Facultad de Humanidades y Ciencias de la Educación–Imprenta López.

[97] Holden, C. (1979). Paul MacLean and the triune brain. *Science*, 204, 1066–1068.

[98] Jakob, C. (1922). Del tropismo a la teoría general de la relatividad: Un capítulo biopsicofiláctico. *Humanidades (La Plata)*, 3, 45–58.

[99] MacLean, P. D. (1997). The brain and subjective experience: question of multilevel role of resonance. *Journal of Mind and Behavior*, 18, 247–268.

[100] Ploog, D. (1970). Social communication among animals. In F. O. Schmitt (Ed.), *The Neurosciences: Second Study Program*. New York: Rockefeller University Press, pp. 349–361.

Brain Struct Funct (2010) 214:319–338
DOI 10.1007/s00429-010-0240-6

REVIEW

Revisiting Christfried Jakob's concept of the dual onto-phylogenetic origin and ubiquitous function of the cerebral cortex: a century of progress

Lazaros C. Triarhou

Received: 22 November 2009 / Accepted: 20 January 2010 / Published online: 11 February 2010
© Springer-Verlag 2010

Abstract This paper revisits a concept combining the evolution, ontogeny and histophysiology of the cerebral cortex, presented, in a quest to explain cognition and behavior, by the neurobiologist Christfried Jakob (1866–1956) at the Second Annual Meeting of the International Society for Medical Psychology and Psychotherapy, organized by Oskar Vogt (1870–1959) in Munich in 1911. Jakob suggested a dual onto-phylogenetic origin and a ubiquitous cortical function, claiming that most receptive pathways end up in an 'outer fundamental layer', which derives from the rhinencephalic apparatus, whereas the 'inner fundamental layer' contains effector elements and derives from the striatum. With advancing evolution, the two fundamental layers become intermingled. By attributing a functional homogeneity to the cortex, Jakob contradicted the theories of Flechsig and Cajal on 'association' and 'mnemonic' areas. The merit of Jakob's concept rests, a century later, with the current resurgence of biological research at the evolutionary–developmental interface and the broadening anticipated from the re-integration of these two fields, especially by adding a functional dimension to the morphological traits.

Keywords Brain development · Brain evolution · Cerebral cortex · Cytoarchitectonics ·

Dorsal ventricular ridge · History of neuroscience · Pallium · Subventricular zone

Introduction

An understanding of neural development and evolution is inevitable for deciphering brain structure and function. The puzzle of the cerebral cortex in particular may become clarified by studying its evolutionary origins (Shimizu and Karten 1991) and the problem of neuronal homologies (Nauta and Karten 1970). As in general biology, so in neurobiology, research on phylogenetic accounts of ontogeny has been gaining momentum (Rakic and Kornack 2001).

Crucial studies on brain evolution and brain development came out from major laboratories between the 1870s and the 1930s (cf. Edinger 1896; Flatau and Jacobsohn 1899; Unger 1906; Brodmann 1909; Obersteiner 1913; Johnston 1915; Vogt and Vogt 1919; Ariëns Kappers et al. 1936). They contain a wealth of knowledge, not fully exploited by modern investigators, as they had been published in languages other than English.

This article revisits a theory proposed by Christfried Jakob (1866–1956), a Bavarian neuropathologist and the founder of Argentinian neurobiology, affiliated with the Universities of Buenos Aires and La Plata. Besides extensive research in comparative and human neuroscience, and numerous publications on cortical ontogeny, phylogeny and neuropathology, Jakob integrated such diversities as general biology, anthropology, paleontology, biogeography, philosophy and music. Half-way through his life, Jakob published in Munich two classic Atlases, called *The human brain* and *From animal brain to human brain* (Triarhou 2008c), and made a key presentation at the

Parts of this work were presented at the 3rd International Conference on Cortical Development, Crete, 22–25 May 2008, and the 38th Annual Meeting of the Society for Neuroscience, Washington, DC, 15–19 November 2008.

L. C. Triarhou (✉)
Economo-Koskinas Wing for Integrative and Evolutionary Neuroscience, University of Macedonia, 156 Egnatia Ave., 540 06 Thessaloniki, Greece
e-mail: triarhou@uom.gr

 Springer

Second Annual Meeting of the International Society for Medical Psychology and Psychotherapy, organized also in Munich by Oskar Vogt on 25–26 September 1911. Jakob's communication was published the following year in two variants (Jakob 1912a, b); he maintained a constant interest in that concept (Jakob 1910b, 1913), and in 1916 published a Spanish version, with minor emendations and an added concluding paragraph (Jakob 1916).

Jakob suggested a dual evolutionary and developmental origin and an ubiquitous function of the cerebral cortex, based on data gathered during his 'first tenure' in Argentina, between 1899 and 1910 (Triarhou and del Cerro 2006a, b; Triarhou 2008b). He conducted hundreds of human neuropathological examinations and carried out comparative studies on over 100 species of the Patagonian fauna, including primates, the legless amphibian *Caecilia lumbricoides*, which resembles a giant earthworm but belongs to the order of the Gymnophiona, and the squamate reptile *Amphisbaena darwini* (blind viper).

Jakob claimed that all cortical regions contain receptive elements and that most sensory pathways end up in the 'outer fundamental cortical layer' (small and medium-size pyramidal cells), which onto-phylogenetically derives from the rhinencephalic apparatus. The 'inner fundamental layer' contains effector elements and derives from the striatum. With advancing evolution, the two fundamental layers become intermingled. Jakob attributes homogeneity to the cortex and contradicts Flechsig (1896, 1898) and Ramón y Cajal (1895, 1899, 1904, 1906) on the existence of dedicated association and memory areas.

The 'triple-synthesis' (evolutionary, developmental and physiological) has been viewed as one of Jakob's prime contributions to science (Moyano 1957; López Pasquali 1965; Orlando 1966; Meyer 1981; Fontana et al. 2002; Lores Arnaiz et al. 2002). Diego Outes, a pupil of Braulio Moyano (Jakob's successor at the Pathology Laboratory of the National Neuropsychiatric Hospital in Buenos Aires), revisited the topic of the dual cortex (*Doppelrinde*) twice (Outes and Benítez 1976; Outes 2006).

The current resurgence of research at the evolutionary–developmental interface of biology witnesses the promising re-integration of these two fields, after almost a century (Raff et al. 1999). With his larger Atlases, Jakob (1911; Jakob and Onelli 1911, 1913) pioneered evolutionary–developmental neurobiology. A cognoscente of 'cortical history' (*Die Geschichte der Hirnrinde*), Jakob promptly applied phylogenetic observations to the ontogeny of the human brain (Moyano 1957; López Pasquali 1965). The originality of his ideas is witnessed by the writings of his contemporaries (Ranke 1911; Siemerling 1911; von Economo and Koskinas 1925; Seldon and Szirko 2005).

Jakob has been compared in caliber to Ramón y Cajal. In their *Cytoarchitectonics*, von Economo and Koskinas

(1925) express the view that future research on the cerebral cortex must be based on the fundamental work of Kaes (1907), Sammet (2006), Ramón y Cajal (1906), and Jakob (1911), Jakob and Onelli (1911). von Economo and Koskinas (1925) stress Jakob's *geniale Ansicht* (ingenious idea) regarding cortical phylo-ontogeny, Cajal's *ganz glänzend* (totally brilliant) use of the Golgi method, and Meynert's *geniale Intuition* (ingenious intuition) in associating the granularity in the area striata of the calcarine sulcus with sensory function (Meynert 1872).

Today, powerful cytochemical and molecular biological methods are available to neurobiologists for studying gene expression, the cascade of molecular pathways, cell–cell interactions, and the phylogeny of embryonic cortical development (Rakic 2007). A century ago, tools at hand were confined to the Nissl and Weigert stains and light microscopy, having to be compensated by the observer's acuity and imagination.

The present study makes Jakob's essay available in English. The discussion that follows re-examines Jakob's concept and the progress made since. The translation of Jakob's communication was rendered from the papers published in *Journal für Psychologie und Neurologie* (vol. 19, no. 1, pp. 379–382, 1912) and *Münchener Medizinische Wochenschrift* (vol. 59, no. 9, pp. 466–468, 1912). The sections added by Jakob in the Spanish version (*La Prensa Médica Argentina*, vol. 2, no. 23, pp. 305–307, 1916) appear in brackets. Figures were supplemented from sources explained in the captions.

Jakob (1918) mentions that he began studying the brains of Argentina's batrachians and reptiles in 1905, reporting his initial findings in an article (Jakob 1910a) and later in his books (Jakob 1911; Jakob and Onelli 1911, 1913). Originally, Jakob (1911; Jakob and Onelli 1911, 1913) wrote that the drawing on the relation of cortical layers depicted the brain of *Caecilia*, an amphibian belonging to the family of Gymnophiona, one of the three orders of modern amphibians (Zardoya and Meyer 2001). As Outes (2006) explains, Jakob (1918) subsequently clarified that the particular specimen had actually been an apod reptile, the *Amphisbaena darwini* (blind viper). The two species looked alike enough in external morphology and shared common habitats; the flaky epidermis and the more advanced brain form eventually led to the distinction of the Amphisbaena from the Gymnophion, whose vesicular brain reminded that of the urodela. The erratum has been remedied in the current translation. A systematic revisionary work on South American amphisbaenids did not appear until well into the 1960s (Gans 1966). Cytoarchitectonic studies have shown that the brain of *Caecilia* markedly differs from other amphibians in having a highly differentiated accessory olfactory bulb and a large overall extension of this brain region (Zilles et al. 1981).

 Springer

On the ubiquity of the dual receptive–effector function of the cerebral cortex as the basis for a new biological conception of the cortical apparatus of mind

by Dr. Chr. Jakob (Buenos Aires) currently in Krailling by Munich

The historical development of our views on the origin and biological bases of mental forces can be divided into the following four phases:

1. The speculative period of antiquity and the middle ages, dominated by the theories of pneuma and ventricular localization.
2. The anatomical period of Vesalius and Varolio (inventor of the fiber teasing method), over to de le Boë Sylvius (1641, first essay on cortical localization) and Gall, characterized by diverse theories attempting to localize mental processes in the cerebral matter.
3. The physiological period of the nineteenth century that led us, experimentally and clinically, from Bouillaud and Flourens to Broca, Fritsch and Hitzig, Wernicke, and finally to Flechsig's theory on the varying importance of the cortex (projection and association centers), which came to a dead end with Ramón y Cajal's theory of 'mnemonic centers'.
4. The current biological–eclectic period, characterized by the systematic application and comparison of all methods, and by its biological trend.

In a systematic series of studies, I examined the regional structure and cortical layers (cyto-, myelo-, fiber-architecture), as well as the comparative histology, histogenesis and histopathogenesis, and their relation to normal and disturbed function. The most important questions toward a synthetic conception of the mental apparatus, which thus far remain unanswered, are:

a. Are there sectors in the human and animal cerebral cortex of an exclusively *receptive* nature and others of a purely *effector* nature or such of a purely *neutral* nature, and how are their components differentiated?
b. What is the evolutionary course of such sectors and what *fundamental differences* characterize the *human cortex*? Are there principal differences compared to animals or is it all about *gradual differentiation*?
c. Is the *origin of the cortex* mono- or poly-phyletic?

I have been addressing these questions for the past decade in collaboration with my students at the Institutes I head at the University of Buenos Aires (Laboratory of Hospital de Las Mercedes, Neurobiological Laboratory of National Women's Psychiatric Hospital, and Neurology Clinic of San Roque Hospital), using a vast collection of human brains, histopathologically (degeneration methods),

experimentally in apes (*Cebus* from Paraguay), and comparatively anatomically in species of the South American fauna (Figs. 1, 2).

The results can be summarized as follows:

(I) *All regions of the human and animal cortex, with no exception, are receptive (the only peculiarity being the status of Ammon's formation, the phylogenetically oldest cortical region).*

This does not have to be discussed for cortical territories (visual, auditory, etc.) classified as 'projection centers'— von Monakow and Déjerine determined the fundamentals of anatomical fiber relations—but it becomes necessary for the remaining parts, i.e. frontal cortex, cingulate gyrus, precuneus, posterior parietal and temporal, temporopolar, and occipito-temporal regions. We systematically examined over 300 cortical areas (mostly softening processes) in serial sections for retrograde thalamic cellular changes with the Nissl–Lenhossék thionin stain. A preliminary study of the human thalamic nuclei was necessary (Jakob 1910a, 1911). We could discern that lesions of areas in all parts of the frontal cortex result in cell loss in most parts of the anterior lateral thalamic nucleus (the only exception being the small fronto-olfactory sector of the frontal base); in a similar manner, degenerative foci in the cingulate gyrus result in retrograde cell loss in the anterior and dorsal nucleus of the thalamus; degenerative foci in the precuneus and posterior parietal cortex result in retrograde cell loss in the dorsal rostral pulvinar (posterior nucleus, dorsal–anterior portion) and posterior lateral nucleus; degeneration of the anterior occipital area results in retrograde cell loss in the dorsal caudal pulvinar; degeneration of the posterior temporal area results in retrograde cell loss in the basal pulvinar (posterior nucleus, basal portion); and degeneration of the anterior temporal cortex in cell loss in the internal basal nucleus of the thalamus.

These data prove that the entire cortical surface has fiber systems receiving sensory stimuli that emanate from distinct thalamic regions, also on directed and descending pathways at the sensory mesencephalon and diencephalon; thus, the cortical surface is completely separated into sectorial receptive fields, each equipped with a specific modality. We obtained similar results in the ape brain. Degeneration areas must have a minimum size to lead to recognizable cell losses.

[One can verify, by translating the anatomical data into physiological facts that the entire cortical surface receives, from each distinct region of the thalamus, pathways which conduct the stimuli toward the cortex (centripetal paths or thalamocortical radiations). It is precisely the cells of origin of those pathways that degenerate secondarily as a consequence of cortical foci. As these pathways are further associated with the lower afferent sensory pathways of the midbrain and the diencephalon, it follows that the pallium

🖄 Springer

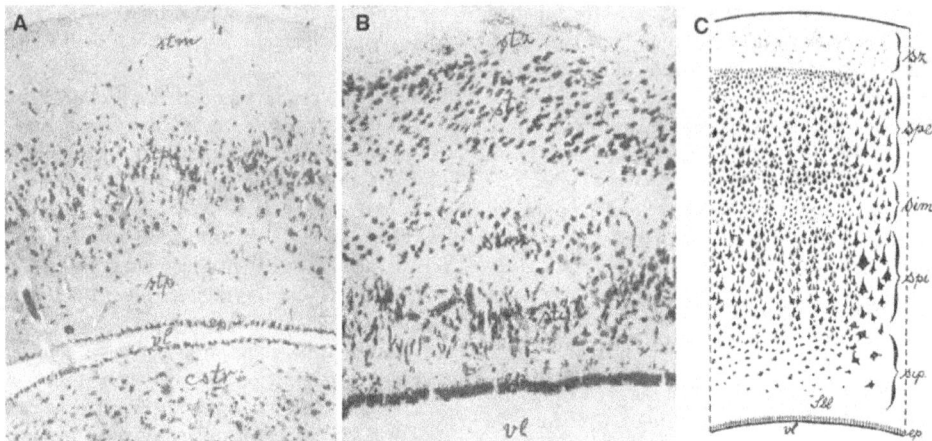

Fig. 1 a Ammon's cortex in the adult *Caiman latirostris* ('yacaré'). *stp* pyramidal cell layer, *stm* molecular layer, *vl* lateral ventricle, *ep* ependyma, *cstr* corpus striatum. Nissl stain ×150. From Jakob (1941, p. 168). **b** Lateral cortex of *Iguana iguana* with the molecular (*stz*), external (*ste*) and internal (*sti*) cortical layers, and an intermediate layer (*stm*). From Jakob (1941, p. 148). **c** Polystratified cortex of a higher mammal; *sz* molecular layer, *spe* external pyramidal layer, *sim* intermediate layer, *spi* internal pyramidal layer, *sip* interpyramidal layer, *sbl* basal layer (Jakob and Onelli 1913, p. 16)

is divided in its entire extent into zones that affect the form of sectors, each with distinct sensory modalities.

Moreover, if one takes into account the relations known to exist between the centers of projection and their respective thalamic nuclei, we conclude that for each sensory modality there are at least two adjacent cortical territories, one of which is *directly* connected with the basal thalamus, and the other *indirectly* connected via afferent pathways with the diencephalon (with the intercalation of the dorsal thalamic nuclei).

Thus, we arrive at the notion that the old 'projection centers' are identical to our direct primary cortical territories; secondary indirect territories (which phylogenetically mature later than primary) occupy parts of the old 'association centers'. The fact that lesions of the latter frequently develop in a latent form can be explained by the intervention of intact primary territories, whereas destruction of primary territories, of consequence to a secondary involvement of the baso-thalamic nuclei, must always disturb the conduction to secondary territories.

I obtained similar results in primates, whose structural plan completely agrees with the human brain. Such foci must always have a certain extent to produce secondary alterations that can be safely discerned.]

(II) *The principal part of these sensory radiations terminates at the external cortical layer.* [Outer layer = molecular layer, small, middle, larger outer pyramids + granular layer (intermediate layer).]

This trend can be proven with the Weigert method only at individual areas (degeneration of Gennari's strip in the calcarine sulcus) connected with the corresponding

thalamic areas; on the other hand, with the Marchi method it is also confirmed in all other areas of both humans and apes, as well as in lower animals. Histological finds (myelination, impregnation) support this view as well. Further, for this purpose, indications are provided by the fact that the outer layer strongly increases always at the expense of the inner layer where ever radiations occur (visual, auditory, tactile cortex). This applies to the entire animal line.

[Studying the engravings of Ramón y Cajal, one can see that he invariably terminates the fibers in what I call the external layer, although he does stress that fact.]

(III) *The inner layer of the cortex has an effector function and exists everywhere.* (Inner layer = deep, large and middle pyramids, deep smaller and polymorphous cellular elements.)

That finding is first a logical result of the previous one, in the sense of a general structural plan of the central nervous system. It is further supported by the fact that everywhere in the cortex, where motor pathways emanate that project to lower centers (pyramidal pathway, bulbar pathways, hypothalamic bundles, pontine pathways, fornix system) this takes place without any exception in the inner layer; and anywhere that an overall effector character stands out, it is accompanied by an increase of the inner layer at the expense of the outer; this happens consistently in the entire animal line (Fig. 3). Definite proof must be furnished not by degeneration studies, but from the phylogeny of the two fundamental layers.

[It has been verified that the cerebral cortex does not in principle depart from the structural plan of lower spinal

 Springer

Fig. 2 Evolution of the pallium
and cortical mantle in five
vertebrate species, from fish
to human. From Jakob (1911,
p. 34); Jakob and Onelli (1913,
p. 12)

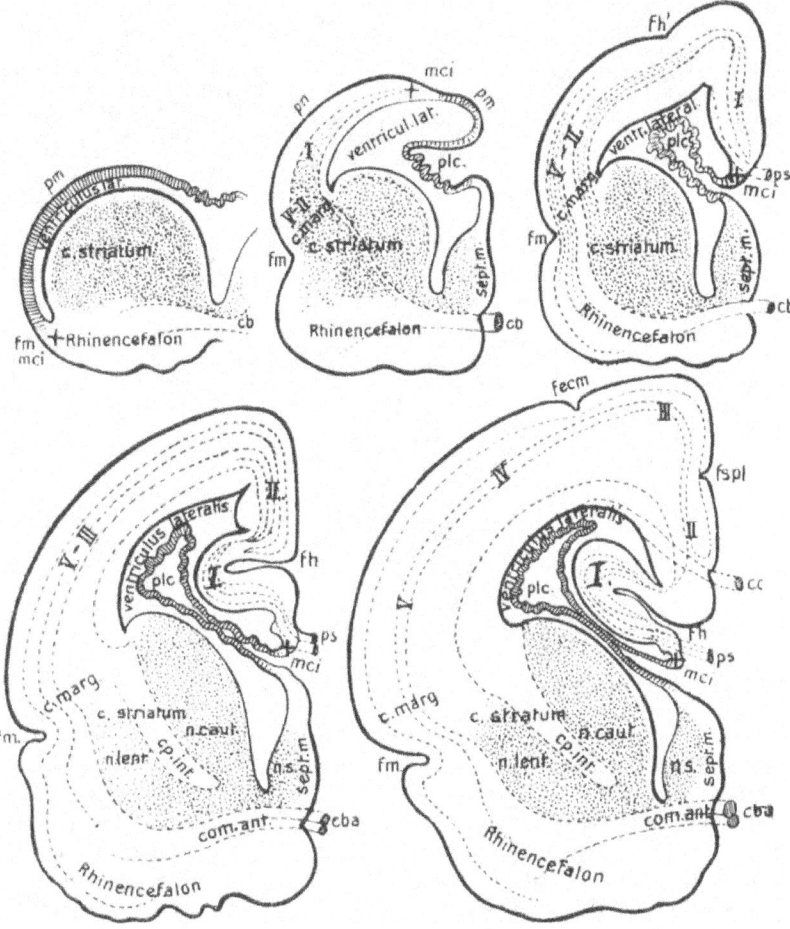

systems, since in them as well, through their entire extent, including the spinal cord, medulla and midbrain, the separate sensory and motor columns are arranged in parallel. The same happens in the cortex, where both fundamental layers run parallel. Similar to these regions, one or the other column may occasionally become overgrown or diminished, according to peripheral exigencies. This fact is repeated in the central apparatus in accordance with homologous laws. I attribute to the concept of such a sensory–motor disposition, for the entire corticality, the same importance for the future, that has been attributed in the history of neurology to the discovery of Bell and Magendie concerning the distinct functional nature of the ventral and dorsal roots.]

(IV) *Both fundamental layers have a uniform dual origin in mammals up to humans (monophyletic evolution) and something similar can be stated for various classes of lower vertebrates (reptiles in particular).* On the contrary,

amphibians do not belong here. According to my studies, the *Amphisbaena* actually represents the fundamental type of evolution of the higher cortical apparatus up to humans (Figs. 4, 5). My comparative histological studies suggest that the two fundamental cortical layers have distinct origins: the receptive outer layer stems from the rhinencephalon (an old sensory part of the brain), whereas the effector inner layer from the corpus striatum (an old motor central nucleus) (Fig. 6). Despite its dual origin, with advancing evolution, the entire cortex reaches a correspondingly more internal intermingling of the two layers. In mammals, this proceeds as a consequence of the augmentation of cellular elements with a highly ramified protoplasm through the formation of axon collaterals, the occurrence of numerous intermediate and intercalated, and the secondary structural fusion of the fundamental cortical layers, which were initially separated in the anlage (Fig. 7); a principle that especially distinguishes the primate and human cerebral cortex.

 Springer

Fig. 3 Cortical evolution from
mono- to poly-stratified. *Left to
right* amphibian, reptilian,
avian, mammalian, and higher
mammalian laminar structure.
From Jakob and Onelli (1911,
p. 23), (1913, p. 16)

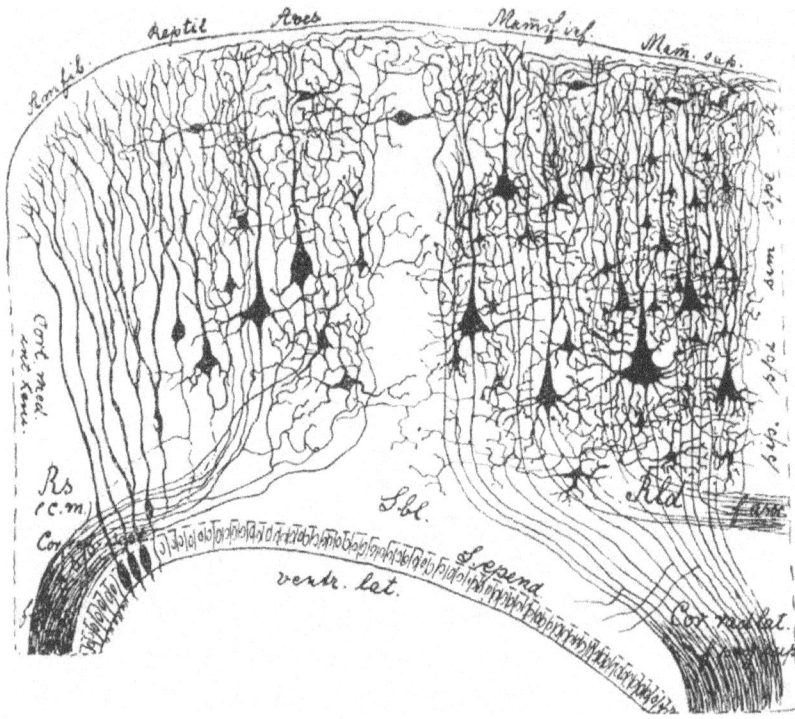

[The bilaminar cortical type can be seen in the brain of *Tapirus terrestris* (Fig. 8), and the union of both outer and inner cortical layers can be demonstrated in the brain of *Erinaceus europaeus* (Fig. 9).]

(V) *A cortex that is neither receptive nor effector (='association cortex') does not exist.*

The thorough study of the cortical surface yields consistent results. The dual-location fundamental layer type only knows one exception: in the entire animal line, the unistriate structure of Ammon's formation (the oldest olfactory cortical portion), which always exhibits only the lower, effector layer. Thus, the entire cortical mantle (pallium) is either sensorimotorily or motor-sensorily active (depending on the dominance of one or the other component). There is no part of the cortex that functions and acts solely as receptive or as effector, as far as the histological and experimental proof is concerned. A cortex that could exclusively claim the descriptor 'association cortex' or 'mnemonic center' does not exist. Such processes are functions of a dynamic nature of both fundamental layers at all localities. Accordingly, I proclaim, against the views of Flechsig and Ramón y Cajal, the equivalence of all cortical zones; with differences being only in gradation.

I thus arrive at the following conclusions:

The cerebral cortex develops monophyletically in the entire mammalian and human spectrum from two originally separate and also functionally different fundamental layers. Both components appear in the cortex in an ascending order into a more intrinsic contact than the case is in lower spinal, bulbar, and other systems. As a result of its structure, the cortex is reckoned as both receptive and effector throughout, the only change being functional qualities and their topographic relationships with lower centers.

As a consequence of the mutual penetration of both layers in mammals, and especially in humans, a process purely effector or exclusively receptive is not plausible at any cortical zone; each state of excitation has to momentarily release the corresponding other phase (receptive or effector). Thus, all cortical acts must a priori be defined as having a 'mixed' nature; an arbitrary separation into the two components seems irreconcilable with the cortical texture. Such a trend is fundamental for the understanding of cortical functions.

Therefore, each fundamental process in the domain of will or perception must carry a similar 'mixed' character; it is erroneous to speak of such processes as distinct (what actually happens is that, in each case, one component prevails over the other). They differ in tendency, not in essence. We thus arrive, from the past 'dualist' views of cortical mental processes, at a concept of 'cortical monism' for the entire functional repertoire. The outcome of those

 Springer

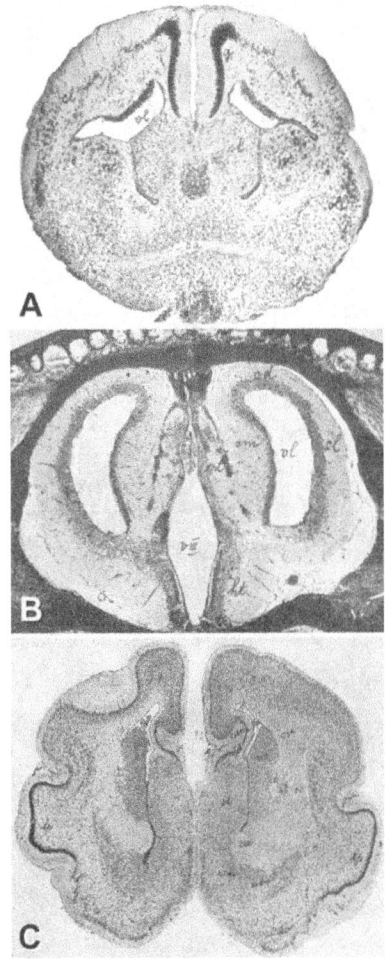

Fig. 4 Coronal sections in typical vertebrates of the Argentinian fauna. **a** *Amphisbaena darwini* initially thought to be *Caecilia lumbricoides*, but later corrected by Jakob (1918), cf. also Outes (2006). *g* Ammon's formation containing the fascia dentata and Ammon's cortex, *cl* lateral cortex, *sl* lateral septum, *olf* olfactory area, *vl* lateral ventricle, *nc* caudate nucleus, *ca* anterior commissure. Nissl stain. From Jakob and Onelli (1913, Plate 2.12). **b** Coronal section of the hemispheres and diencephalon in an amphibian (*Gymnophion* or *Chthonherpeton*). *cl*, *cd*, *cm*, lateral, dorsal, medial cortex; *vl*, *vIII*, lateral and third ventricle; *ht*, hypothalamus; *ept*, habenula or epithalamus. From Jakob (1941, p. 122). **c** Anterior coronal section in a marsupial ('comadreja' or *Didelphis paraguayensis*), Nissl stain. Gray paleoneuronal (Ammonic) and neoneuronal (dorsolateral) cortical formations. *fr* rhinal sulcus, *ao* olfactory area, *lp* piriform lobe, *sl* septum, *nc* caudate nucleus, *nl* lenticular nucleus. From Jakob (1941, p. 113)

views, which represent a new important biological basis for clinical, psychological (Jakob 1913) and physiological studies, contribute to bridging the gap between biological and mental phenomena.

I intend to revisit these new views on the nature and function of our mind apparatus. I would like to point out that these results, which derive from modern biological brain research, closely match in certain points the views of the newer philosophy (cf. Wundt's apperception theory, the doctrines of the subconscious, etc.). Thus, regarding the future of cortical biology, I would particularly emphasize the important point that the findings of mental research no more contradict those of brain research, as has been the case until now.

Discussion

Jakob (1895, 1898) had laid out a plan to study the brain in the early Atlases, which constituted a testimony to his versatility in neurohistology (Meyer 1981). The approaches considered meaningful to better understand the nervous system were (1) histological staining and serial section reconstruction of the adult human brain; (2) neuropathological changes and their sequels; (3) comparative neuroanatomy and neuroembryology; (4) human brain development and myelination; and (5) experimentally induced lesions in animals.

Within a dozen years of working in Buenos Aires, Jakob collected the tissues and was able to materialize his original plan of integrating data gathered from clinico-pathological correlations (e.g. focal cortical degeneration in neurosyphilis), lesion-induced retrograde degeneration (to track thalamocortical projections in monkeys), and comparative neurohistological studies (in diverse species of the Patagonian fauna). As Moyano (1957) relates, five decades later and a continent apart, Jakob had materialized his plans in the lavishly illustrated Atlas volumes of *Folia Neurobiológica Argentina*, which, in 1,200 pages, 1,000 figures and 650 macrophotographic plates, covered (I) human topographic neuroanatomy, (II) human neuropathology, and (III) cerebral ontogeny and phylogeny (Jakob 1939a, b, 1941).

Jakob and Onelli (1911, 1913) intended to establish a biological basis of mental phenomena, convinced that the comparative study of central nervous system structure and function would yield psychological clues, otherwise psychology is doomed to the limitations of its descriptive methods (Papini 1988). Jakob viewed the 'cortical apparatus of mind' as a mere quantitative evolution from animals to humans (Jakob 1914, 1945): 'Motivated by a study of animal language, I probed the varying production of affective language by galliformes: although humans surpass all animals through the free and extensive use of symbols, which only they know how to separate from the ensuing emotions, thus being able to intellectually rise to higher ideative constructions. Such an economy of thought facilitates a mental life freed from momentary emotional

🍀 Springer

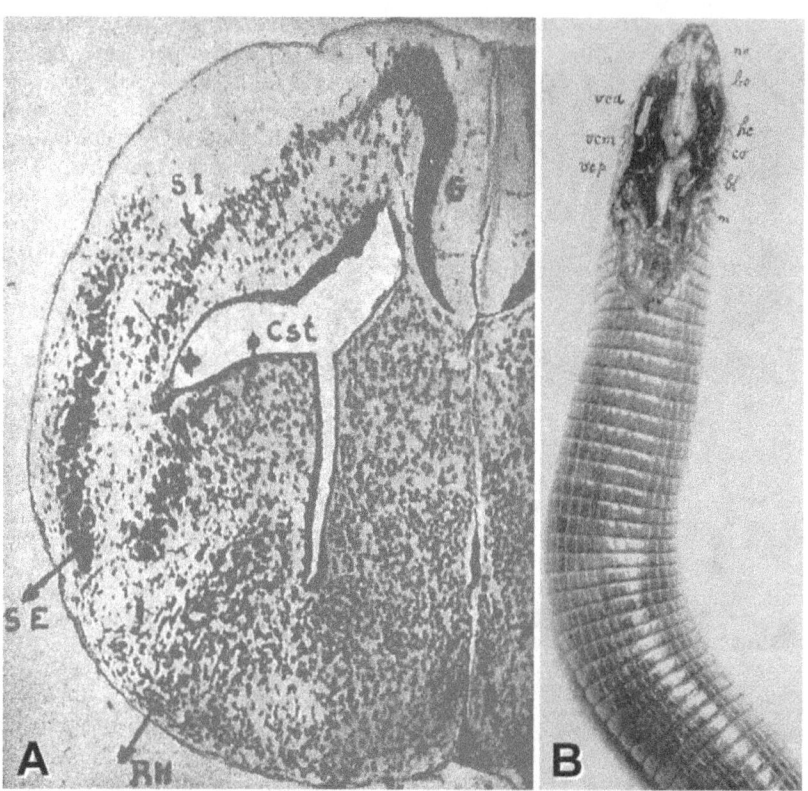

Fig. 5 a Photomicrograph of *Amphisbaena darwini* (Jakob 1918). The inner layer (*si*) of the lateral cortex continues dorsally toward Ammon's horn and ventrally toward the striatum (*Cst*). Jakob emphasized the ventral continuity, since the striatum is associated with motor function, although in this particular photomicrograph, the continuity with Ammon's horn is more conspicuous than that with the striatum. *se* outer layer, *Rh* rhinencephalon. From Outes (2006). **b** The brain of *Amphisbaena darwini* ('víbora ciega') dissected in situ, showing the olfactory bulb (*bo*), cerebral hemisphere (*hc*), rudimentary optic body (*co*), cerebellum, medulla oblongata (*bl*) and beginning of spinal cord (*m*). Magnification ×2. From Jakob (1941, p. 65)

responses, to which the animal psyche remains bound, but deep inside the psychogenetic vocabulary, numbering 8–10 in the hens, 500 in primitive man, and 50,000 or more in civilized peoples, lies an essentially quantitative question of intensity grade and extent, and not of qualitative differences concerning its biological basis.'

The originality of Jakob's ideas is reiterated by contemporary authors (Ranke 1911; Siemerling 1911). The lifelong interest of Jakob in fetal development is echoed in the fact that his first publication from Argentina was on cortical ontogeny during the first trimester of gestation (Jakob 1899), and one of his last books (Jakob et al. 1945) dealt with human embryology.

Flechsig (1896, 1898) had argued that 'illness of association areas is what, above all, produces mental diseases and thus forms the true object of psychiatry', agreeing with an earlier theoretical conjecture by Wernicke that mental diseases constitute alterations of association systems

(Keegan 2003). Ramón y Cajal (1904, 1906, 1995) describes, following the groundwork by von Monakow, Déjerine, Siemerling, Vogt and others, that all the areas claimed by Flechsig to be association are connected with lower centers by way of projection fibers, a fact later acknowledged by Flechsig as well. Cajal hypothesizes the existence of three varieties of cortical areas (bilateral *perception*, and unilateral *primary* and *secondary mnemonic* areas), all of which issue centrifugal fibers: perception areas receive afferents from the thalamus, and mnemonic areas receive input from cortical perception areas. Cajal stands by Vogt's stance that the association area theory cannot adequately explain psychological mechanisms, and emphasizes that attempts to localize cognitive phenomena at 'privileged areas' amount to pursuing a chimera. In Cajal's view, intellectual operations result from the combined activity of multiple primary and secondary mnemonic areas (Ramón y Cajal 1904, 1906, 1995).

 Springer

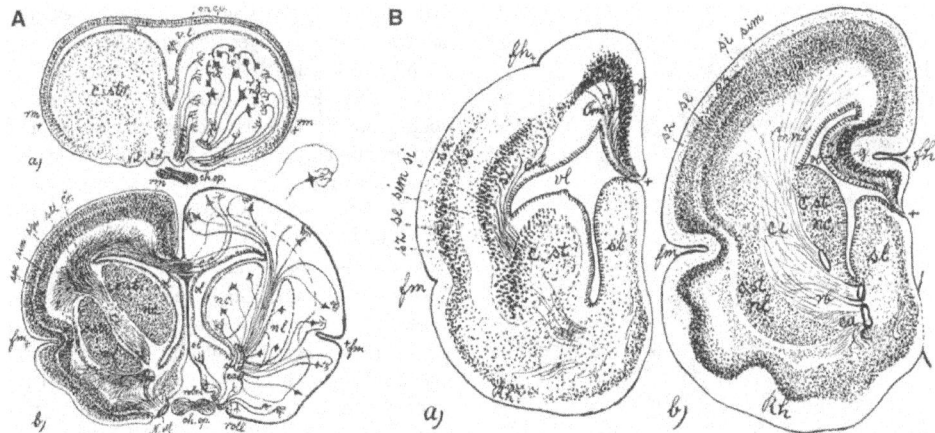

Fig. 6 a Coronal sections in a fish (*a*) and a mammal (*b*). The corpus striatum (*c str*) is robust, whereas the hemispheric mantle (pallium) appears as a thin ependymal membrane (*mep*) over the lateral ventricles (*vl*). From Jakob and Onelli (1911, p. 6), (1913, p. 15); field (A)/(*a*) also used by von Economo and Koskinas (1925). **b** Coronal sections of the cerebral hemispheres in the apod reptile *Amphisbaena darwini* (*a*) and the marsupial *Didelphis azarae* (*b*). The striatum (*c st*) appears robust. The thin pallium encloses the lateral ventricles (*vl*) dorsally; neurons are seen in the archipallium (*g*). These cells originate in the lateral band of the striatum and form the inner fundamental layer (*si*). The outer fundamental layer (*se*) grows over *si*. Therefore, *se* derives from the rhinencephalon (*Rh*) and later merges with *si*, which is derived from the striatum. *fm* marginal sulcus, *sz* molecular layer, *sim* intermediate layer, *fh* hippocampal sulcus, *g* fascia dentata, *ci* internal capsule, *sl* septum pellucidum, *nc* caudate nucleus, *nl* lenticular nucleus, *cor rad* corona radiata, *rb* radiatio basalis. From Jakob (1911, p. 38), (1911, p. 12), (1913, p. 14); also used by von Economo and Koskinas (1925)

In discoursing his evolutionary postulate of *progressive cerebration*, von Economo (1929) credits Jakob [and Onelli (1911)] for rightfully pointing out that the human organism 'is freed from the primitive and coarse law of the simple reflex; external stimuli no longer lead to a simple reflex, but, depending on past experiences, a different ingredient is obtained, whereby that simple reflex is finally converted to a process bearing the individual characters of the personality.'

A perspective from Jakob's intellectual progeny

Outes (2006) notes that Luys (1876), Nissl (1908) and Ariëns Kappers (1909) were forerunners of the idea that individual cortical layers may have discrete functions: Luys (1876) could be the first anatomist to think that each cortical layer has an individual and specific function. Thalamocortical connections were studied with the method of retrograde degeneration for the first time by Nissl (1908), who reflected on the meaning of each layer rather than cytoarchitectonics. Nissl (1908) suggested the existence of two parallel and superimposed cortical layers and attributed a distinct function to each one, the lower layer being concomitantly effector and receptive, and the upper fundamental psychic.

Jakob claimed that the entire cortex was receptive, against the views of Flechsig (1896, 1898), who, on the basis of his myelogenetic studies, had concluded that the primary association areas of the cortex did not project or receive any fibers.

Studying phylogeny, Ariëns Kappers (1909) suggested that the cortex is divided into an upper layer, with a receptive–psychic function, and a lower layer, with effector function, based on the finds that (1) the reptilian archicortex contains massive granule cells, almost double in number than pyramidal cells; (2) the forebrain of lower mammals shows a substantial increase of pyramidal cells in the hippocampus (archicortex), an area assuming projection and association functions; (3) the formation of an olfactory-mental field (subicular zone) in the mammalian hippocampus coincides with the appearance of a large number of pyramidal cells in the supragranular layer.

Ariëns Kappers (1909) further cited neuropathological data, such as those of Siebenmann and Bing (1907) who had reported reduced thickness of the lower layers in the auditory cortex of a deaf patient, and von Monakow (1905), who had made a similar observation in a case of deafness with lesions of the internal capsule. Thus, Ariëns Kappers (1909) concludes that 'the subgranular layers depend on the local subcortical system, whereas supragranular layers mostly depend on interregional associations with neighboring and distant cortical zones.' Supragranular pyramidal cells are the last to appear ontogenetically (Mott 1907) and would have association functions of a higher order (Ariëns Kappers 1909). That idea was later

 Springer

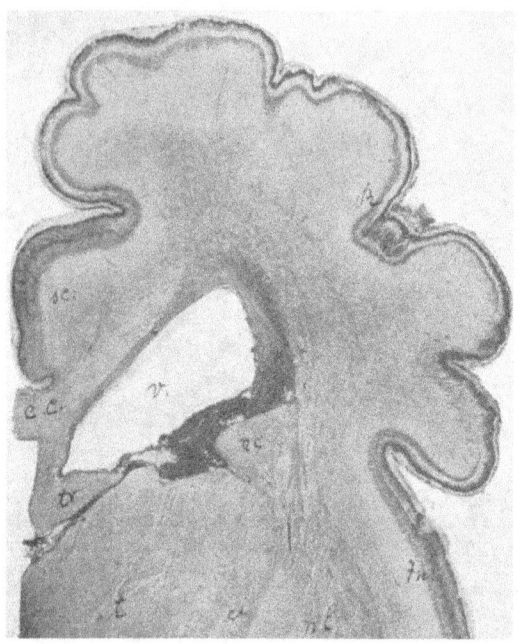

Fig. 7 Cerebral anlage of a human embryo at the beginning of the fifth month of gestation. von Economo and Koskinas (1925) discuss the cortical cell arrangement in a double layer, explaining that the 'fetal bilaminar type' corresponds to the two 'fundamental layers' of Jakob (1911), a trend seen in the human fetal brain and in the animal line, whereby the mixed 'sensorimotor function' of the entire human cortex is rooted. *cc* corpus callosum, *sc* supracallosal gyrus, *R* central sulcus, *ci* internal capsule, *nc* caudate nucleus, *nl* lentiform nucleus, *t* thalamus, *v* lateral ventricle. From Jakob (1911, p. 38)

developed by Kleist (1926, 1934), for whom the robust growth of layer III is typical of 'psychic' centers.

In a later schematic depiction of the cortex of *Lacerta* by Kuhlenbeck (1922), the neocortex or lateral cortex does not appear as bilaminar, and its lower parts seem to be in continuation with the corpus striatum (epistriatum); the lateral part of Ammon's horn is placed beneath the neocortex, at a position that could correspond to Jakob's inner layer (Outes and Benítez 1976; Outes 2006). Bilamination of the lateral cortex or primitive neocortex is more marked in the Amphisbaena (Jakob 1918) (Fig. 5).

Neuroanatomical and morphofunctional considerations

von Economo and Koskinas (1923) discovered that cortical areas known to be involved in sensory functions exhibit a homogeneous layering and a robust layer IV. The idea that sensory areas can be structurally distinguished from motor and association areas by the 'dusty' appearance of *koniocortex* is considered as one of Economo's major contributions to neuroscience (Marburg 1932). von Economo and

Koskinas (1925) relied on Jakob's views, who had defined the inner cortical layers as effector and the outer as receptive (Marburg 1932). von Economo (1926) further showed that receptive areas only occupy the wall and sulcus floor of a gyrus and not the dome, a notion he subsequently used to explain the pattern formation of gyri.

von Economo and Koskinas (1925) devote a segment of their *Textband* to discussing the ideas of Jakob (1911; Jakob and Onelli 1911) and conclude that 'the future will show whether Jakob's novel and basic ideas on the fundamental layers and sector development will prove correct' (Figs. 6, 10, 11). An English translation of the corresponding section can be found in Seldon and Szirko (2005).

According to Meyer (1982), the concept of the 'sensorimotor' cortex, which has remained a problem for a century, is attributed to Munk (1881) by Mott (1894), with an early contribution by Bechterew (Bechterew and Meyer 1978). Flechsig (1896) understood that idea as an almost reflex-like unity between sensory and motor cortical function. Poliak (1932), Foerster (1936), Penfield and Boldrey (1937) and Penfield and Rasmussen (1950) resumed the idea of the sensorimotor cortex on the grounds of motor and somatosensory responses from both the postcentral and precentral gyrus. Another relevant point is that isolated giant cells of Betz, typically associated with motor function, spread over layer V from the giant pyramidal precentral area (FA_γ) to the giant pyramidal (PA) and intermediate (PC) postcentral areas, which are normally associated with sensorimotor function (von Economo and Koskinas 2008; von Economo 2009). The overwhelming majority of neurons that make up the brain and spinal cord, in fact all but motor neurons, cannot be classified as either sensory or motor in the strict sense (Nauta and Karten 1970).

A vindication of Jakob's views is echoed in *Gray's Anatomy* (Warwick et al. 1973): 'Even the simpler differentiation of the cortex into *sensory* areas receiving afferent projection fibres and *motor* areas projecting efferents—the remainder being regarded as 'silent' or *associational*—can no longer be considered appropriate, being itself an inaccurate over-simplification. Evidence has accumulated during the last three decades to show that the areas receiving or originating projection fibres are much more extensive than the initial classical studies indicated. Furthermore, the division into *receiving* and *originating* projection areas is by no means so distinct as first appeared… It is hence more appropriate to speak of the pre- and post-central areas as being *sensory–motor*; and since a mixture of afferent and efferent connections has been shown to exist also in respect to the projection fibres of the acoustic and visual *sensory* areas, they also are more accurately described as sensory–motor in character … It is this afferent–efferent character of most, and probably all, the sensory–motor areas which

 Springer

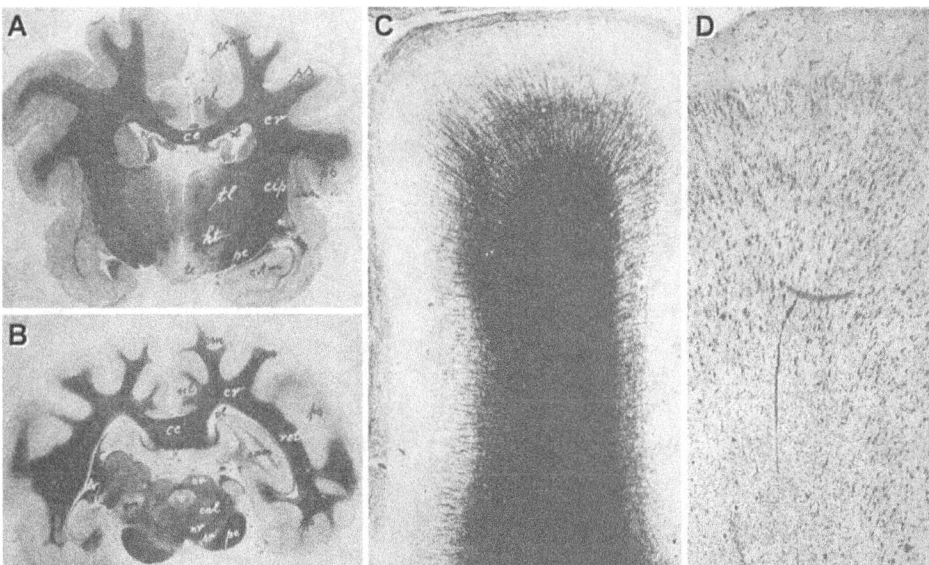

Fig. 8 Cerebrum of *Tapirus terrestris* ('tapir americano'). **a** Posterocapsular section of anterior cerebrum. Weigert method ×3/4. From Jakob (1941, p. 263). **b** Section through the knee of the corpus callosum. Weigert method ×2/3. From Jakob (1941, p. 263).

c Cortical fibers. Prefrontal gyrus. Weigert method ×100. From Jakob and Onelli (1913, Plate 47.3). **d** Insular cortex. Nissl stain ×75. From Jakob and Onelli (1913, Plate 25.3)

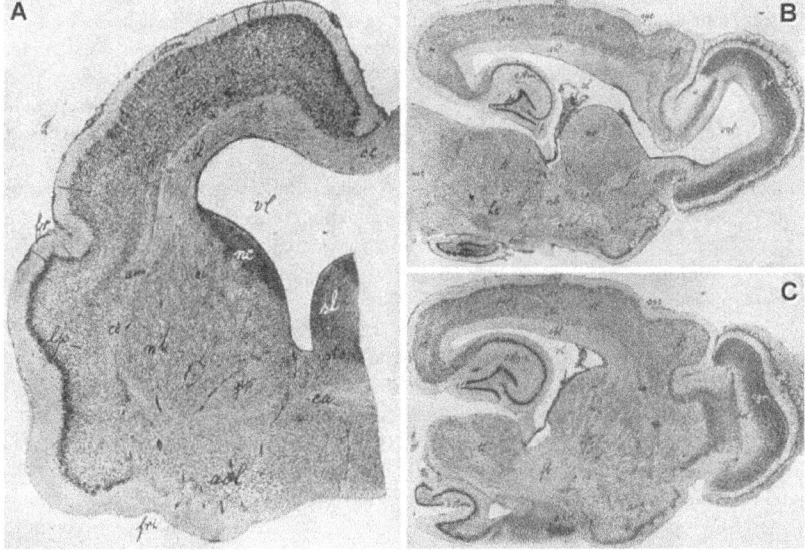

Fig. 9 Cerebrum of the hedgehog or *Erinaceus europaeus* ('erizo'). **a** Left hemisphere with neocortical divisions; the claustrum (*am*) continues upwards into the internal pyramidal layer. Nissl stain ×40. From Jakob (1941, p. 129). **b** Midsagittal section through the cerebral hemisphere, with the olfactory lobe. The neopallium continues toward Ammon's horn (*cAm*). At the base one distinguishes the olfactory (*aol*) and parolfactory (*apol*) areas, caudate nucleus (*nc*), fasciculus

(*fb*) of nucleus basalis (*nb*), thalamus (*t*) and hypothalamus (*ht*). Nissl stain ×25. From Jakob (1941, p. 131). **c** A more lateral section. The capsular radiation (*rfb*) traverses the striatum in disseminated fascicles. The Ammonic zone (*zam*) is continuous with the inner cortical layer of Ammon's horn (*cAm*). Nissl stain ×25. From Jakob (1941, p. 131)

 Springer

Brain Struct Funct (2010) 214:319–338

Fig. 10 Septal projection pathways (reptile) and lateral projection pathways (mammal) (Jakob and Onelli 1913, p. 26)

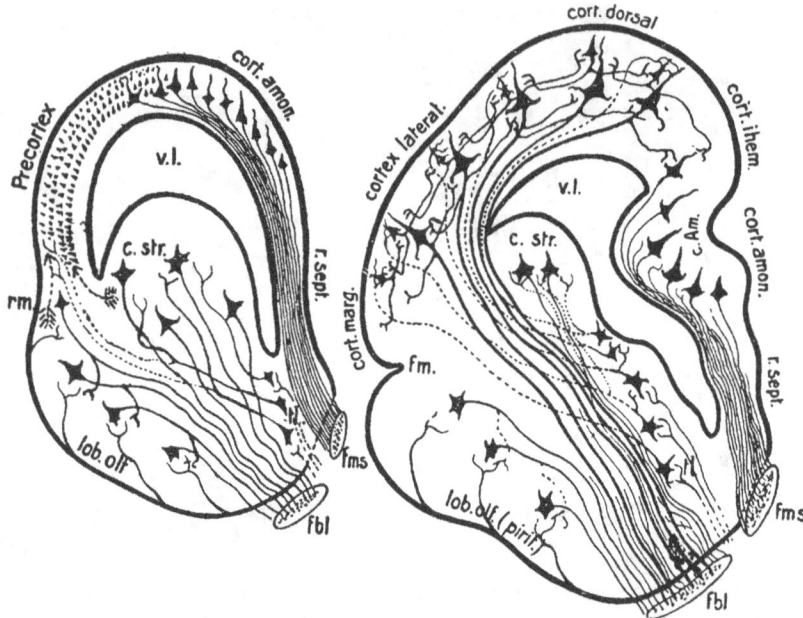

makes the concept of distinguishable motor and sensory parts of the cortex anatomically invalid and functionally misleading. It is clear from the above remarks that much less of the cerebral cortex remains to be dubbed as *associational*, in the vague but well-established meaning of the term.'

The term 'motor' is understood today as a function controlling movements via corticobulbar and corticospinal tracts, and it cannot be accepted, e.g. for the function of layer V neurons in V1 projecting to other cortical or subcortical areas (e.g., the lateral geniculate nucleus). The use of the term 'sensory–motor' by Jakob is more restricted and focused, otherwise all output systems are motoric and all input systems sensoric. Such an inflation of terminology might make the content of Jakob's idea cloudy in a modern context. Therefore, the terms 'effector' and 'receptive' are used to convey what he probably had in mind.

Tracing studies in mice indicate an overlap of neuron populations with simultaneous projections to both somatosensory and frontal association cortical areas, suggesting a widespread pattern of sensory–motor and premotor integration (Mitchell and Macklis 2005). The transcription factor *COUP-TFI* (or *Nr2f1*) appears to be necessary for balancing neocortical patterning, through a repression of motor (frontal) area identities and a specification of sensory (parietal and occipital) area identities (Armentano et al. 2007).

Kaas (1999) suggests that complex brains evolved from simpler brains not by adding vast amounts of general-purpose association cortex, but by expanding hierarchies to include additional areas for sensory processing and motor control; he bases his reasoning on the detection of multiple sensory and motor areas in what was once thought to be 'association cortex' territory in the cat, and the finding that sensory and motor representations occupy most of the neocortex in small-brain mammals.

Mounting evidence on the neural underpinnings of cognition supports the notion that neocortical operations are essentially multisensory at all levels of cortical processing, against an older view that senses operate independently (Ghazanfar and Schroeder 2006). Traditionally, it has been assumed that the integration of different kinds of sensory information in the cortical parenchyma was the task of specialized, higher-order association areas to produce a unified, coherent representation of the outside world. In contrast to such an assumption, recent data (reviewed by Ghazanfar and Schroeder 2006) suggest that much, if not all, of the neocortex is multisensory, which deviates us from the validity of probing the brain unimodally in addressing cognitive aspects, from development to social cognition.

The modern synthesis of functional evolution–development

The disciplines of evolutionary and developmental biology have operated separately for most of the twentieth century;

 Springer

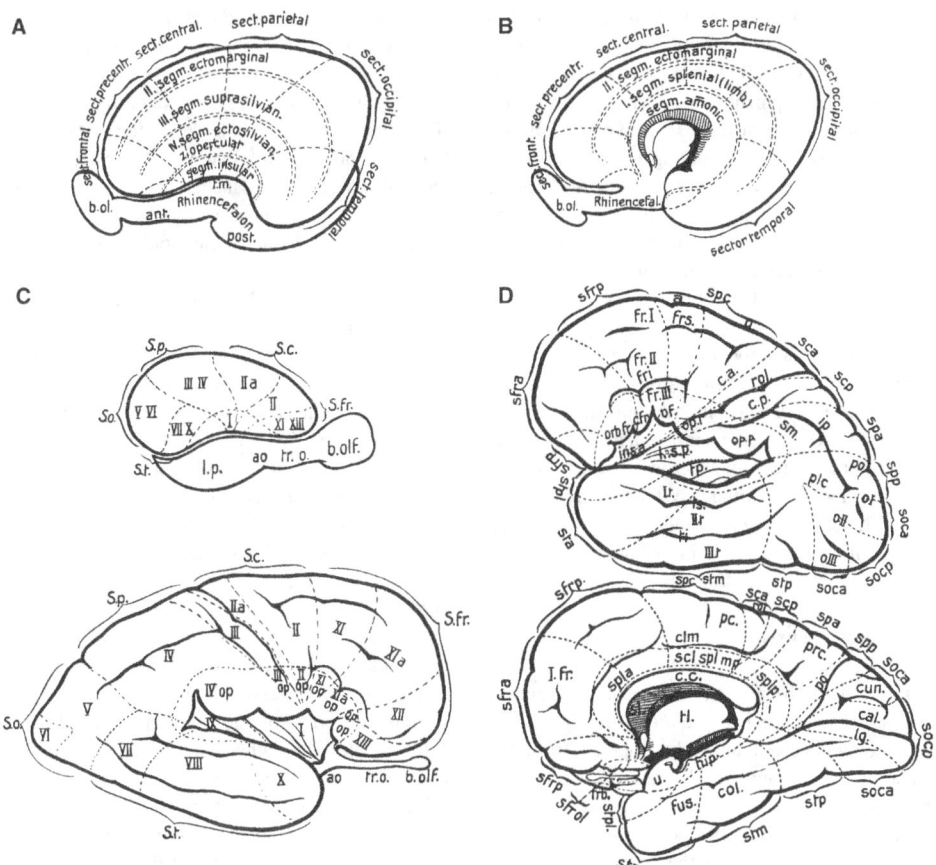

Fig. 11 **a** Lateral and **b** midsagittal facies depicting the hemispheric sector formation in the cerebral surface of a lissencephalic animal. The fan-shaped development of sectors and segments is shown in an anteroposterior direction. The insula forms the rotation point of such a growth. Jakob suggested the development of four sagittal cortical 'pre-gyri', laterally to Ammon's formation. He designated them as (*I*) gyrus splenialis or limbicus, where he places the *visceral cortex*, (*II*) the *bodily axis–hindlimb zone* located between the splenial and ectomarginal sulci, (*III*) the *forelimb zone* between the ectomarginal and suprasylvian sulci, and (*IV*) the *facio-mandibulo-lingual zone* between the suprasylvian and marginal sulci. The formation of these 'segments' has its origin in the base of the marginal sulcus, the insular area of higher mammals. From Jakob (1911, p. 33); Jakob and Onelli (1911, p. 19), (1913, p. 18); field (A) also used by von Economo and Koskinas (1925). **c** Sector formation in the cerebral hemispheric surface of a lissencephalic mammal (*upper*) and the gyrencephalic human brain (*lower*). The temporal lobe is pushed downwards and forward with the fan-shaped growth, whereas the occipital lobe is displaced caudally. From Jakob (1911, p. 34); Jakob and Onelli (1911, p. 20); Jakob and Onelli (1913, p. 18); figures also used by von Economo and Koskinas (1925). **d** Jakob's concept of sector formation

in the human cerebral hemispheres (lateral and midsagittal views) in greater detail. From Jakob and Onelli (1911, p. 9), (1913, p. 36). Jakob considered the 'development of sectors' as the most important principle in the organization of the cerebral cortex, already noted in the brains of lower vertebrates such as edentates. He thus explained regional variations in cortical cytoarchitectonics, which he ascribed to five 'pre-sectors' and their sector partitions—frontal (with five partitions), central (three), parietal (three), occipital (two), and temporal (five)—as well as a robust 'subsector conformation'. The entire cortical mantle was viewed as a system of similarly constructed radiating sectors in a fan-shaped form, with their tip oriented towards the insula, and their expansions towards the upper hemispheric edge. Based on the pattern of fiber growth, he reckoned that sectors possess centripetal virgate parts in their coronae, with centrifugal segments consistently appearing only in certain areas across species. Jakob suggested that all sectors are receptively active, serving simultaneously both projection and association functions, and rejected the separation of the cortex into independent projection and association areas. For detailed reviews of these concepts, see Triarhou and del Cerro (2006a, b) and Triarhou (2008a, 2009)

some visionaries recognized this fact, but lacked the means for a conceptual and experimental synthesis (Raff et al. 1999).

The role of development as an evolutionary factor is currently studied in the context of 'Evolutionary Developmental Biology', with the discovery of the conservation

 Springer

of genes with prominent roles in development, and with a focus on developmental mechanisms that generate new variation (Raff 1996; Carroll 2005). A main goal is to understand how developmental mechanisms influence evolution and how such mechanisms evolved (Butler 1999). A comprehensive understanding of crucial evolutionary modifications of development with features of complete ontogenies, as opposed to static adult morphogenesis, is necessary, whereby multiple fields must be brought together into an interdisciplinary synthesis, seeking an integrated science of biological form (Raff et al. 1999).

An understanding of adaptive evolution requires the use of the entire conceptual spectrum, particularly the fusion of functional aspects with evolution and development (Breuker et al. 2006). One argument is that in many instances, evolution–development has emphasized a structural and partly historical perspective, without systematically addressing function. Therefore, Breuker et al. (2006) argue that the link to function is essential in gaining an integrated view of the role of development in evolution.

Phylogenetic considerations

Brain organization has been studied more in mammalian than in non-mammalian species, one reason being the fact that neuroanatomy received an early impetus from the neuropsychiatric clinic (Nauta and Karten 1970).

Brodmann (1909, 2006) described the comparative anatomy and cytoarchitecture of the cerebral cortex in numerous mammalian orders, from the hedgehog—with its unusually large archipallium—up to non-human primate and human brains; he introduced terms such as *homogenetic* and *heterogenetic formations* to denote two different basic cortical patterns, which, respectively, are either derived from the basic six-layer type or do not demonstrate the six-layer stage. Brodmann was intrigued by the phylogenetic increase in the number of cytoarchitectonic cortical areas in primates, and was astute in pointing out the phenomenon of phylogenetic regression as well (Striedter 2005). Vogt and Vogt (1919) laid the foundations of fiber pathway architecture; they defined the structural features of allocortex, proisocortex, and isocortex, and extensively discussed the differences between paleo-, archi-, and neo-cortical regions (Vogt and Vogt 1919; Vogt 1927; Zilles 2006).

The growth of the cortex by intercalation has been considered fairly well established (von Bonin 1963), in line with the neocortical evolutionary idea of Dart (1934) on a dual origin from hippocampal and prepiriform regions and an intercalation of newer parts between phylogenetically older parts.

Combining cyto- and myelo-architectonics, Sanides (1962, 1964) placed emphasis on transition regions

(*Gradationen*) that accompany 'streams' of neocortical regions coming from paleo- and archi-cortical sources (Pandya and Sanides 1973). [Vogt and Vogt (1919) had already spoken of 'areal gradations'.] The idea of a 'koniocortex core' and 'prokoniocortex belt areas' in the temporal operculum (Pandya and Sanides 1973) was modified by Kaas and Hackett (1998, 2000), who speak of histologically and functionally distinct 'core', 'belt' and 'parabelt' subdivisions in the monkey auditory cortex, with specified connections.

While there are still open questions regarding the evolutionary origin of the mammalian neocortex, comparative studies indicate that at least three pallial subdivisions—lateral (olfactory), dorsal and medial (hippocampal)—characterized the roof of the cerebral hemispheres of earliest vertebrates (Northcutt and Kaas 1995). The forebrain of early mammals was dominated by an olfactory bulb, and the olfactory (piriform) cortex appeared large relative to the small amount of neocortex; in reptiles, the dorsal cortex—homolog of the mammalian neocortex—is proportionally almost as large in surface as the neocortex of early mammals, but thinner, comprising mainly a single pyramidal layer (Kaas 2008).

To adequately explain the evolution of the mammalian neocortex, an understanding of correlative changes in surrounding areas of the telencephalic pallium and subpallium, close neighbors in a common morphogenetic field and postulated sources of certain cortical neuron subsets, is deemed necessary (Puelles 2001). The developmental evidence that cells originating in a compartment corresponding to the dorsal ventricular ridge (DVR) become included in structures generated in a compartment corresponding to isocortex could reconcile the proposed evolutionary homology between the reptilian/avian DVR and mammalian isocortex (Aboitiz 1999).

The evidence seems to favor a correspondence of isocortex with the dorsal cortex of reptiles (Fig. 12b): sensory projections that terminate in the ventral pallium of reptiles end in the dorsal pallium (isocortex) of mammals, possibly owing to their phylogenetic participation in associative networks between dorsal, olfactory, and hippocampal cortices subserving spatial or episodic memory in early mammals (Aboitiz et al. 2003).

The mammalian neocortex is characterized by an inside-out developmental neurogenetic gradient, with deep layer neurons being born earlier than superficial layer neurons, whereas the reptilian cortex originates in a reverse, outside-in gradient (Goffinet et al. 1986; Aboitiz 1993; Aboitiz et al. 2002). The older, inferior layers (V–VI) of the mammalian isocortex resemble reptilian cortical cells in morphology, neurotransmitter signatures and subcortical connections, while the younger superficial layers (II–IV)

 Springer

exhibit their own local and corticocortical connections (Aboitiz et al. 2002).

It was suggested that the neocortex has resulted from a translocation of large neuronal masses, which in ancestral forms occupied subcortical stations, in particular the region of the external striatum (Nauta and Karten 1970). The superior and temporal neocortices in particular appear to resemble, from a phylogenetic viewpoint, the dorsal cortex and the DVR in reptiles, and the Wulst (a territory at the transition between the dorsalmost zone of the external striatum, or hyperstriatum, and the pallial mantle, characterized by a clearly defined layer of granule cells) and the DVR in the avian forebrain, respectively (Nauta and Karten 1970; Shimizu and Karten 1991; Reiner 2000). The lateral pallium of amphibians, reptiles and birds may be homologous to the hexalaminar cortex of mammals (Northcutt 1981); its expansion has been viewed as associated with a displacement of the hippocampal formation (archicortex) and the piriform cortex (paleocortex) toward the medial telencephalic wall (Rakic and Kornack 2001). The cytoarchitectonically continuous cortical plate (with a medial/dorsomedial hippocampal portion and a more dorsal

portion) and the subcortical DVR form two portions of the pallium that are relevant to neocortical evolution in modern turtles (Reiner 2000).

The term 'dorsal ventricular ridge' was originally introduced by Johnston (1915) as a descriptive label (Lohman and Smeets 1991). Half a century earlier, Hunter (1861) had described in the lateral ventricle of reptiles the prominent eminence that would eventually bear his name: Hunter's eminence was viewed by many neuroanatomists during the early part of the twentieth century, perhaps including Jakob, as homologous to the basal ganglia of the mammalian brain (Fig. 12), thence the general application of the term 'striatum' to this structure (Lohman and Smeets 1991). Elliot Smith (1919) divided it into hypopallium, paleostriatum and amygdaloid complex; Ariëns Kappers et al. (1936) into neostriatum, paleostriatum and archistriatum; in modern usage, it comprises the DVR, the striatum and the amygdaloid complex (Lohman and Smeets 1991).

On the other hand, the three longitudinal cortical zones comprised mediodorsal, dorsal and lateral cortex (Edinger 1896), *Ammonsrinde*, dorsal and lateral cortex (Unger 1906), hippocampal, dorsal and piriform cortex (Goldby

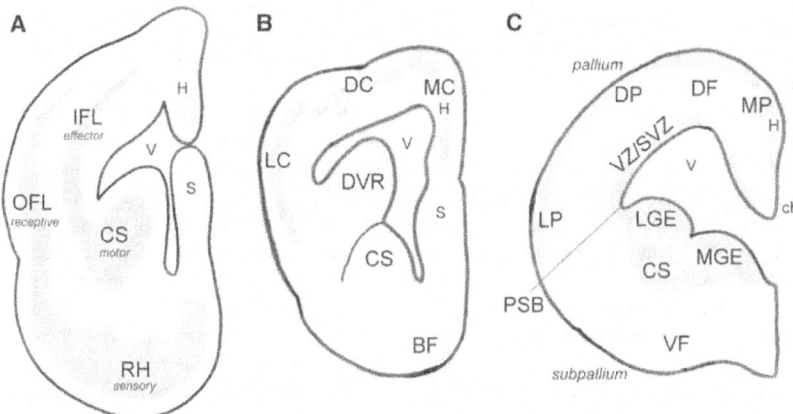

Fig. 12 **a** A summary of Jakob's idea on the dual onto-phylogenetic origin and ubiquitous function of the cerebral cortex, adapted from López Pasquali (1965, p. 30). Jakob mentions that the outer and inner cortical layers, respectively 'derive' from the rhinencephalon and the striatum (Jakob and Onelli 1911, p. 37); although the word 'migration' is not mentioned explicitly, the verb 'derive' logically implies that this is what he had in mind (Outes 2006). *CS* corpus striatum, *H* hippocampus (Ammon's horn), *IFL* inner fundamental layer, *OFL* outer fundamental layer, *RH* rhinencephalon, *S* septal nucleus, *V* lateral cerebral ventricle. **b** Modern view of forebrain phylogeny, shown in the reptilian cerebral hemisphere. The thin lateral cortex (*LC*) arises by radial migration from the lateral ependyma, whereas the dorsal ventricular ridge (*DVR*)—whose marked protrusion into the lateral ventricle led to being erroneously thought for a long time as homologous to the basal ganglia of the human brain—arises through the ventrolateral migration of neurons

from the ependyma of the ventrolateral wall of the cerebral ventricle (Karten 1997). *BF* basal forebrain, *DC* dorsal cortex, *MC* medial cortex. Drawing based on Northcutt and Kaas (1995), Karten (1997), Tissir et al. (2002), and Aboitiz and Montiel (2007). **c** Modern view of mammalian forebrain ontogeny. The lateral ganglionic eminence (*LGE*) gives rise to the striatum (*CS*), and the medial ganglionic eminence to the globus pallidus; the medial pallium (*MP*) gives rise to the hippocampus, the lateral pallium (*LP*) to olfactory cortex, and the ventral pallium to the claustro-amygdaloid complex (Aboitiz and Montiel 2007). In the human forebrain, interneurons originate both in the ganglionic eminence and the ventricular (*VZ*) and subventricular (*SVZ*) zones. *ch* cortical hem, *DF* dorsal forebrain, *DP* dorsal pallium, *PSB* pallial-subpallial boundary, *VF* ventral forebrain. Drawings based on Northcutt and Kaas (1995), Karten (1997), Tissir et al. (2002), Aboitiz and Montiel (2007) and Rakic (2009)

🕸 Springer

1934), fascia dentata, Ammon's formation/general cortex and piriform cortex (Curwen 1937), and medial, dorsal and lateral cortex in modern usage (Lohman and Smeets 1991).

Although the consideration that the neocortex is elaborated from the rhinencephalon (Ariëns Kappers et al. 1936) is no longer tenable, it is conjectured that in primitive mammals the origin of the cerebral cortex was triggered, in a context of adaptation to nocturnal life, by the development of the olfactory system (Aboitiz 1992).

Herrick (1924), who acknowledges the contribution of Jakob and Onelli (1911) to the anatomy of the olfactory region in the opossum, emphasizes the large olfactory bulb that is connected with the remainder of the cerebral hemisphere by a distinct olfactory crus; in the latter, superficial cells assume a cortical cellular type, namely, piriform cortex laterally and anterior hippocampal cortex medially, which in turn extend forward in contact with the bulbar formation.

One puzzle was the striking discrepancy between the relatively small pallium and large striatum in sauropsids, and the apparent inverse ratios of large areas of pallium and only moderate amounts of striatum in mammals (Karten 1969). Nauta and Karten (1970) hypothesized that two distinct populations of founder cells contribute to cortical formation in mammals. One possibility is that DVR neuroblasts of ancient reptiles shifted into the lateral pallium and provided additional cells for the hexalaminar neocortex; the 'external striatum' of reptiles and birds is absent as such from mammals, which, conversely, feature a 'neocortex' in the pallial mantle (Nauta and Karten 1970; Rakic and Kornack 2001). Experiments in mouse chimeras support a dual origin, and further suggest that the two phylogenetically distinct populations stay segregated during development as the separate laminar and radial clones in the mammalian cerebral cortex (Kuan et al. 1997).

The resolution of the evolutionary origins of the neocortex, which has long puzzled anatomists, requires analyses of both developing and adult brains (Karten 1991). Studies in the mature reptilian and avian brain have clarified some basic questions on the origin of the neocortex, which can be viewed as consequent to two events: (1) The elaboration of constituent neuronal populations and their associated connections that are common to the telencephalon of non-mammalian and mammalian amniotes: such populations are found within the neocortex in mammals, whereas most of them are seen within the DVR and the dorsolateral ventricular ridge (DLVR) in reptiles and birds. (2) In mammals, the components of the DVR and DLVR are incorporated into the thin overlying pallium to form a laminated 'neocortex' (Karten 1991).

An onto-phylogenetic comparison of pallial organization indicates that the lateral cortex of reptiles is homologous to the avian and mammalian piriform cortex; the anterior DVR of reptiles is probably homologous to the neostriatum and ventral hyperstriatum of birds and to the endopiriform nucleus of mammals, whereas the posterior DVR of reptiles is most likely homologous to the archistriatum of birds and the mammalian amygdala; the dorsal cortex of reptiles is probably homologous to mammalian isocortex (Striedter 1997).

Multiple evolutionary origins of the neocortex have been proposed, based on its separation into the precursors of non-laminar and laminar regions; ancestral reptiles and proto-mammals have been thought of possessing discrete populations that were the precursors of cells of the SVZ of mammals and the DVR of non-mammalian amniotes (Shimizu and Karten 1991). Migratory routes may follow a mediolateral course from the DVR–SVZ in reptiles, and a lateral and dorsolateral course in mammals (Shimizu and Karten 1991).

In all, two major modern hypotheses have been proposed to explain the origin of the mammalian cortex (reviewed by Northcutt and Kaas 1995): the 'out-group hypothesis' holds that the DVR of living reptiles and birds and the isocortex of mammals are 'homoplastic' structures, representing independent transformations of the ancestral pallium. The 'recapitulation hypothesis' rejects the validity of the cerebral hemispheres of living amphibians as an appropriate out-group, and postulates that an additional pallial subdivision (the DVR) existed in the common ancestor of terrestrial vertebrates; thus, the presence of a DVR represents the primitive or ancestral condition, which has been retained in living reptiles and birds, while the mammalian isocortex arose by a dual differentiation of the dorsal cortex and a migration of the cells of the DVR. In that case, the DVR and the dorsal cortex of reptiles are viewed as homologous to the mammalian isocortex owing to the single transformation in one radiation.

Sauropsida (reptiles and birds) have a trilaminar dorsal cortex (corresponding to layers I, V and VI in mammals), whereas metatheria (marsupials) possess six-layer cortices; a SVZ—which appears to have emerged prior to the eutherian–metatherian split—may not be required for the generation of a hexalaminar cortex in all mammals, a clue supported by the absence of an organized SVZ in the South American gray short-tailed opossum (*Monodelphis domestica*) (Cheung et al. 2009).

Ontogenetic considerations

A dichotomy exists concerning the embryogenesis of the upper and the lower cortical layer in mammals (Fig. 12c), based on cytoarchitectonic and functional criteria, as well as axonal projections and gene expression.

Distinct neocortical neuron populations are generated within two spatially and molecularly segregated proliferative

Springer

domains (Lai et al. 2008; Azim et al. 2009): progenitors of the dorsally located pallium positive for the transcription factor SOX6 give rise to excitatory projection neurons, whereas SOX5 positive progenitors of the ventrally located subpallium give rise to inhibitory interneurons (such an expression pattern becomes reversed in postmitotic neurons).

The deep cortical or infragranular layers (V–VI) are formed by early-generated neurons, born from dividing neuronal precursors at the ventricular zone (VZ). The upper cortical or supragranular layers (II–IV) derive from late-generated neurons, born through mitotic divisions of intermediate progenitor cells at the subventricular zone (SVZ) (Parnavelas et al. 2000; Molnár et al. 2006; Cheung et al. 2007; Noctor et al. 2007; Abdel-Mannan et al. 2008; Cubelos et al. 2008). Descending projection pathways to subcortical targets mainly arise from pyramidal neurons of the deep cortical layers, whereas projection neurons of the upper cortical layers do not extend long corticofugal axons (Kwan et al. 2008). Corticofugal axons (including corticothalamic and corticospinal fibers) that extend from neurons in the deep cortical layers are guided through the internal capsule, possibly through the specification of neurons expressing the transcription factors *Nfla* and *Nflb* (Plachez et al. 2008).

Specific genes, such as the homeodomain transcription factor *Cux-2*, selectively control the proliferation rates of SVZ precursors and therefore the number of upper cortical neurons (Cubelos et al. 2008). On the other hand, the nuclear factor I gene product NFIB is predominantly expressed in corticofugal projection neurons of the deep cortical layers, V and VI (Plachez et al. 2008).

Cells from the lateral/ventral pallium migrate to the lateral ganglionic eminence (LGE) and a number of cells cortically derived from the *Emx1* progenitor lineage persist in the adult striatum, thus being a putative source of neural diversity in the ventral telencephalon (Cocas and Corbin 2008).

Conclusion

It would be conjectural, even anachronistic, to attempt to guess to which structures, based on our current knowledge and nomenclature, correspond the areas construed by Jakob as the elements of his onto-phylogenetic and morphofunctional ideas (Fig. 12). Could what he described as 'striatum' in the reptilian forebrain be, at least in part, the dorsal ventricular ridge? Or could the ganglionic eminence of the embryonic human brain be included in what he viewed as striatal domains?

From the point of view of Outes (2006), it appears logical to think that the inner cell layer is the last migratory stream toward the pia, overlapping with the striatal mass insofar as this is also the last stream of a sector more inferior to the telencephalic vesicle; all this might indicate a kinship between these two sectors, both originating by the same secondary migratory wave.

The new foci of comparative neurobiology (Striedter 1998) comprise (1) the integration of comparative and developmental neurobiology and genetics to test phylogenetic hypotheses at a mechanistic level; (2) the comparative morphology of neural circuits and their relation to physiology and behavior; and (3) phylogenetic analyses of independently evolved similarities to discover general rules on how neural systems operate and become modified in the course of evolution. Jakob's functional evolutionary–developmental integration seems to endure, a century later. Concepts from the classical bibliography may thus complement the sophisticated modern means in deciphering the structural–functional workings of the brain and its mind.

Acknowledgments The author gratefully acknowledges Michael C. Triarhou, LL.M., for invaluable help with the German language, Professor Karl Zilles for constructive criticism, Noelia Fiorentino of *La Prensa Médica Argentina*, Buenos Aires, for a copy of Jakob's 1916 article, and Dr. Daniel S. Margulies for Jakob's biography by López Pasquali.

References

Abdel-Mannan O, Cheung AFP, Molnár Z (2008) Evolution of cortical neurogenesis. Brain Res Bull 75:398–404

Aboitiz F (1992) The evolutionary origin of the mammalian cerebral cortex. Biol Res 25:41–49

Aboitiz F (1993) Further comments on the evolutionary origin of the mammalian brain. Med Hypotheses 41:409–418

Aboitiz F (1999) Comparative development of the mammalian isocortex and the reptilian dorsal ventricular ridge: evolutionary considerations. Cereb Cortex 9:783–791

Aboitiz F, Montiel J (2007) Origin and evolution of the vertebrate telencephalon, with special reference to the mammalian neocortex. Adv Anat Embryol Cell Biol 193:1–112

Aboitiz F, Montiel J, Morales D, Concha M (2002) Evolutionary divergence of the reptilian and the mammalian brains: considerations on connectivity and development. Brain Res Rev 39:141–153

Aboitiz F, Morales D, Montiel J (2003) The evolutionary origin of the mammalian isocortex: towards an integrated developmental and functional approach. Behav Brain Sci 26:535–552

Ariëns Kappers CU (1909) The phylogenesis of the paleocortex and archicortex compared with the evolution of the visual neocortex. Arch Neurol Psychiatry (Lond) 4:161–173

Ariëns Kappers CU, Huber GC, Crosby EC (1936) The comparative anatomy of the nervous system of vertebrates, including man, 2 vols. Hafner, New York

Armentano M, Chou SJ, Tomassy GS, Leingärtner A, O'Leary DD, Studer M (2007) *COUP-TFI* regulates the balance of cortical patterning between frontal/motor and sensory areas. Nat Neurosci 10:1277–1286

Azim E, Jabaudon D, Fame RM, Macklis JD (2009) SOX6 controls dorsal progenitor identity and interneuron diversity during neocortical development. Nat Neurosci 12:1238–1247

Bechterew W, Meyer A (1978) The concept of a sensorimotor cortex: its early history, with especial emphasis on two early experimental contributions by W. Bechterew. Brain 101:673–685

Breuker CJ, Debat V, Klingenberg CP (2006) Functional evo-devo. Trends Ecol Evol 21:488–492

Brodmann K (1909) Vergleichende Lokalisationslehre der Großhirnrinde. J. A. Barth, Leipzig

Brodmann K (2006) Localisation in the cerebral cortex. Springer Science, New York (translated by L. J. Garey)

Butler AB (1999) Whence and whither cortex? Trends Neurosci 22:332–334

Carroll SB (2005) Endless forms most beautiful: the new science of evo devo. Norton, New York

Cheung AFP, Pollen AA, Tavare A, De Proto J, Molnár Z (2007) Comparative aspects of cortical neurogenesis in vertebrates. J Anat 211:164–176

Cheung AF, Kondo S, Abdel-Mannan O, Chodroff RA, Sirey TM, Bluy LE, Webber N, Deproto J, Karlen SJ, Krubitzer L, Stolp HB, Saunders NR, Molnár Z (2009) The subventricular zone is the developmental milestone of a 6-layered neocortex: comparisons in metatherian and eutherian mammals. Cereb Cortex Sep 2. (Epub ahead of print). doi:10.1093/cercor/bhp168

Cocas LA, Corbin JG (2008) Embryonic Emx1+ progenitor cells generate diverse neural subtypes in the mature striatum. In: Fishell G, Kriegstein AR, Parnavelas JG (eds) Cortical development: stem cells, neurogenesis, migration, circuit formation, cortical disorders. Mediterranean Agronomic Institute, Crete, pp 43–44 (abstract)

Cubelos B, Sebastián-Serrano A, Kim S, Moreno-Ortiz C, Redondo JM, Walsh CA, Nieto M (2008) Cux-2 controls the proliferation of neuronal intermediate precursors of the cortical subventricular zone. Cereb Cortex 18:1758–1770

Curwen AO (1937) The telencephalon of Tupinambis nigropunctatus. I. Medial and cortical areas. J Comp Neurol 66:375–404

Dart RA (1934) The dual structure of the neopallium: its history and its significance. Anat Rec 69:1–19

Edinger L (1896) Untersuchungen über die vergleichende Anatomie des Gehirns. III. Neue Studien über das Vorderhirn der Reptilien. Abh Senckenb Naturforsch Gesch 19:313–388

Elliot Smith G (1919) A preliminary note on the morphology of the corpus striatum and the origin of the neopallium. J Anat (Lond) 53:271–291

Flatau E, Jacobsohn L (1899) Handbuch der Anatomie und vergleichenden Anatomie des Centralnervensystems der Säugetiere. Karger, Berlin

Flechsig P (1896) Gehirn und Seele. Veit and Comp, Leipzig

Flechsig P (1898) Études sur le cerveau. Vigot Frères, Paris (trad. L. Levi)

Foerster O (1936) Motorische Felder und Bahnen: sensible corticale Felder. In: Bumke O, Foerster O (eds) Handbuch der Neurologie, vol 6. Springer, Berlin, pp 1–448

Fontana H, Belziti H, Requejo F (2002) El espacio perforado anterior y zonas aledañas. Consideraciones funcionales. Parte I. Rev Argent Neurocirug 16:1–11

Gans C (1966) Studies on amphisbaenids (Amphisbaenia, Reptilia). 3. The small species from Southern South America commonly identified as Amphisbaena darwini. Bull Am Mus Nat Hist 134:185–260

Ghazanfar AA, Schroeder CE (2006) Is neocortex essentially multisensory? Trends Cogn Sci 10:278–285

Goffinet AM, Daumerie C, Langerwerf B, Pieau C (1986) Neurogenesis in reptilian cortical structures: [³H]-thymidine autoradiographic analysis. J Comp Neurol 243:106–116

Goldby F (1934) The cerebral hemispheres of Lacerta viridis. J Anat (Lond) 68:157–215

Herrick CJ (1924) The nucleus olfactorius anterior of the opossum. J Comp Neurol 37:317–359

Hunter J (1861) Essays and observations on natural history, anatomy, physiology, psychology and geology. van Voorst, London

Jakob C (1895) Atlas des gesunden und kranken Nervensystems nebst Grundriss der Anatomie, Pathologie und Therapie desselben. Lehmann, München

Jakob C (1898) An atlas of the normal and pathological nervous systems, together with a sketch of the anatomy, pathology, and therapy of the same. Baillière Tindall and Cox, London (translated by J. Collins)

Jakob C (1899) Sobre el desarrollo de la corteza cerebral. Rev Soc Méd Argent 7:397–403

Jakob C (1910a) La célula cortical en la locura (Estudios histopatológicos sobre las células piramidales en las enfermedades mentales). Anales de la Administración Sanitaria y Asistencia Pública: Ediciones de La Semana Médica. Imprenta de E. Spinelli, Buenos Aires, pp 263–267

Jakob C (1910b) La histoarquitectura comparada de la corteza cerebral y su significación para la psicología moderna. Argent Méd (B Aires) 8:437–438

Jakob C (1911) Das Menschenhirn: Eine Studie über den Aufbau und die Bedeutung seiner grauen Kerne und Rinde. Lehmann, München

Jakob C (1912a) Über die Ubiquität der senso-motorischen Doppelfunktion der Hirnrinde als Grundlage einer neuen, biologischen Auffassung des corticalen Seelenorgans. J Psychol Neurol (Leipz) 19:379–382

Jakob C (1912b) Ueber die Ubiquität der senso-motorischen Doppelfunktion der Hirnrinde als Grundlage einer neuen biologischen Auffassung des kortikalen Seelenorgans. Münch Med Wochenschr 59:466–468

Jakob C (1913) La psicología orgánica y su relación con la biología cortical. Arch Psiquiatr Criminol (B Aires) 12:680–698

Jakob C (1914) El lenguaje de los animales. Rev Jard Zool B Aires (Época II) 10:129–135

Jakob C (1916) Sobre la existencia simultánea de una doble función sensomotriz de la corteza cerebral como base de una nueva concepción biológica del órgano psíquico cortical. Prensa Méd Argent 2:305–307

Jakob C (1918) La filogenia cortical: sobre la corteza cerebral de gimnofiones y anfisbenas argentinas (abstract). Actas Trab 1er Congr Nacl Med (B Aires) 1:82

Jakob C (1939a) El cerebro humano: su anatomía sistemática y topográfica (Folia Neurobiológica Argentina, Atlas I). López, Buenos Aires

Jakob C (1939b) El cerebro humano: su anatomía patológica en relación a la clínica (Folia Neurobiológica Argentina, Atlas II). López, Buenos Aires

Jakob C (1941) El cerebro humano: su ontogenia y filogenia (Folia Neurobiológica Argentina, Atlas III). López, Buenos Aires

Jakob C (1945) El yacaré (Caimán latirostris) y el origén del neocortex: estudios neurobiológicos y folklóricos del reptil más grande de la Argentina (Folia Neurobiologica Argentina, Tomo IV). López, Buenos Aires

Jakob C, Onelli C (1911) Vom Tierhirn zum Menschenhirn: vergleichend morphologische, histologische und biologische Studien zur Entwicklung der Grosshirnhemisphären und ihrer Rinde. Lehmann, München

Jakob C, Onelli C (1913) Atlas del cerebro de los mamíferos de la República Argentina: estudios anatómicos, histológicos y biológicos comparados sobre la evolución de los hemisferios y de la corteza cerebral. Kraft, Buenos Aires

Jakob C, Jakob A, Pedace EA (1945) El embrión humano, folleto III: el proceso real de la gastrulación en un embrión con dos somitos. López, Buenos Aires

🕮 Springer

Johnston JB (1915) The cell masses in the forebrain of the turtle *Cistudo carolina*. J Comp Neurol 25:393–468

Kaas JH (1999) The transformation of association cortex into sensory cortex. Brain Res Bull 50:425

Kaas JH (2008) The evolution of the complex sensory and motor systems of the human brain. Brain Res Bull 75:384–390

Kaas JH, Hackett TA (1998) Subdivisions of auditory cortex and levels of processing in primates. Audiol Neurootol (Basel) 3:73–85

Kaas JH, Hackett TA (2000) Subdivisions of auditory cortex and processing streams in primates. Proc Natl Acad Sci USA 97:11793–11799

Kaes T (1907) Die Grosshirnrinde des Menschen in ihren Massen und in ihrem Fasergehalt: ein gehirnanatomischer Atlas mit erläuterndem Text und schematische Zeichnung. Fischer, Jena

Karten HJ (1969) The organization of the avian telencephalon and some speculations on the phylogeny of the amniote telencephalon. Ann NY Acad Sci 167:164–179

Karten HJ (1991) Homology and evolutionary origins of the 'neocortex'. Brain Behav Evol 38:264–272

Karten HJ (1997) Evolutionary developmental biology meets the brain: the origins of mammalian cortex. Proc Natl Acad Sci USA 94:2800–2804

Keegan E (2003) Flechsig and Freud: late 19th-century neurology and the emergence of psychoanalysis. Hist Psychol 6:52–69

Kleist K (1926) Die einzeläugigen Gesichtsfelder und ihre Vertretung in den beiden Lagen der verdoppelten inneren Körnerschicht der Sehrinde. Klin Wochenschr 5:3–10

Kleist K (1934) Gehirnpathologie. Barth, Leipzig

Kuan C, Elliott EA, Flavell RA, Rakic P (1997) Restrictive clonal allocation in the chimeric mouse brain. Proc Natl Acad Sci USA 94:3374–3379

Kuhlenbeck H (1922) Über den Ursprung der Großhirnrinde: eine phylogenetische und neurobiotaktische Studie. Anat Anz (Jena) 55:337–365

Kwan KY, Lam MM, Krsnik Ž, Kawasawa YI, Lefebvre V, Šestan N (2008) SOX5 postmitotically regulates migration, postmigratory differentiation, and projections of subplate and deep-layer neocortical neurons. Proc Natl Acad Sci USA 105:16021–16026

Lai T, Jabaudon D, Molyneaux BJ, Azim E, Arlotta P, Menezes JR, Macklis JD (2008) SOX5 controls the sequential generation of distinct corticofugal neuron subtypes. Neuron 57:232–247

Lohman AHM, Smeets WJA (1991) The dorsal ventricular ridge and cortex of reptiles in historical and phylogenetic perspective. In: Finlay BL, Innocenti G, Scheich H (eds) The neocortex: ontogeny and phylogeny. Plenum Press, New York, pp 59–74

López Pasquali L (1965) Christfried Jakob: su obra neurológica, su pensamiento psicológico y filosófico. López Libreros Editores S.R.L.–Talleres Gráficos de La Prensa Médica Argentina, Buenos Aires

Lores Arnaiz MR, Borrego Maturana F, Azzara S (2002) Las ideas de Christofredo Jakob sobre mapa cortical y functiones superiores. Rev Hist Psicol 23:9–36

Luys J-B (1876) Le cerveau et ses fonctions. Baillière, Paris

Marburg O (1932) Konstantin Economo Freiherr von San Serff. Dtsch Z Nervenheilk 123:219–229

Meyer L (1981) Cristofredo Jakob: a veinticinco años de su muerte. Acta Psiquiátr Psicol Amér Lat 27:13–14

Meyer A (1982) The concept of a sensorimotor cortex: its later history during the twentieth century. Neuropathol Appl Neurobiol 8:81–93

Meynert T (1872) Der Bau der Gross-Hirnrinde und seine örtlichen Verschiedenheiten, nebst einem pathologisch-anatomischen Corollarium. Heuser, Leipzig

Mitchell BD, Macklis JD (2005) Large-scale maintenance of dual projections by callosal and frontal cortical projection neurons in adult mice. J Comp Neurol 482:17–32

Molnár Z, Métin C, Stoykova A, Tarabykin V, Price DJ, Francis F, Meyer G, Dehay C, Kennedy H (2006) Comparative aspects of cerebral cortical development. Eur J Neurosci 23:921–934

Mott FW (1894) The sensory motor functions of the central convolutions of the cerebral cortex. J Physiol 15:464–487

Mott FW (1907) The progressive evolution of the structure and functions of the visual cortex in mammalia. Arch Neurol Pathol Lab Lond 3:1–48

Moyano BA (1957) Christfried Jakob (25/12/1866–6/5/1956). Acta Neuropsiquiátr Argent 3:109–123

Munk H (1881) Über die Funktionen der Großhirnrinde. Gesammelte Mitteilungen aus den Jahren 1877–1880, mit Einleitung und Anmerkungen. Hirschwald, Berlin

Nauta WJH, Karten HJ (1970) A general profile of the vertebrate brain, with sidelights on the ancestry of cerebral cortex. In: Schmidt FO (ed) The neurosciences second study program. Rockefeller University Press, New York, pp 7–26

Nissl F (1908) Experimentalergebnisse zur Frage der Hirnrindeschichtung (38. Versammlung der Südwestdeutschen Irrenärzte, Heidelberg, 2.–3. November 1907). Mschr Psychiatr Neurol 23:186–188

Noctor SC, Martínez-Cerdeño V, Kriegstein AR (2007) Contribution of intermediate progenitor cells to cortical histogenesis. Arch Neurol 64:639–642

Northcutt RG (1981) Evolution of the telencephalon in nonmammals. Annu Rev Neurosci 4:301–350

Northcutt RG, Kaas JH (1995) The emergence and evolution of the mammalian neocortex. Trends Neurosci 18:373–379

Obersteiner H (1913) Die Kleinhirnrinde vom Elephas und Balaenoptera. Arb Neurol Inst (Wien) 20:145–154

Orlando JC (1966) Christofredo Jakob: su vida y obra. Editorial Mundi, Buenos Aires

Outes DL (2006) A medio siglo de la muerte de Christofredo Jakob, 1956–2006: fuentes de la concepción biológica de la doble corteza. Electroneurobiología (B Aires) 14:3–28

Outes DL, Benítez I (1976) Sobre el origen de la concepción biológica de la doble corteza: a veinte años de la muerte de Christofredo Jakob, 1956–1976. Rev Neurol Argent 1:220–228

Pandya DN, Sanides F (1973) Architectonic parcellation of the temporal operculum in rhesus monkey and its projection pattern. Z Anat Entwicklungsgesch 139:127–161

Papini MR (1988) Influence of evolutionary biology in the early development of experimental psychology in Argentina (1891–1930). Int J Exp Psychol 2:131–138

Parnavelas JG, Anderson SA, Lavdas AA, Grigoriou M, Pachnis V, Rubenstein JL (2000) The contribution of the ganglionic eminence to the neuronal cell types of the cerebral cortex. In: Bock GR, Cardew G (eds) Evolutionary developmental biology of the cerebral cortex. Wiley and Sons, Chichester, pp 129–139

Penfield W, Boldrey E (1937) Somatic motor and sensory representation in the cerebral cortex of man as studied by electrical stimulation. Brain 60:389–443

Penfield W, Rasmussen TB (1950) The cerebral cortex of man: a clinical study of localization of function. MacMillan, New York

Plachez C, Lindwall C, Sunn N, Piper M, Moldrich RX, Campbell CE, Osinski JM, Gronostajski RM, Richards LJ (2008) Nuclear factor I gene expression in the developing forebrain. J Comp Neurol 508:385–401

Poliak SL (1932) The main afferent fiber systems of the cerebral cortex in primates. University of California Press, Berkeley, pp 107–207

Puelles L (2001) Thoughts on the development, structure and evolution of the mammalian and avian telencephalic pallium. Phil Trans R Soc Lond (Biol) 356:1583–1598

Raff RA (1996) The shape of life: genes, development, and the evolution of animal form. University of Chicago Press, Chicago

Raff RA, Arthur W, Carroll SB, Coates MI, Wray G (1999) Chronicling the birth of a discipline. Evol Dev 1:1–2

Rakic P (2007) The radial edifice of cortical architecture: from neuronal silhouettes to genetic engineering. Brain Res Rev 55:204–219

Rakic P (2009) Evolution of the neocortex: a perspective from developmental biology. Nat Rev Neurosci 10:724–735

Rakic P, Kornack DR (2001) Neocortical expansion and elaboration during primate evolution: a view from neuroembryology. In: Falk D, Gibson KR (eds) Evolutionary anatomy of the primate cerebral cortex. Cambridge University Press, Cambridge, pp 30–56

Ramón y Cajal S (1895) Algunas conjeturas sobre el mecanismo anatómico de la ideación, asociación y atención. Rev Med Cirug Práct (Madr) 19:497–508

Ramón y Cajal S (1899) The sensori-motor cortex. In: Story WE, Wilson LN (eds) Clark University 1889–1899 decennial celebration. Norwood Press, Norwood, MA, pp 311–382

Ramón y Cajal S (1904) Textura del sistema nervioso del hombre y de los vertebrados, tomo II, secunda parte. Moya, Madrid, pp 1121–1152

Ramón y Cajal S (1906) Studien über die Hirnrinde des Menschen, 5. Heft. Barth, Leipzig, pp 41–79 (übers. von J Bresler)

Ramón y Cajal S (1995) Histology of the nervous system of man and vertebrates, vol. II. Oxford University Press, New York, pp 707–729 (translated by N. Swanson and L.W. Swanson)

Ranke O (1911) Bücheranzeigen und Referate: Chr. Jakob und Cl. Onelli, Vom Tierhirn zum Menschenhirn; Chr. Jakob, Das Menschenhirn. Münch Med Wochenschr 58:2510–2512

Reiner AJ (2000) A hypothesis as to the organization of cerebral cortex in the common amniote ancestor of modern reptiles and mammals. In: Bock GR, Cardew G (eds) Evolutionary developmental biology of the cerebral cortex. Wiley and Sons, Chichester, pp 83–108

Sammet K (2006) Wilhelminian myelinated fibers—Theodor Kaes, myeloarchitectonics and the asylum Hamburg-Friedrichsberg 1890–1910. J Hist Neurosci 15:56–72

Sanides F (1962) Die Architektonik des menschlichen Stirnhirns. Springer, Berlin

Sanides F (1964) The cyto-myeloarchitecture of the human frontal lobe and its relation to phylogenetic differentiation of the cerebral cortex. J Hirnforsch 47:269–282

Seldon HL, Szirko M (2005) The comments on Professor Christfried Jakob's contributions made in *Die Cytoarchitektonik der Hirnrinde des erwachsenen Menschen* by Constantin von Economo and Georg N. Koskinas (1925). Electroneurobiología (B Aires) 13:46–73

Shimizu T, Karten HJ (1991) Multiple origins of neocortex: contributions of the dorsal ventricular ridge. In: Finlay BL, Innocenti G, Scheich H (eds) The neocortex: ontogeny and phylogeny. Plenum Press, New York, pp 75–86

Siebenmann F, Bing H (1907) Über den Labyrinth- und Hirnbefund bei einem an Retinitis pigmentosa erblindeten Angeboren-Taubstummen. Z Ohrenheilk (Wiesb) 54:265–280

Siemerling E (1911) Referat: Chr. Jakob und Cl. Onelli, Vom Tierhirn zum Menschenhirn; Chr. Jakob, Das Menschenhirn. Arch Psychiatr Nervenkrankh (Berl) 49:353–355

Striedter GF (1997) The telencephalon of tetrapods in evolution. Brain Behav Evol 49:179–213

Striedter GF (1998) Progress in the study of brain evolution: from speculative theories to testable hypotheses. Anat Rec (New Anat) 253:105–112

Striedter GF (2005) Principles of brain evolution. Sinauer Associates, Sunderland, MA

Tissir F, Lambert de Rouvroit C, Goffinet AM (2002) The role of reelin in the development and evolution of the cerebral cortex. Braz J Med Biol Res 35:1473–1484

Triarhou LC (2008a) Centenary of Christfried Jakob's discovery of the visceral brain: an unheeded precedence in affective neuroscience. Neurosci Biobehav Rev 32:984–1000

Triarhou LC (2008b) Christfried Jakob's 1911 proposition on the dual onto-phylogenetic origin and ubiquitous sensory-motor function of the cerebral cortex (abstract). In: Fishell G, Kriegstein AR, Parnavelas JG (eds) Cortical development: stem cells, neurogenesis, migration, circuit formation, cortical disorders. Mediterranean Agronomic Institute, Crete, pp 89–90

Triarhou LC (2008c) The books of Christofredo Jakob: lasting treasures of evolutionary neuroscience (abstract). Soc Neurosci Abstr 38:221.16

Triarhou LC (2009) Tripartite concepts of mind and brain, with special emphasis on the neuroevolutionary postulates of Christfried Jakob and Paul MacLean. In: Weingarten SP, Penat HO (eds) Cognitive psychology research developments. Nova Science Publishers, Hauppauge, NY, pp 183–208

Triarhou LC, del Cerro M (2006a) Semicentennial tribute to the ingenious neurobiologist Christfried Jakob (1866–1956). 1. Works from Germany and the first Argentina period, 1891–1913. Eur Neurol 56:176–188

Triarhou LC, del Cerro M (2006b) Semicentennial tribute to the ingenious neurobiologist Christfried Jakob (1866–1956). 2. Publications from the second Argentina period, 1913–1949. Eur Neurol 56:189–198

Unger L (1906) Untersuchungen über die Morphologie und Faserung des Reptiliengehirns. Anat Hefte 31:271–341

Vogt O (1927) Architektonik der menschlichen Hirnrinde. Zbl Gesamte Neurol Psychiatr 45:510–512

Vogt C, Vogt O (1919) Allgemeinere Ergebnisse unserer Hirnforschung. J Psychol Neurol (Leipz) 25:279–461

von Bonin G (1963) The evolution of the human brain. University of Chicago Press, Chicago

von Economo C (1926) Die Bedeutung der Hirnwindungen. Allg Z Psychiatr Psych-Gerichtl Med 84:123–132

von Economo C (1929) Der Zellaufbau der Grosshirnrinde und die progressive Cerebration. Ergebn Physiol 29:83–128

von Economo C (2009) Cellular structure of the human cerebral cortex. Karger, Basel (translated and edited by L. C. Triarhou)

von Economo C, Koskinas GN (1923) Die sensiblen Zonen des Großhirns. Klin Wochenschr 2:905

von Economo C, Koskinas GN (1925) Die Cytoarchitektonik der Hirnrinde des erwachsenen Menschen. Textband und Atlas. Springer, Wien

von Economo C, Koskinas GN (2008) Atlas of cytoarchitectonics of the adult human cerebral cortex. Karger, Basel (translated, revised and edited by L. C. Triarhou)

von Monakow C (1905) Gehirnpathologie, zweite Aufl. Hölder, Wien

Warwick R, Williams PL, Bannister LH (1973) Neurology. In: Warwick R, Williams PL (eds) Grays' Anatomy, 35th edn. Longman, Edinburgh, pp 745–1169

Zardoya R, Meyer A (2001) On the origin of and phylogenetic relationships among living amphibians. Proc Natl Acad Sci USA 98:7380–7383

Zilles K (2006) Architektonik und funktionelle Neuroanatomie der Hirnrinde des Menschen. In: Förstl H, Hautzinger M, Roth G (eds) Neurobiologie psychischer Störungen. Springer Medizin, Heidelberg, pp 75–140

Zilles K, Welsch U, Schleicher A (1981) The telencephalon of *Ichthyophis paucisulcus* (Amphibia, Gymnophiona [= *Caecilia*]): a quantitative cytoarchitectonic study. Z Mikrosk-Anat Forsch (Leipz) 95:943–962

In: *Neuroanatomy Research Advances*
Editors: C.E. Flynn and B.R. Callaghan

ISBN 978–1–60741–610–4
© 2010 Nova Science Publishers

Final Publications of Christfried Jakob: On the Frontal Lobe and the Limbic Region

Lazaros C. Triarhou[*]

Economo–Koskinas Wing for Integrative and Evolutionary Neuroscience,
University of Macedonia, Thessaloniki, Greece;

Abstract

One of the foremost neuroanatomists of the twentieth century, Christfried (Christofredo) Jakob (1866–1956) left a legacy of over 30 monographs and 200 papers, now becoming appreciated in the English biomedical literature. Born in Germany, he was summoned in 1899 to Buenos Aires by the Argentinian psychiatric academia. He spent the rest of his professional life (save for a brief return to Germany between 1910–1912) in affiliation with the National Universities of La Plata and Buenos Aires. The writings of Jakob cover a wide spectrum of topics, from the pathology of neuropsychiatric disorders to the phylogeny, ontogeny and dynamics of the cerebral cortex and their mental corollaries, and ultimately some of the most fundamental neurophilosophical questions. Although in many respects his innovative ideas opened up new ways of thinking in brain and behavior research, they still remain largely unheeded, most likely owing to their exclusive appearance in German or Spanish. The present study revisits Jakob's last two formal publications, dating to 1949. These are entitled 'The task of the frontal lobe in connection with a synthetic quantification of its constitutive elements' and 'The neuronal quantification of the limbic region in its relation to the endogenous affective sphere' (co-authored with his pupils Eduardo A. Pedace and Andrés R. Copello, respectively), and represent the culmination of Jakob's thought, integrating morphofunctional concepts in his quest for understanding the neuroanatomical fundamentals of the human mind. Cognitive function and emotional processing are at the core of current neurobiological research, and Jakob's pioneering concepts remain worthy of consideration six decades later.

* E-mail address: triarhou@uom.gr, phone +30 2310 891-387, fax +30 2310 891-388 (Corresponding author)

Introduction

Christfried (Christofredo) Jakob (1866–1956) was a German-born neuropathologist, neurobiologist and neurophilosopher, who spent most of his professional life in Argentina. Having worked under Friedrich Albert von Zenker (1825–1898) and Adolf von Strümpell (1853–1925) at Erlangen and Bamberg, Germany, Jakob subsequently became affiliated with the Universities of La Plata and Buenos Aires, and established one of the most important neuropathological laboratories in South America. Jakob is considered the father of Argentinian neurobiology [Orlando, 1966; Outes, 2006; Triarhou & del Cerro, 2006a, 2006b].

Brief Exposé of Jakob's Work

Jakob has left an invaluable treasure of over 30 monographs and 200 articles, written in German or Spanish, spanning over a wide range of diverse scientific topics [Orlando, 1966; Outes, 2006; Triarhou & del Cerro, 2006a, 2006b; Triarhou, 2007]. He had already reached international renown through early successful atlases of human [Jakob, 1911a] and comparative neuroanatomy [Jakob & Onelli, 1911, 1913]. He later produced landmark works on cortical and evolutionary neurobiology, studying in detail the normal and pathological human nervous system, as well as dozens of the autochthonous species – some extinct today – found in the Patagonian fauna, including the broad-snouted 'yacaré' (*Caiman latirostris*), a reptile of the Alligatoridae family [Jakob, 1945], and the 'pichiciego' or fairy armadillo (*Chlamyphorus truncatus*), a mammal of the Dasypodidae family [Jakob, 1943a].

On a coarse recounting, Jakob beginnings have a positivistic Virchowian view, mixed with Herbertian philosophy. His positivism is refined until 1912, when some general ideas of his mentor Theodor Ziehen are provisorily adopted; these, in turn, become gradually rejected around 1930, when a mystic, yet positivistic, *Weltanschaung* from his travels gives him a less Kantian, more admirative stance in front of the cosmos. Such a dominant *leitmotiv* leads him to a more Pythagorean worldview, introducing even proportions such as the *section aurea* in neuronal counting, which becomes very clear in his articles of the late 1940s. From 1949 to 1953 Jakob sketched his interests in unpublished anthropological notes [Crocco, 2008].

Jakob's Final Publications

Among Jakob's strongest interests were the human frontal lobes [Pedace, 1949] and the limbic system, particularly the anatomical bases of emotion [Orlando, 1964; Triarhou, 2008a, 2008b].

The two articles that follow are the first English translations of Jakob's last two published papers [Jakob & Pedace, 1949; Jakob & Copello, 1949]. This endeavour forms part of an ongoing effort to make available select landmark works by Jakob not hitherto available in the English biomedical literature. Besides their historical interest, these documents contain valuable scientific information, especially in view of the current interest in the frontal lobe and the limbic region, as well their structure and function under normal and pathological circumstances.

Reflecting a life's culmination in the thought of the 83-year-old neurobiologist, the papers were presented at the Third South American Congress of Neurosurgery in Buenos Aires, on April 3–9, 1949, with Professor Ramón Carrillo (1906–1956), the great Argentinian neurosurgeon and social policy officer – and a Jakob alumnus [Crocco, 2006; Ordóñez, 2004] – as General Secretary of the Congress.

The two papers give a summary of mostly quantitative neuroanatomical data on the cellular and axonal components of the human frontal lobe, cingulate (supracallosal) and hippocampal (inferior limbic) gyrus. For the sake of numerical comparisons, it is herein reiterated that Economo & Koskinas [1925] had estimated the total number of neurons in the cerebral cortex of both hemispheres at about 14×10^9 (6×10^9 being the smaller granule cells and 8×10^9 all the remaining larger neurons); the current estimate of the number of nerve cells in the human cerebral cortex stands at 20×10^9 [Pakkenberg & Gundersen, 1997].

Conclusion

Jakob's views on frontal lobe function evolved over half a century, from his early anatomical and neuropsychological papers [Jakob, 1906a, 1906b, 1910, 1911b], through the 'middle period' [Jakob, 1913a, 1913b, 1921, 1923], all the way to the late neurobiological [Jakob, 1939a, 1939b, 1941a, 1941b, 1941c, 1943b] and neurophilosophical [Jakob, 1946] synthetic treatises.

A detailed discussion of the older views and the current state of affairs regarding the anatomical components of the 'visceral brain' has been given elsewhere [Triarhou, 2008a], as has the evolutionary context of Jakob's ideas [Triarhou, 2008b].

Acknowledgements

The author gratefully acknowledges the courtesy of the National Library of Medicine, Bethesda, MD, and Staatsbibliothek Berlin, Germany.

References

Crocco, M. (2006). Breve biografía de Ramón Carrillo (1906–1956). *Electroneurobiología (Buenos Aires)*, **14**, 173–186.

Crocco, M. (2008). Personal communication, December 17, 2008.

Economo, C. von, & Koskinas, G.N. (1925). *Die Cytoarchitektonik der Hirnrinde des Erwachsenen Menschen.* Wien–Berlin: Julius Springer.

Jakob, C. (1906a). Estudios biológicos sobre los lóbulos frontales cerebrales. *La Semana Médica (Buenos Aires)*, **13**, 1375–1381.

Jakob, C. (1906b). La leyenda de los lóbulos frontales cerebrales como centros supremos psíquicos del hombre. *Archivos de Psiquiatría y Criminología (Buenos Aires)*, **5**, 678–699.

Jakob, C. (1910). La significación de la histoarquitectura comparada para la psicología moderna. *Revista del Jardín Zoológico de Buenos Aires*, **6**, 159–162.

Jakob, C. (1911a). *Das Menschenhirn: Eine Studie über den Aufbau und die Bedeutung seiner Grauen Kerne und Rinde. I. Teil. Tafelwerk nebst Einführung in den Organisationsplan der Menschlichen Zentralnervensystems.* München: J. F. Lehmann's Verlag.

Jakob, C. (1911b). La histoarquitectura comparada de la corteza cerebral y su significación para la psicología moderna. *Archivos de Psiquiatría y Criminología (Buenos Aires)*, **10**, 385–387.

Jakob, C. (1913a). La biología en el sistema de las ciencias filosóficas y naturales. *Anales de la Academia de Filosofía y Letras*, **2**, 55–67.

Jakob, C. (1913b). La psicología orgánica y su relación con la biología cortical. *Archivos de Psiquiatría y Criminología (Buenos Aires)*, **12**, 680–698.

Jakob, C. (1921). La teoría actual de las gnosias y praxias como factores fundamentales en el dinamismo de la corteza cerebral. *Crónica Médica (Lima)*, **38**, 17–24.

Jakob, C. (1923). *Elementos de Neurobiología, vol. I: Parte Teórica.* La Plata, Argentina: Facultad de Humanidades y Ciencias de la Educación de la Universidad Nacional de La Plata.

Jakob, C. (1939a). *Folia Neurobiológica Argentina, Atlas I – El Cerebro Humano: Su Anatomía Sistemática y Topográfica.* Buenos Aires: Aniceto López.

Jakob, C. (1939b). *Folia Neurobiológica Argentina, Atlas II – El Cerebro Humano: Su Anatomía Patológica en Relación a la Clínica.* Buenos Aires: Aniceto López.

Jakob, C. (1941a). *Folia Neurobiológica Argentina, Atlas III – El Cerebro Humano: Su Ontogenía y Filogenía.* Buenos Aires: Aniceto López.

Jakob, C. (1941b). *Folia Neurobiológica Argentina, Tomo I. Neurobiología General.* Buenos Aires: Aniceto López.

Jakob, C. (1941c). La función psicogenética de la corteza cerebral y su posible localización. *Anales del Instituto de Psicología de la Facultad de Filosofía y Letras de la Universidad de Buenos Aires*, **3**, 63–80.

Jakob, C. (1943a). *Folia Neurobiológica Argentina, Tomo II. El Pichiciego (Chlamydophorus Truncatus): Estudios Neurobiológicos de un Mamífero Misterioso de la Argentina.* Buenos Aires: Aniceto López.

Jakob, C. (1943b). *Folia Neurobiológica Argentina, Tomo III. El Lóbulo Frontal: Un Estudio Monográfico Anatomoclínico sobre Base Neurobiológica.* Buenos Aires: Aniceto López.

Jakob, C. (1945). *Folia Neurobiológica Argentina, Tomo IV. El Yacaré (Caimán latirostris) y el Origen del Neocortex: Estudios Neurobiológicos y Folklóricos del Reptil más Grande de la Argentina.* Buenos Aires: Aniceto López.

Jakob, C. (1946). *Folia Neurobiológica Argentina, Tomo V. Documenta Biofilosófica, Folleto I. Biología y Filosofía.* Buenos Aires: López y Etchegoyen.

Jakob, C., & Copello, A.R. (1949). La cuantificación neuronal de la región limbica en su relación con la esfera introyental afectiva. *Archivos de Neurocirugía*, **6**, 475–481.

Jakob, C., & Onelli, C. (1911). *Vom Tierhirn zum Menschenhirn: Vergleichend Morphologische, Histologische und Biologische Studien zur Entwicklung der Grosshirnhemisphären und ihrer Rinde.* München: J. F. Lehmann's Verlag.

Jakob, C., & Onelli, C. (1913). *Atlas del Cerebro de los Mamíferos de la República Argentina: Estudios Anatómicos, Histológicos y Biológicos Comparados sobre la Evolución de los Hemisferios y de la Corteza Cerebral*. Buenos Aires: Guillermo Kraft.

Jakob, C., & Pedace, E.A. (1949). La misión del lóbulo frontal frente a una cuantificación sintética de sus elementos productores. *Archivos de Neurocirugía*, **6**, 467–474.

Ordóñez, M.A. (2004). Ramón Carrillo, el gran sanitarista Argentino. *Electroneurobiología (Buenos Aires)*, **12**, 144–147.

Orlando, J.C. (1964). Sobre el cerebro visceral. Documentación histórica de una prioridad científica. *Revista Argentina de Neurología y Psiquiatría*, **1**, 197–201.

Orlando, J.C. (1966). *Christofredo Jakob: Su Vida y Obra*. Buenos Aires: Editorial Mundi.

Outes, D.L. (2006). A medio siglo de la muerte de Christofredo Jakob, 1956–2006: Fuentes de la concepción biológica de la doble corteza. *Revista Electroneurobiología (Buenos Aires)*, **14**, 3–35.

Pakkenberg, B., & Gundersen, H.J.G. (1997). Neocortical neuron number in humans: effect of sex and age. *Journal of Comparative Neurology*, **384**, 312–320.

Pedace, E.A. (1949). Contribución de la Escuela Neurobiológica Argentina del Profesor Chr. Jakob en el estudio del lóbulo frontal. *Archivos de Neurocirugía*, **6**, 464–466.

Triarhou, L.C. (2007). Christofredo Jakob as a naturalist: the 1923 scientific voyage aboard HSDG *Cap Polonio* to La Tierra del Fuego. *Electroneurobiología (Buenos Aires)*, **15**, 61–116.

Triarhou, L.C. (2008a). Centenary of Christfried Jakob's discovery of the visceral brain: an unheeded precedence in affective neuroscience. *Neuroscience and Biobehavioral Reviews*, **32**, 984–1000.

Triarhou, L.C. (2008b). Tripartite concepts of mind and brain, with special emphasis on the neuroevolutionary postulates of Christfried Jakob and Paul MacLean. In: S. P. Weingarten, & H. O. Penat (Eds.), *Cognitive Psychology Research Developments*. Hauppauge, NY: Nova Science Publishers (in press).

Triarhou, L.C., & del Cerro, M. (2006a). Semicentennial tribute to the ingenious neurobiologist Christfried Jakob (1866–1956). 1. Works from Germany and the first Argentina period, 1891–1913. *European Neurology*, **56**, 176–188.

Triarhou, L.C., & del Cerro, M. (2006b). Semicentennial tribute to the ingenious neurobiologist Christfried Jakob (1866–1956). 2. Publications from the second Argentina period, 1913–1949. *European Neurology*, **56**, 189–198.

In: *Neuroanatomy Research Advances*
Editors: C.E. Flynn and B.R. Callaghan

ISBN 978–1–60741–610–4
© 2010 Nova Science Publishers

The Task of the Frontal Lobe in Connection with a Synthetic Quantification of its Constitutive Elements[1]

Christofredo Jakob and Eduardo A. Pedace

Service of Pathological Anatomy, Hospital Nacional de Alienadas, Buenos Aires

Introduction

The frontal lobes represent 25% of the human brain, i.e. about 350–370 g of cerebral mass, but reflect the latest acquisitions in the ascending neurophylogeny. These facts, solely considered from a quantitative standpoint, must alone testify to the higher task of their cortical functions. In them, contrary to the *cognitive orientation* of the experiences achieved by the individual and reserved for the other lobes of the cerebral hemispheres, we recognize the centers of experiential accumulation resulting from the personal *"intervention"*, progressively elaborated for the elemental and highest human skills, stimulated by the corresponding emotional manifestations.

Both zones, in close gnosio-praxic collaboration, execute the conscious activation of human mentality in its creative labor from the concrete to the abstract in an intimate synthesis between their endogenous and exogenous domains, i.e. from their affectivity and intellectuality, reciprocally.

It now becomes imperative, given the complexity of its structure, to extend in detail the radius of action of our quantitative knowledge directly towards the neuronal elements that intervene in its game of cortico-intercortical and subcortical focalization, projection and

---

[1] An English translation of 'La misión del lóbulo frontal frente a una cuantificación sintética de sus elementos productores,' originally published in *Actas del Tercer Congreso Sudamericano de Neurocirugía* (Buenos Aires, April 3–9, 1949), *Archivos de Neurocirugía*, vol. VI, no. 1–4, pp. 467–474 [Jakob & Pedace, 1949]; translated and edited from the Spanish text by L.C. Triarhou* with the help of A.B. Vivas.

* E-mail address: triarhou@uom.gr, phone +30 2310 891-387, fax +30 2310 891-388 (Corresponding author)

association (Figs. 1–5). Only in this way can we penetrate into the intimacy of its real psychogenetic dynamics. These data will speak clearer than what any graphic diagram or reproduction is capable of achieving: if vegetative vital phenomena rest on the principle of labor division among numerous collaborating units, this must be even more so in the case of cerebral operations, whose high degree of differentiation increasingly necessitates a greater affluence of cellular and axonal elements. The quantitative factor is the basis for every qualitative process resulting from the creations of such multiplying collaborations, since without capital of work there are no benefits, neither in the material nor in the ideal. Like all the other lobes, the frontal lobe has 'autochthonous' [intrinsic] focal elements (capital proper) as well; in addition, it requires the 'transfocal' [intermediary] systems of correlation (circulating capital), whose numerical capacity we will address in the following order:

(1) *Afferent and efferent axonal systems* of cortico-subcortical charge and discharge.

(2) *Cortico-pyramidal focal systems* of commemorative accumulation in its afferent and efferent layers.

(3) Short, semi-long, and long ipsilateral and contralateral *transcortical association systems.*

Afferent and Efferent Axonal Quantification

(a) The frontal afferent systems derive in their majority (more than 95%) from the thalamus, originating in its anterior one-third; the rest being of olfactory origin (medial frontobasal olfactory pathways), do not present any major interest, as they have been dealt with in various publications [Jakob, 1943].

The thalamofrontal systems pass through the genu and anterior segment, better to say the anterior radiation of the internal capsule, splitting up as the frontal corona radiata between the four frontal gyri: the superior limbic (supracallosal at the median facies), the superior, middle and inferior frontal gyri, at the dorsolateral hemispheric facies); a remaining segment of the radiation reaches the anterior insular pole.

In total these axons represent the considerable amount of 4 million among a total of 19 million axons in the capsula, as compared to the 15 million axons for the remaining cortico-retrofrontal; thus the ratio of the frontal contingent to the other centro-parieto-occipito-temporal lobes is of 4:15. (As can be seen, in our study on the frontal lobe we exclude the anterior Rolandic region and its axons, which belong functionally to the praxic centers as opposed to the remaining cognitive.)

The task of the frontal thalamic avalanche is distributed in two portions: a medial limbic of 400,000 axons, and a lateral of 3,600,000, in round numbers. (The counts were effected in magnified microphotographs of sections treated with the axon silver impregnation technique.)

The medial or limbic portion is formed with the continuation of the mamillothalamic system (bundle of Vicq d'Azyr) and carries according to our studies commenced over 30 years ago (that little by little appear to be confirmed) viscerosympathetic sensitivity from the respective thoracic organs until the pelvic, towards the supracallosal cortex, creating in this way the basis for the elemental notions, so variable physiologically, of the vegetative malaise and well-being.

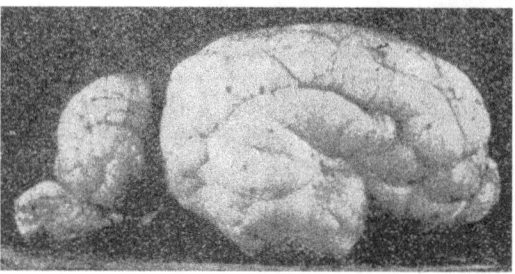

Fig. 1. Hemisphere of a sloth (*Bradypus tridactylus*, order *Edentata*) with its characteristic rotation and primordial segmentation system [Jakob & Pedace, 1949].

The lateral avalanche relays essentially the stimuli of the cerebello-rubro-thalamic radiation (superior hypothalamic radiation) and, with it, especially the cerebellar muscular sensitivity, further comprising other intercalated thalamic categories (its physiology represents still a very poor chapter in research).

The repartition of the afferent systems over the four frontal gyri is established approximately as follows (these numbers we consider nothing but an approximate orientation; their value is not absolute, but relative):

Superior limbic gyrus:	400,000 axons
Superior frontal gyrus:	1,600,000 axons
Middle frontal gyrus:	1,200,000 axons
Inferior frontal gyrus:	800,000 axons
Total:	4,000,000 axons

These afferent systems terminate as it is well known in the external pyramidal layer of the frontal cortex, a certain amount getting in through the tangential layer (possibly the phylogenetically oldest form according to the comparison with animal brains).

(b) The *frontal efferent* systems, corresponding to the first, can also be divided into two fascicles: medial and lateral. Both originate in the internal pyramidal layer, representing the medial portion, of superior limbic (supracallosal) origin via the limbic-tuberian tract with a total of 5,000 axons, that crossing through the corona radiata and its medial portion accompany the internal capsule, terminating little by little in the lateral paraependymal and tuberian sympathetic nuclei of the diencephalon (paleoneuronal sympathetic subcortical centers for the smooth-viscero-vascular musculature).

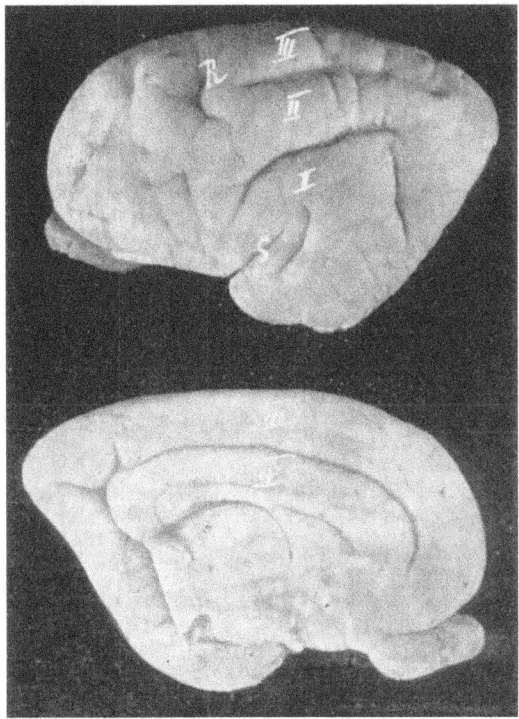

Fig. 2. Left hemisphere of a prosimian (*Cheiromys*) with its primordial gyral system [Jakob, 1943].

Along with them but coursing more laterally, emanate from the lateral frontal cortex 25,000 axons, which according to earlier studies formed the fronto-pontine tract; our own studies confirmed the concept of Meynert that part of them arrive at the substantia nigra (about 10,000 axons) and whose total represents, like the limbic-tuberian tract, the so-called *fronto-nigral* tract connected to the striated normokinesia and the frontal praxias.

The remaining 15,000 axons continue from the cerebral peduncle towards the dorsomedial pontine ganglia, forming the *fronto-pontine* tract, thus half of the frontal efferent pathways, together with the lateral thalamofrontal, close the dynamic afferent frontocerebellar loop, connected by the cerebellar pontine systems to the contralateral cerebellar hemisphere; thus result the two crossed pathways: the afferent cerebello-rubro-thalamic frontal system and the efferent cerebellar fronto-pontine system, i.e. the circuit of the neokinesic system superimposed on the paleokinesic (instinctive) and archikinesic (reflex).

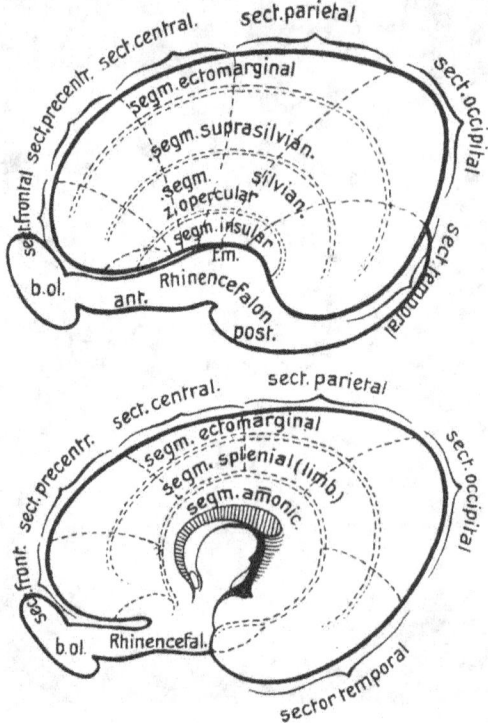

Fig. 3. Scheme of the segmentation and sectorization of the mammalian hemisphere [Jakob, 1911; Jakob & Onelli, 1911, 1913].

Its task connects evidently the frontal praxic dynamics to the cerebellar normokinesia for the appropriate execution of the volitive acts learned individually, from the most elemental skills to the mobilization of symbolic thought; such is the task of the lateral corticality, as opposed to the medial which is destined to contribute the vegetative affectivity oriented at the regulatory influence that the viscero-vasomotor endogenous functions, through the frontal limbic cortex, exercise in the reactive, biophylactic, neoneuronal phylogeny.

Then, by means of such medial afferent frontal radiations originates the notion of the *"anxiety"* of viscero-sympathetic origin in the frontal limbic cortex in pathological states; its elements represent, according to what has been exposed, only 10% of the thalamofrontal radiation.

The Frontal Pyramidal Elements

The frontal cortex presents as a site of contact between the afferent and efferent pathways, mentioned above, the system of transformation of the specific stimulations and reactions (endogenous-exogenous frontalization) and with it the final accumulation of its elaborations (frontalized commemorative function).

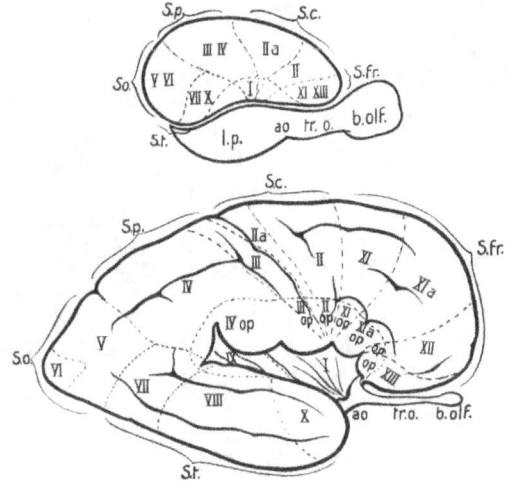

Fig. 4. Scheme of the cytoarchitectonic areas in the *Didelphis* (*upper*) and human brain (*lower*). Note the striking frontal phylogenetic development in the human [Jakob, 1911; Jakob & Onelli, 1911, 1913].

The first is effected by the neurodynamic exchanges of its receptive external pyramidal layer with the effector internal by the macrodynamic waves of charge and discharge. The focal accumulation, that constitutes the respective commemorative representation, is in charge of its minor elements (microdynamic waves); thus, from both layers this focal microdynamic charge is the one that encloses the outcomes of the personal praxic experience in the form of neurodynamic energy of the past, of increasing and available tension; for all acts of conscious intervention in the future (neokinesias) its latent game represents, consequently, the praxic capital of the individual in both domains: medial affective and lateral external environmental.

It is understood that the extension of possible interventions in humans requires an increased number of collaborative elements, without getting here into greater detail; in total we estimate for the frontal lobe (excluding always the anterior Rolandic) in the vicinity of 1,200 million pyramidal cells over a cell total of 5,000 million for each hemisphere (excluding the fusiform elements of the supra- and infrapyramidal layer). More than half (about 60%) belong to microdynamisms, the remaining part to systems of associative waves, as we shall see, and a small remnant to systems of discharge already studied (30,000).

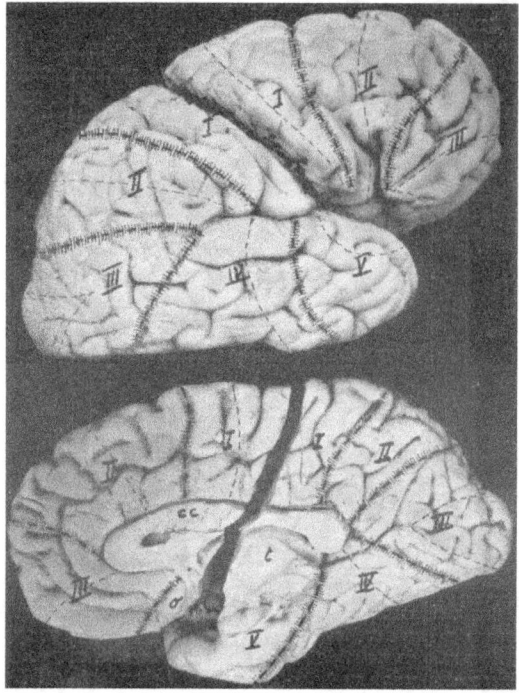

Fig. 5. The 'golden section' of the human cerebral hemisphere. External (*upper*) and midsagittal facies (*lower*). Note the importance of the frontal 'praxic' sectors [Jakob, 1943].

As far as the repartition of the pyramidal cells over the gyri are concerned, we only have an approximate orientation; taking into account the extent in surface, thickness and cell wealth, we calculate the following:

Cortex of the supracallosal (superior limbic) gyrus:	150 million
Cortex of the superior frontal gyrus:	450 million
Cortex of the middle frontal gyrus:	350 million
Cortex of the inferior frontal gyrus:	250 million
Total:	1,200 million

It will be interesting to establish next the coefficients between cells and projection axons; in general, to each afferent axonal element correspond about 300 pyramidal cells, whereas for an efferent neuron it is concentrated into the collaboration of more than 100 times (i.e. about 40,000 cells). Naturally, great regional differences exist in that respect.

Frontal Associative Systems

Although our histological knowledge offers sufficient information, one must confess that the functional relations are almost totally unknown; thus we will deal with them superficially, indicating only that the frontal lobe appears to possess the greater number of such special elements in its gyral association as much as contralaterally.

We know in the endogenous domain, as it was shown graphically by Burdach, the *cingulum* as an associative pathway composed of semi-long accumulated elements; their count has given us at different levels, from 25,000 to 35,000 axonal elements. In the lateral exogenous zone, in its dorsal portion, we have the arcuate fasciculus with 35,000 axons and in the base the uncinate gyrus with 12,000.

One should add an unknown but very high number of short intergyral systems (U-fibers), both in the limbic zone and the lateral gyri.

As far as the commissural system is concerned, we have in the genu of the corpus callosum a potent interfrontal pathway that we estimate at more than 1,500,000 axons split into afferent and efferent.

It is not possible to indicate a distribution of the corresponding gyri yet.

Conclusion

Having by this time finished our modest contribution to the numerical knowledge provisional of the frontal functions, we think that the presented numbers, without insisting on their absolute fidelity, will increase in the mind of our neurosurgeon colleagues somewhat more their medical responsibility when they cut so confidently axons and cells that will never be able to regenerate.

It is meaningful to note that our theory of the endogenous-exogenous cortical centers and systems, already conceived by one of us more than 30 years ago [Jakob & Onelli, 1913], anticipated a scientific basis accessible for the current psychosomatic clinical doctrines, confirming, through these cortical centers, the physio-psychological synthesis indispensable for the mutual collaboration and control between the human mind and body.

On the other hand, this short exposition in enough to understand the higher frontal task, both in its endogenous affective contribution and its exogenous praxic intervention; the real fact that the frontal lobe gathers so closely both domains of our conscious mentality, is a living indication for the current psychological concept, already established from the philosophy of Schopenhauer and Wundt, that in the volitive domain amalgamate intimately afferent affective intonations and efferent praxic skills as a creative frontal tribute, besides the postfrontal cognitive orientation for the total human mentality.

References

Jakob, C. (1911). *Das Menschenhirn: Eine Studie über den Aufbau und die Bedeutung seiner Grauen Kerne und Rinde. I. Teil. Tafelwerk nebst Einführung in den Organisationsplan der Menschlichen Zentralnervensystems.* München: J. F. Lehmann's Verlag.

Jakob, C. (1943). *Folia Neurobiológica Argentina, Tomo III. El Lóbulo Frontal: Un Estudio Monográfico Anatomoclínico sobre Base Neurobiológica.* Buenos Aires: Aniceto López.

Jakob, C. (1946). El trígono cerebral. Su significación neurobiológica. (Vía central eferente para la musculatura lisa víscero-vascular de la esfera gnósica-emotiva). *Revista Neurológica de Buenos Aires,* **11**, 2–36.

Jakob, C., & Onelli, C. (1911). *Vom Tierhirn zum Menschenhirn: Vergleichend Morphologische, Histologische und Biologische Studien zur Entwicklung der Grosshirnhemisphären und ihrer Rinde.* München: J. F. Lehmann's Verlag.

Jakob, C., & Onelli, C. (1913). *Atlas del Cerebro de los Mamíferos de la República Argentina: Estudios Anatómicos, Histológicos y Biológicos Comparados sobre la Evolución de los Hemisferios y de la Corteza Cerebral.* Buenos Aires: Guillermo Kraft.

Jakob, C., & Pedace, E.A. (1949). La misión del lóbulo frontal frente a una cuantificación sintética de sus elementos productores. *Archivos de Neurocirugía,* **6**, 467–474.

In: *Neuroanatomy Research Advances*
Editors: C.E. Flynn and B.R. Callaghan

ISBN 978-1-60741-610-4
© 2010 Nova Science Publishers

The Neuronal Quantification of the Limbic Region in its Relation to the Endogenous Affective Sphere[1]

Christofredo Jakob and Andrés R. Copello

Laboratory of Neuropathology, Hospital Nacional de Alienadas, Buenos Aires

Abstract

In the times of Broca one already separated the limbic gyrus from the rest of the hemisphere, an anatomical fact verified by phylogeny and ontogeny, and later by psychopathology, especially by the limbic form of Pick disease. This medial cortex is related to the periependymal tuberomamillary diencephalic paleoneuronal sympathetic centers (vasomotility and endocriny). Of the 5,000 million cortical pyramidal elements that a hemisphere has, 325 million correspond to the supracallosal gyrus and 75 million to the hippocampus; the former receives 600,000 axons from the anterior thalamic nucleus and emits the limbicotuberian pathway that is composed of 5,000 axons. The hippocampus, which is related to olfaction, hosts the central olfactory radiation (250,000 axons) and gives origin to the fornix (65,000 axons) to take stimuli to ipsilateral and contralateral periependymal and tuberian nuclei, to the peri-raphé nuclei of the mesencephalon and rhombencephalon from which it stimulates glands, vasomotility, etc. The supracallosal account as an associative path to the cingulum (25,000–35,000 axons) and U-fibers, and the hippocampus with only U-fibers and as an interhemispheric association to one-tenth of the genu of the corpus callosum, also to one-tenth of the anterior and interammonic commissures. In summary, the limbic gyrus maintains a functional independence from the rest of the hemisphere and is linked to viscero-vasomotor stimuli; as a phylogenetically older cortex, it provides us with the notion of well-being or vegetative malaise, getting to oscillate between euphoria and anxiety in a more pronounced manner. Thus, we deduce that here reside the bases for the affectively intoned feeling of endogenous ('introyental') mental life, as opposed to that which the lateral cortex carries out, toward the associative contribution for the total conscious mentality.

[1] An English translation of 'La cuantificación neuronal de la región limbica en su relación con la esfera introyental afectiva,' originally published in *Actas del Tercer Congreso Sudamericano de Neurocirugía* (Buenos Aires, April 3–9, 1949), *Archivos de Neurocirugía*, vol. VI, no. 1–4, pp. 475–481 [Jakob & Copello, 1949]; translated and edited from the Spanish text by L.C. Triarhou* with the help of A.B. Vivas.

* E-mail address: triarhou@uom.gr, phone +30 2310 891-387, fax +30 2310 891-388 (Corresponding author)

Introduction

Since the times of Broca, comparative neurobiology has separated the marginal cortical circle (supra and infracallosal) of the median hemispheric facies from the rest of the lateral hemispheric facies. Such a limbic zone essentially consisted of the supracallosal (superior limbic) and the hippocampal (inferior limbic) gyrus at the base. This purely anatomical concept has been confirmed in its phylogenetic existence and ontogeny, although its physiological study remained almost in darknesses, and it is just lately that psychopathology, especially the limbic form of Pick disease [Jakob, 1946b] opened up the possibility of elaborating clearer concepts with respect to its functions. In addition, normal and pathological histotopography have contributed to the clarification of certain limbic axonal relations with the periependymal and tuberomamillary diencephalic centers [Jakob, 1946a] that connect their cortical functions with those sympathetic paleoneuronal centers (vasomotility and endocriny). From it a fundamental concept emerges on the existence of distinct functions in two cortical spheres: the lateral cortex, represented by afferent and efferent pathways, is linked to the external environment; on the other hand, the internal ('introyental') medial limbic domain is related to sensitivity and viscerovascular motility. The former acts on striated musculature (pyramidal tract, etc.); the latter, on smooth musculature (fornix, etc.). As the latter is intimately associated with our emotive sphere (circulation, nutrition, inner secretions, etc.), it would represent our inner affective cortex; both together generate, intimately combined, the individual consciousness of the self (neopsychisms).

Anatomically, a large part of the limbic system belongs to the frontal lobe (supracallosal), and the remaining to the temporal lobe (hippocampus), but this one is also related to the frontal lobe through the intermediation of the olfactory area. Although altogether the limbic zone in humans only represents 10% of the cerebrum (in the descending phylogeny in animals it is inversed) one cannot speak in humans of a substantial and less functional reduction, without denying, of course, the superiority in lateral gain. It is for that reason of interest to know with greater detail, the quantity of functional elements that enter into its histoneuronal organization, because on the total of its elements evidently depends the efficacy of its impulses that dominate within the set of mental functions.

We divided its neuronal elements into the following three groups:

(1) Those corresponding to its cortical centers of neurodynamic accumulation (focalization of its commemorative work).

(2) Its elements of charge and discharge (afferent and efferent) in its projection toward subcortical regions.

(3) Those destined to its associative and commissural transcortical correlations with the rest of the corticality.

Pyramidal Cortical Elements

Of the total number of *pyramidal cortical elements* that in one hemisphere reach approximately 5,000 million [the count was obtained from magnified microphotographs of sections subjected to silver impregnation of cells and axons], correspond to the limbic zone around 400 million that we divide in the following manner: the superior limbic (supracallosal) cortex has 325 million, and the inferior limbic (hippocampus) 75 million. Each one represents the fundamental type of two superimposed layers. Thus, in the supracallosal gyrus, the receiving external layer of stimuli contains 225 million cells, and the effector internal layer 100 million cells. Between both layers the most numerous microdynamic elements form the stationary waves of sensory-motor consolidation. In the hippocampus, the receiving external layer is represented by the elements of the dentate gyrus (near 50 million) and the efferent external by the pyramidal cells of Ammon's horn (25 million). All these elements are the ones that store the acquired experience individually and transform it into volitional motor realizations, according to the mode of the received and transformed stimuli.

Afferent and Efferent Limbic Axons

To orient oneself really on the specific function of a cortical center, a knowledge of its pathways of charge and discharge is indispensable, because it is based on the category of those stimulations and reactions in its correlated projection that one must determine on the function of the focal center as it occurs with the two limbic zones, that in spite of being topographically and associatively reunited present essential differences.

(a) The *superior limbic gyrus* receives as afferent pathway (Fig. 1) to the *anterior dorsothalamic* radiation formed by a contingent of approximately 600,000 axons. Its anterior frontal portion, the greater of the two, participates with 65% (near 400,000 axons) and it is precisely about this portion that we have more precise functional information. In effect, the bundle of Vicq d'Azyr, which contains around 15,000 microaxons, terminates in the anterior thalamic nucleus; this bundle of very old phylogeny originates in the medial portion of the mamillary body and takes – according to confirmed personal observations a long time ago – viscerosympathetic stimuli of all thoraco-abdomino-pelvic organs, a fact so old in its phylogeny that it confirms in its relations to the individual and generic trophic functions (nutrition, sexuality) from the inferior mammals, although lacking to still penetrate into its special physiologic interpretation. In humans the observations of cases with tissue softening, such as progressive general paresis and Pick disease, confirm such facts.

In the same site of termination of the mamillothalamic fasciculus (Vicq d'Azyr) the frontal anterior radiation of the supracallosal gyrus originates precisely from the thalamic cells; the thalamus functions in general as an enormous multiplicator system, with an increase from the 15,000 axons of the bundle of Vicq d'Azyr to 500,000 of the thalamolimbic radiation, i.e. more than 30 times. These radiations extend in the form of a large fan in front of the genu of the internal capsule, from their dorsal portion to the base, entering the external pyramidal layer of the supracallosal cortex; their thicker fibers arrive at the tangential layer; they are reflected towards the external layer, a fact particularly demonstrative in mammals: the external tangential layer has nothing to do with association, as it was once considered, which is in fact evident in the inferior limbic gyrus. However, the pathway of discharge of the limbic is much more limited, also according to a general law that applies to all efferent radiations; it is represented by the *limbico-tuberian* pathway, formed by a contingent of around 5,000 axons and which, originating in the internal pyramidal layer of the cortex, continues crossing the radiation and genu of the internal capsule towards its base, where it occupies its more medial portion on the inside of the frontonigral and pontine systems [Jakob, 1947] to terminate in the lateral juxtaependymal and tuberian sympathetic nuclei of the diencephalon, and to conduct the stimuli from here by pathways little known towards the viscerovasomotor smooth musculature by means of the bulbospinal sympathetic reflex centers.

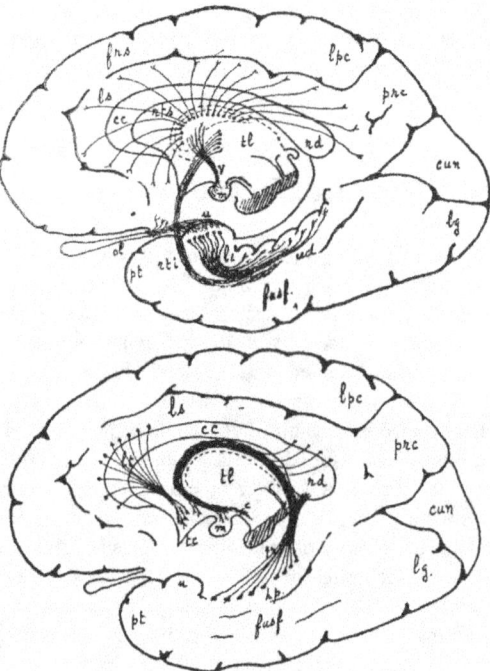

Fig. 1. Scheme of the afferent projection pathways to the superior and inferior limbic gyrus (*upper*) and of efferent projection systems (*lower*). Abbreviations: *c*, crus fornix; *hp*, hippocampus; *li*, inferior limbic; *ls*, superior limbic; *lt*, limbico-tuberian pahtway; *m*, mamillary body; *ol*, olfactory pathways; *rts*, superior thalamic radiation; *tc*, tuber cinereum; *tl*, thalamus; *tr*, fornical system; *u*, hippocampal uncus; *uc*, uncodentate system; *V*, bundle of Vicq d'Azyr [Jakob & Copello, 1949].

Thus closes the *limbic internal ('introyental')* functional circuit through which visceral affective states, from the euphoric to the anxious, are mediated, accumulated and discharging in the personal experience during normal individual life, biologically relating the limbic cortex to visceroendocrine functions. This is evidently a complex psychophysiological problem still not resolved, and common to the entire animal series, to which belong, among others, the care for personal hygiene and the young; cave and nest formation in animals; elimination of organic wastes via sphincters and regulation of nutritive and sexual functions.

The training for such functions begins in early childhood and never ceases; it is the evoked inner affect, which forms the biological basis of being, and which we connect foremost with cortical functions of the superior limbic gyrus. In these studies, human pathology has a most important part with its clinico-anatomical correlations.

(b) *Inferior limbic gyrus* (hippocampus). More has been known since earlier times about this gyrus, because comparative anatomy provided evidence as to the relation with the chemical sense of olfaction, a fact that already in itself links its cortical functions to the experience related to nutrition and sexuality, where the sense of olfaction dominates. The olfactory afferent central axons originate in the uncus of the hippocampus, where the lateral radiations also arrive from the olfactory area; that central radiation, that is the one of the hippocampus, integrated by near 250,000 axons (very approximately) crosses the subiculum (perforant fibers) and becomes distributed on a superficial surface; it arrives at the subicular external tangential layer, coursing soon, from that marginal zone to the hippocampal dentate gyrus, a pathway that Cajal called sphenodentate and which to our understanding is more appropriate to the dentate perforant radiation, a pathway that in animals originates in the piriform lobe (the uncus); so that topographically one should designate to it the name of *uncodentate* system. To this pathway become aggregated axons of the inferior thalamic sort/mode (gustative pathway?). We have therefore a second source of thalamo-temporal stimuli, whose true mission is in discussion; anyway, olfaction as a unique Ammonic function is debatable. Its pathway of discharge is far better known, originating in the Ammonic layer (effector layer of the hippocampus) integrated by an approximated contingent of 65,000 axons forming the well-known fornix (*cerebral trigonum*) that, as we have shown in a special work [Jakob & Copello, 1948] takes most of its axons towards the ipsilateral and contralateral periependymal and tuberian medial nuclei, and whose longer fibers arrive at the zone of the hypothalamic commissure, at the peri-raphé nuclei of the mesencephalon and the rhombencephalon. Again we have an efferent projection towards viscerosympathetic nuclei from which are stimulated the secretion and motility of the digestive and vascular systems.

Considering the mission of the superior limbic dynamic circuit associated here, we confirm its similar function in the visceral sympathetic inferior limbic circuit of the total internal environment ('introyente'). "Hunger and love emit from those cortices their categorical imperatives in the animal and human organism" [Jakob & Onelli, 1913].

It has functions already reflected and instinctively regulated by subcortical systems, but the evidence increasingly suggests the role of the limbic cortex as a neoneuronal superposed center individually dominant in relation to the lower archineuronal and paleoneuronal centers.

Associative Limbic Systems

Given the limited knowledge on the function of those systems that we know better from their histotopography, we shall only treat them briefly. In the area of the superior limbic gyrus is known in the first place for a long time the cingulum, a longitudinal system jointly formed by short and semi-long pathways that unite at distinct regions of the supracallosal gyrus. This path in front of the genu of the corpus callosum begin at the base, and taking a parallel course to the great interhemispheric commissure, it extends backwards, it reaches the splenium and following its curvature, the longer axons arrive all the way to the posterior parts of the hippocampus. In the entire trajectory, a part of its fibers irradiates to the supracallosal cortex, uniting in a double arc to its different portions. It is formed by a large amount of very fine axons (microaxons) which are disposed in a fascicular grouping; their count has given us as many as 25,000–35,000 axons, reaching the maximum in the frontal zone directly in front of the genu. In addition the supracallosal gyrus is connected by an appreciable amount with intergyral systems (U-fibers) to the entire medial facies of the hemisphere, being especially numerous in its anterior half (association with the superior frontal and paracentral lobules). The relations with the indusium griseum (*taenia tecta*) are doubtful.

In the inferior limbic gyrus, the associative existence is poorer; only some intergyral systems appear (U-fibers) that connect the hippocampus with the temporal internal facies. A quantification of these systems has not been possible for us until now. We shall not consider here the ancient occipitotemporal pathway, in the past interpreted as associative, since it belongs to the optic projection.

As far as the commissural interhemispheric association is concerned, it contributes by the anterior supracallosal gyrus a certain part of the genu of the corpus callosum, which we were able to calculate as one-tenth of its constitution; the larger amount of fibers is directed toward the frontolateral cortex. In the hippocampus we know the anterior and interammonic commissures; a count of the former has given us nearly 30,000 axons, but we must consider that part of it constitute crossed olfactory projection pathways, perhaps one-tenth as well.

Altogether we can affirm that the limbic endogenous zone is in ample associative contact with the ipsilateral and contralateral, especially as far as the superior limbic gyrus is concerned; in reality, numerous details still lack.

Conclusion

The following conclusion is drawn from the above exposition: the limbic gyrus maintains, as far as its afferent-efferent projections are concerned, a functional independence from the remaining gyri of the lateral neocortex, which are related to the outer environment. This important circuit is not fully understood yet; being phylogenetically older cortex, it provides us with viscerovasomotor signals and the feeling of well-being or of vegetative malaise, oscillating between euphoria and anxiety. Thus, we can deduce that here reside the bases for the affectively colored feelings of inner mental life, as opposed to those subserved by the lateral cortex, thus providing the association for integrated conscious mentality. In spite of these centers existing in lower vertebrates, it is in mammals that they successively become associated with the lateral outer effects, intensifying the intellectualization of states in higher consciousness in relation to the elementary nature of the others. Only in primates, and especially in humans with their verbal symbolization, one rises to a supreme mental degree, where the sphere of inner feelings of visceral origin becomes amalgamated with the environmental gnosio-praxic experience; in that superior synthesis arises the neopsychic sphere, from the concrete to the most abstract. But it consists, according to the most ideal reasoning that invariably accompanies it, although in a latent form, the affective endogenous stimulation forming the indispensable biophylactic individual-genetic basis, a priori, of all human existential life, as popular psychology already guesses when it says that the head needs the help of the heart if it wants to produce lasting values.

In this study on the limbic-endogenous cortical centers, we attempted to achieve a constructive synthesis for the creation of those inner affective mental phenomena in their relation to external cognitive phenomena. The associative process between both systems creates the conscious and total personality: affect and cognition, respectively located at the medial fornical-limbic and at the lateral pyramidal systems. The effects of the involuntary smooth musculature in combination with the voluntary striated muscles present an indisputable and fruitful basis for current doctrines in the psychosomatic sphere, because the internal-environmental interactions form the basis for this psychosomatic dynamic unity that represents the old concept of body and mind.

References

Jakob, C. (1946a). El trígono cerebral: su significación neurobiológica (vía central eferente para la musculatura lisa víscero-vascular de la esfera gnósica-emotiva). *Revista Neurológica de Buenos Aires,* **11**, 2–36.

Jakob, C. (1946b). La demencia progresiva: un análisis neurobiológico de la enfermedad de Pick. *Revista Neurológica de Buenos Aires,* **11**, 81–94.

Jakob, C. (1947). La significación neurobiológica y clínica de la cuantificación de los sistemas cerebrales. *Revista Neurológica de Buenos Aires,* **12**, 229–245.

Jakob, C., & Copello, A.R. (1948). La psicointegración introyento-ambiental orgánica y sus problemas para la neuropsiquiatría y psicología (primera parte: su filogenía constructiva). *Revista Neurológica de Buenos Aires,* **13**, 63–79.

Jakob, C., & Copello, A.R. (1949). La cuantificación neuronal de la región limbica en su relación con la esfera introyental afectiva. *Archivos de Neurocirugía,* **6**, 475–481.

Jakob, C., & Onelli, C. (1913). *Atlas del Cerebro de los Mamíferos de la República Argentina: Estudios Anatómicos, Histológicos y Biológicos Comparados sobre la Evolución de los Hemisferios y de la Corteza Cerebral.* Buenos Aires: Guillermo Kraft.

Available online at www.sciencedirect.com

SciVerse ScienceDirect

Journal homepage: www.elsevier.com/locate/cortex

ELSEVIER

Special issue: Historical paper

Challenging the supremacy of the frontal lobe: Early views (1906–1909) of Christfried Jakob on the human cerebral cortex

Zoë D. Théodoridou and Lazaros C. Triarhou[*]

Economo-Koskinas Wing for Integrative and Evolutionary Neuroscience, University of Macedonia, Thessaloniki, Greece

ARTICLE INFO

Article history:
Received 14 July 2010
Revised 6 September 2010
Accepted 15 December 2010
Published online 21 January 2011

Keywords:
Christfried Jakob
cerebral cortex
history of neuroscience
frontal lobe
localization of cognitive functions

ABSTRACT

This article focuses on a series of six studies that address functional localization in the frontal lobe; they were published in Argentina between 1906 and 1909 by Christfried Jakob (1866–1956), one of the great thinkers in early 20th century neuropathology and neuro-philosophy. At that time, the localization-holism controversy was at a peak, having been triggered by the historic Marie-Déjerine aphasiology debate. Jakob held the view that constitutive physiological elements of cognition are localized. Nonetheless, he cast doubt on phrenological approaches that considered the frontal lobe as 'superior' to the other cortical regions. Jakob studied the human frontal lobe from fetal life through senility, in normality and pathology, including tumors, injuries, softening, general paralysis and dementia. Based on those finds, he considered strict localization theories a dead-end. Taking a critical look at Flechsig's ideas on the parallel ontogenies of frontal association centers and intellect, Jakob argued that the frontal lobe does not carry any selective advantage over the remaining human cerebral lobes or even over the frontal lobe in non-human primates. Regarding lesion experiments in laboratory animals, he pointed to methodological caveats, such as insufficient recovery time, that may lead to disorientating conclusions, and rejected élite brain research, calling it superficial and inexact. Jakob was convinced that the verification of the anatomical connections of the frontal lobe would elucidate its functions. Thus, he viewed the frontal lobe as a central station receiving input via olfactory pathways and thalamic radiations, pertinent to muscular and cutaneous senses, and attributed a perceptive character to a brain region traditionally associated with productive functions. Modern neuroscience seems to support Jakob's rejection of distin-guishable motor and sensory regions and to adopt a cautious stance concerning over-simplified localization views.

© 2011 Elsevier Srl. All rights reserved.

1. Introduction

After more than a century of cortical research, frontal lobe function still poses challenges. The complexity of the cerebral cortex has led authors to consider it anything from 'the apparatus of civilization' to an organ, the removal of which may not always lead to behavioral deficits (Teuber, 2009). The fact that the human frontal cortex occupies one-third of the

[*] Corresponding author. Neuroscience Wing, University of Macedonia, 156 Egnatia Ave., Bldg. Z-312, 54006 Thessaloniki, Greece.
E-mail addresses: ztheodoridou@hotmail.com (Z.D. Théodoridou), triarhou@uom.gr (L.C. Triarhou).
0010-9452/$ – see front matter © 2011 Elsevier Srl. All rights reserved.
doi:10.1016/j.cortex.2011.01.001

total cortical surface has instilled in researchers the expectation that the unveiling of frontal lobe function might explain the uniqueness of human behavior (Raichle, 2002).

A long debate has been taking place with regard to the functional localization of higher neurocognitive processes in the frontal lobe. Modern theoretical stances fall into a continuum that ranges from fractionated approaches to central concepts; at the same time, attempts are being made to reconcile contrasting views. The common denominator of fractionated approaches (cf. Koechlin et al., 2003; Shallice, 2002; Shallice and Burgess, 1996; Stuss et al., 2002) is the belief that there is no unitary frontal lobe process. The anterior part of the brain rather subserves multiple distinct control processes that underpin executive functions (Godefroy et al., 1999). Within such a framework, modularity and fractionation may even pertain to higher human abilities (Baddeley, 1996; Stuss et al., 2002). A more central concept has been put forth by Duncan and Miller (2002), who reject a fixed functional specialization and highlight the adaptability of select regions of the prefrontal cortex in order to complete a goal-directed activity. Finally, Stuss (2006) argues that the debate between fractionation and adaptability is a false debate and suggests that brain networks may be both locally segregated and functionally integrated (Yeterian et al., 2012, this issue; Catani et al., in press). Marshaled evidence showing the recruitment of the same frontal regions for different cognitive demands (Duncan and Owen, 2000) suggests that in spite of the fractionation, frontal processes are applicable to many domain-specific modules, and therefore are domain-general (Stuss, 2006).

However, the issue of functional localization has been at the core of neuropsychological research, as well as of philosophical delving, since the 19th century (Catani and Stuss, 2012, this issue). Although the idea of specific cerebral localizations of physiological functions was adopted before 1861 by several researchers including Gall and Bouillaud (cf. Finger, 2000), it was Broca's (1861) lecture to the Paris Anthropological Society that brought it forcefully to the scientific world (Lorch, 2008). The second of the two liveliest debates in the history of aphasiology took place when Marie questioned Broca's views, while Déjerine defended localization at a special joint meeting of the New York and the Philadelphia Neurological Societies and at the Neurological Society in Paris two years later, triggering a debate that spread internationally (cf. Tsapkini et al., 2008). At the same time, Jakob (1866–1956), a neurobiologist working in Buenos Aires, would adopt an integrative approach in his attempt to elucidate cortical function.

Born and educated in Germany, Jakob (Fig. 1) went to Argentina in July 1899. At that time, he had already made an international name for himself through his early brain Atlas (Figs. 2 and 3). Zülch (1975) credits Jakob (1899, 1901, Plate 15.5) for demonstrating that, at the direct corticospinal level, the pyramidal pathway is not yet myelinated in the newborn human; as Flechsig (1927) had described, only pathways that pass from motor cortex to the midbrain are myelinated at birth. In all, Jakob left 30 books and 250 articles that cover developmental, evolutionary, anatomical, pathological and philosophical issues in neurobiology (Barutta et al., 2011; Moyano, 1957; Triarhou and del Cerro, 2006a, 2006b; 2007).

Fig. 1 – A sketch of Christfried Jakob by his student and biographer López Pasquali (1965). Signature from Orlando (1966).

The frontal lobe occupied Jakob's thought constantly in a path of enquiry spanning over five decades. Having studied the frontal lobe in its various developmental stages, and in neuropathological conditions, Jakob (1906a, 1907c) cast doubt on its 'supremacy' (Fig. 4). He pointed to potential historical reasons—linked to classical Greek philosophy—that might explain the importance attached to it. Jakob noted that physical characteristics, such as the upright posture, the extremities, and the extended forehead, distinguish humans from animals. In particular, Jakob (1943) considered the 'Olympian forehead', artistically expressed in the sculptures of Zeus, as the symbol of 'humanization'.

Jakob's contributions, written in German and Spanish, have been largely neglected in the English scientific literature. The present study aims at highlighting key concepts from his 'early' period. In that context, we provide selected translated passages from six papers, published between 1906 and 1909 (Jakob, 1906a, 1906b, 1906c, 1907a, 1907b, 1909), that address biological, anatomo-clinical and pathophysiological aspects of the frontal lobe. A psychobiological theory on the gnoses and praxes that culminated during Jakob's 'middle' period is presented elsewhere (Théodoridou and Triarhou, 2010).

2.   Neuroanatomical studies

Jakob constantly viewed morphology in a functional context (Tsapkini et al., 2008). He believed that the elucidation of the

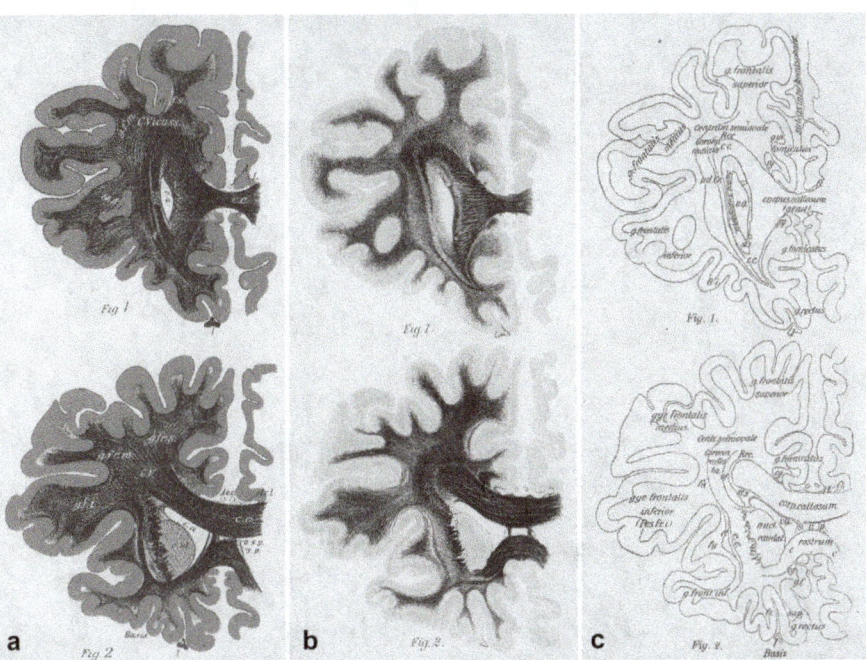

Fig. 2 − Drawings of coronal sections of the frontal lobe by Jakob for his early brain Atlases. (a) Plate 24 from the first edition (Jakob, 1895, 1896): frontal sections through the knee of the corpus callosum and the anterior segment of the frontal lobes (upper) and through the head of the caudate (lower). Abbreviations: *g.f.s.*, superior frontal gyrus; *g.f.m.*, middle frontal gyrus; *g.f.i.*, inferior frontal gyrus; *c.a.*, anterior horn of lateral ventricle; *f.a.*, lateral association bundles; *C.Vieuss.*, centrum semiovale; *I*, olfactory bulb; *s.p.*, septum pellucidum; *c.st.*, head of the caudate nucleus; *c.i.*, anterior limb of the internal capsule. (b, c) Plate 28 from the second edition and explanatory diagram (Jakob, 1899, 1901). Abbreviations: *Rcc.*, radiation of corpus callosum; *pd.Cr.*, base of corona radiata; *ft*, tangential fibers; *st.s*, central gray matter of the ventricle; *v*, ventricle; *of*, occipitofrontal fasciculus; *cg*, cingulum; *fa*, arcuate fasciculus; *fu*, uncinate fasciculus; *ce*, external capsule; *cl*, claustrum; *pes fr.i.*, foot of inferior frontal gyrus; *sL*, nerves of Lancisi; *st.a.*, central gray matter.

anatomical connectivity of the frontal lobe would decipher its functions. Therefore, the anatomo-clinical approach was taken as the safest way in reaching conclusions on function. Jakob emphasized the importance of studying connections, an idea consistent with the current hodological trend (cf. Catani and ffytche, 2005; ffytche and Catani, 2005; Thiebaut de Schotten et al., 2012, this issue). Furthermore, his writings on connections seem attuned to more recent theories of frontal systems and neural networks, such as Alexander et al.'s (1986) influential concept of parallel but segregated frontal-subcortical circuits that has been put into a clinical framework. An in-depth discussion of the association between frontal-subcortical circuits and neurobehavioral disorders can be found in Chow and Cummings (1999) and other papers of the special issue (Krause et al., 2012, this issue; Cubillo in press, 2012; Langen et al., in press).

In studying the structure of the frontal lobe, Jakob did not see any substantial differences from the remaining lobes of the cerebral hemispheres: "The frontal lobe has three categories of fibers just like the other lobes: afferent and efferent projection fibers, association fibers and commissural fibers... Through the study of the afferent pathways we understand that in the major part of the frontal lobe, covering the whole of

its convexity lies the great center of the muscular senses of a higher order" (Jakob, 1906b).

Concerning the connections between the frontal gyri and the Rolandic motor areas via 'U' fibers, Jakob (1906b) wrote: "These fibers join the superior frontal gyrus with motor foci that innervate the lower extremities, relate the middle frontal gyrus with the foci of the arms, and the inferior frontal gyrus with facial-lingual movements... Moreover, there exist short association fibers that connect the three gyri among them, and commissural fibers that, passing through the corpus callosum, enable the communication between the frontal gyri of the two sides... Thus, we come across the existence of an apparatus inserted between the muscles of the periphery and the cerebellum on one side and the Rolandic centers on the other" (Fig. 5).

Jakob (1906c) described the sensory-muscular pathways that arrive at the frontal lobe via the cerebellum, the red nucleus and the thalamus, concluding: "Although it is doubtful whether tactile senses arrive at the frontal lobe, it is true that numerous muscular sensory inputs enter the frontal lobe." In concordance with such an argument, Cappe et al. (2009) demonstrated the existence of thalamic projections to the cerebral cortex using neuroanatomical track-tracing

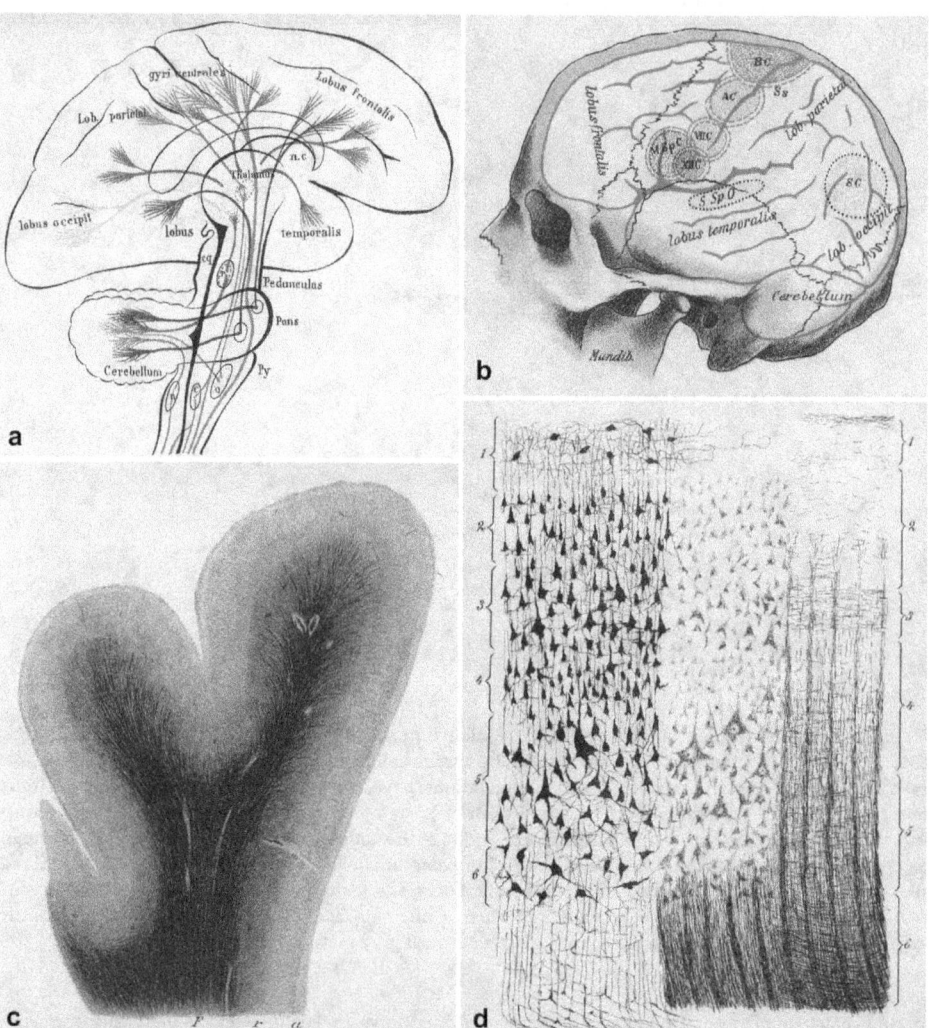

Fig. 3 – Additional drawings by Jakob from the second edition of his early brain Atlas (Jakob, 1899, 1901). (a) Plate 56.1 showing a general view of projection paths. Fibers forming the corona radiata enter the optic thalamus (brown). The frontal and temporal pontine pathway reaches the cerebellum through the contralateral middle cerebellar peduncle (blue). The pyramidal tract appears red, the sensory tract green, the cerebello-olivary tract violet, the optic radiation yellow, and the brachia brown. (b) Plate 21.3 depicting the position of psychomotor and psychosensory cortical centers in the cavity of the skull. Abbreviations: BC, motor center for lower extremities; AC, motor center for upper extremities; VIIC, XIIC, centers for muscles innervated by the facial and hypoglossal nerves; MSpC, SSpO, motor and sensory speech centers; SC, visual center. (c) Plate 20.1 showing a section from the center of the anterior central gyrus (carmin myelin sheath stain). White matter (F) appears blue-black; radial bundles (r) radiate in all directions and end in the cortex; a, outermost subpial layer. (d) Plate 19 showing the arrangement of cells (left and middle, stained with silver and methylene blue, respectively) and fibers (right) in the cerebral cortex. Cytoarchitectonic layer nomenclature: (1) Stratum zonale; (2) first layer of small pyramidal cells; (3) layer of medium-sized and large pyramidal cells; (4) second layer of small closely packed pyramidal cells; (5) second layer of medium-sized and large pyramidal cells with a few giant pyramidal cells; (6) layer of polymorphous cells. Myeloarchitectonic layer nomenclature: (1) stratum zonale with superficial layer of tangential fibers; (2) superradial reticulum and Bekhterev-Kaes stripe; (3) coarser tangential fibers (stripes of Baillarger, Gennari, Vicq d'Azyr); (4) interradial reticulum of tangential fibers; (5) closely packed radial bundles; (6) medullary layer with radiating white fibers (projection, commissural, and long association tracts) and transverse short association bundles (arcuate fibers of Meynert). For the most part, nerve fibers pass from the cerebral white matter into the cortex; collected in bundles, they enter the second layer of cells, where their terminal fibrils end. These radial bundles (radii) therefore have a vertical arrangement. They are crossed at right angles by other fibers running parallel with the cortical surface and forming the plexus of tangential fibers – the superradial reticulum above the radii, and the interradial reticulum with the radii.

ANALYSES   595

(1053) **La légende des Lobes Frontaux en tant que Centres supérieurs du Psychisme de l'Homme,** par Cristofredo Jakob (de Buenos-Aires). *Archivos de Psiquiatria y Criminologia,* Buenos-Ayres, an V, p. 679-698, novembre-décembre 1906.

L'auteur donne plusieurs observations de lésions des lobes frontaux sans déficit psychique d'aucune sorte. La conclusion de son travail est que les lobes frontaux n'exercent aucune hégémonie sur le reste du cerveau; ce qui est perdu de la personnalité psychique à la suite des lésions étendues des lobes frontaux n'est qualitivement, ni quantitativement différent de ce qui est perdu à la suite de la destruction étendue de tout autre lobe cérébral.   F. Déléni.

Fig. 4 – Jakob's 1906 paper abstracted in French in the prestigious *Revue Neurologique* (Jakob, 1907c). The summary reiterates Jakob's conclusion, based on observations that lesions in the frontal lobes do not lead to any substantial mental deficit: "The frontal lobes do not exert any hegemony over the rest of the brain. Any mental deterioration after damage to the frontal lobe does not differ qualitatively or quantitatively from that seen after damage to any other cerebral lobe".

markers. Furthermore, Goldman-Rakic and Porrino (1985) showed that the prefrontal cortex is defined by multiple specific relationships with the thalamus. Performing retrograde tracing experiments, Mitchell and Cauller (2001) examined the corticocortical and thalamocortical afferents to layer I of the rat frontal cortex and affirmed the existence of afferent projections from thalamic nuclei to the frontal lobe.

Based on his anatomical observations, Jakob (1906c) viewed the major part of the frontal lobe as a central station with multiplier and combinatorial characteristics, constantly receiving stimuli from all the motility organs via multiple pathways (Fig. 6). According to Jakob (1911), the various centripetal pathways course into all sectors; thus, the cortex has a perceptive activity over its entire extent (Triarhou, 2010). Jakob's position is compatible with modern views on the function of the anterior parts of the human brain: the prefrontal cortex is considered a locus of synthesis of the outputs of various brain systems which provides the basis for the orchestration of complex behavior (Duncan & Miller, 2002). Furthermore, the role of the frontal lobe in integrating

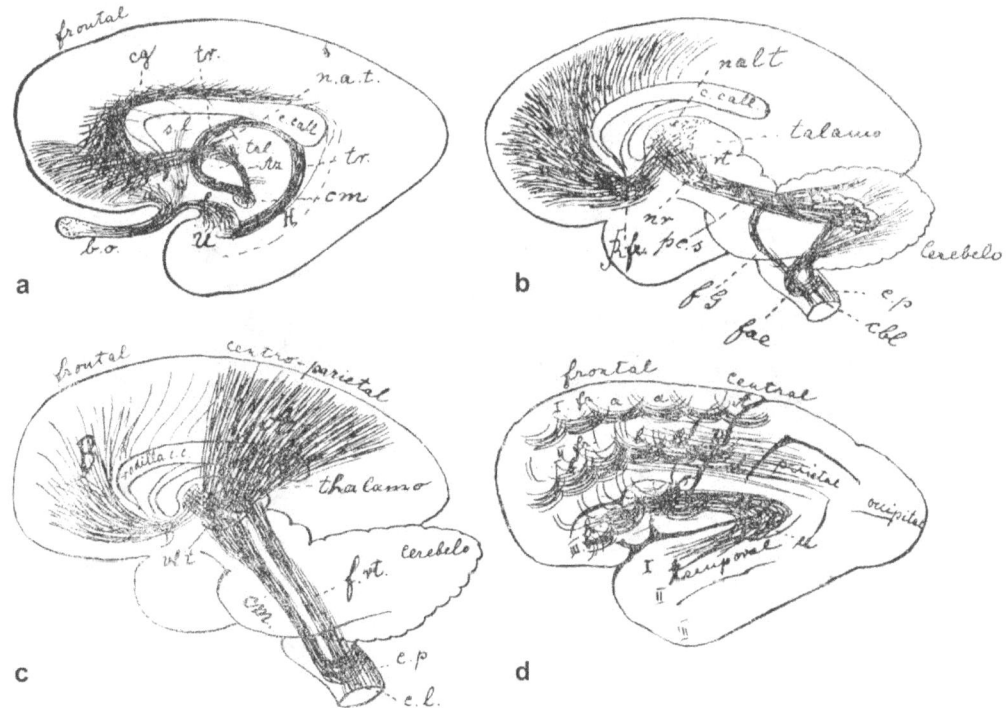

Fig. 5 – Schematic drawings by Jakob (1906b) showing: (a) Olfactory pathways in the frontal lobe. Abbreviations: b.o., olfactory bulb; *a,* internal root; *b,* lateral root; *sl,* septum pellucidum; *u,* uncus; *h,* hippocampus; *tr,* trigonum; *cm,* mamillary body; *Az,* bundle of Vicq d'Azyr; *nat,* anterior nucleus of thalamus; *Rf,* thalamo-frontal radiation; *cg,* cingulum. (b) Cerebello-frontal pathways. Abbreviations: *cp,* posterior spinal fasciculus; *cbl,* lateral cerebellar bundle; *fG,* bundle of Gowers; *pcs,* superior cerebellar peduncle; *nr,* red nucleus; *rt,* rubro-thalamic pathway; *nalt,* anterior lateral thalamic nucleus; *Rfr* frontal radiation. (c) Direct medullo-thalamo-frontal pathways. Abbreviations: *cp,* posterior spinal bundle; *cl,* lateral bundle; *frt,* reticular formation; *cm,* median band of Reil; *vlt,* ventral nuclei of thalamus; *A,* thalamo-Rolandic pathway; *B,* thalamo-frontal pathway. (d) Association pathways in the frontal lobe. Abbreviations: *A,* Rolandic center of crus; *a,* U-fibers of superior frontal gyrus (I); *B,* brachial center; *b,* U-fibers of middle frontal gyrus (II); *C,* facio-lingual center; *c,* U-fibers of inferior frontal gyrus (III); *u,* uncinate fasciculus; *d,* superior longitudinal fasciculus.

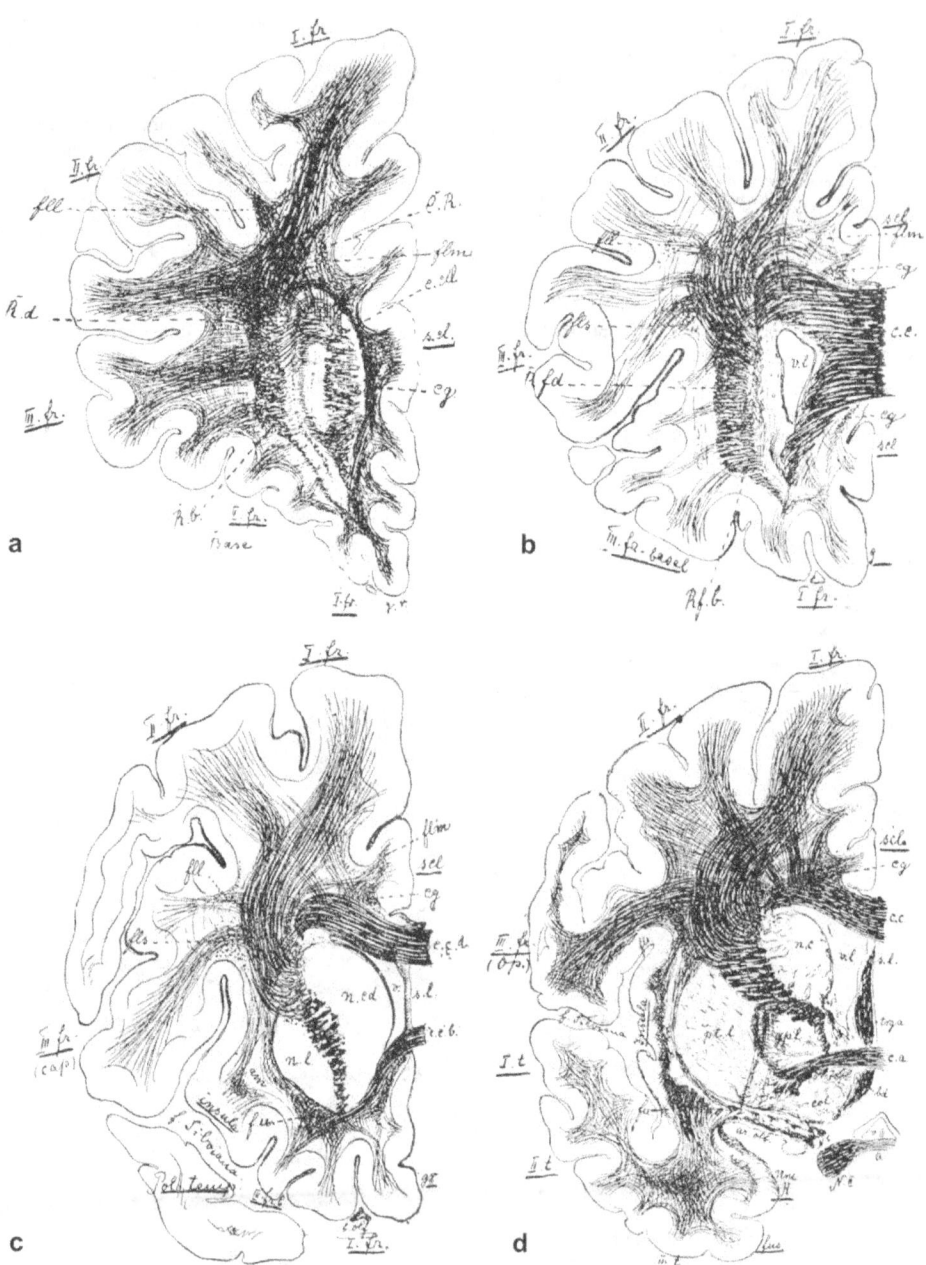

Fig. 6 – Schematic drawings by Jakob (1906c) based on a complete series of serial sections through the frontal lobe; Weigert method to depict fiber pathways. (a) Frontal section (no. 1154) in front of the corpus callosum. (b) Long projection, commissural and association pathways; section (no. 1076) through the knee of the corpus callosum. (c) Projection and commissural pathways and the formation of the internal capsule (frontal radiation); section (no. 948) through the corpus striatum. (d) Section (no. 882) through the posterior region of the frontal lobe with all its long frontal pathways. Abbreviations: fr, frontal; t, temporal; v, ventricle; l, lateral; nc, caudate nucleus; ptl, putamen-lenticular nucleus; gpl, globus pallidus; CR, corona radiata; RD, dorsal radiation of internal capsule; Rb, basal radiation of internal capsule; gr, rectal gyrus; cg, cingulum; flm, fll, fls, medial, lateral and superior longitudinal fasciculus; fu, uncinate fasciculus; cc, corpus callosum; sl, septum pellucidum; am, claustrum; scl, supracallosal gyrus; nl, lenticular nucleus; ca, anterior commissure; vIII, third ventricle; NI, olfactory nerve; NII, optic nerve; ar.olf, olfactory area; trga, anterior pillars of trigonum; col, coliculi of corpus striatum.

information from multiple brain areas supports its crucial involvement in learning, comprehension and reasoning (Baddeley, 2002). Frontal and prefrontal regions have been linked to visual, auditory and somatosensory inputs (Fogassi et al., 1996; Graziano et al., 1994, 1999; Wallace et al., 1992). Sensory, mnemonic and response signals that a single neuron displays provide strong evidence that prefrontal neurons behave as sensorimotor integrators (Goldman-Rakic, 2000). Prefrontal cortical neurons are considered to be a part of integrative neural systems that subserve cross-modal interactions across time (Fuster et al., 2000). According to Fuster's (2006) theorizing, actions related to human behavior, reasoning, and language are organized by means of interactions between prefrontal and posterior networks at the top of the 'perception–action cycle.' In non-human primates, multisensory integration takes place in frontal, parietal and temporal areas (Avillac et al., 2005). Thus, mounting evidence shows that much if not all of the neocortex is involved in multisensory integration (Ghazanfar and Schroeder, 2006).

3. Histological studies

3.1. Cytoarchitectonics

Based on the argument that structural differences signal functional specialization, Jakob studied human brain cytoarchitecture (Fig. 3d). Jakob (1906c) summed up his research as follows: "The frontal cortex is organized in the same cell layers, in the same associations of pyramidal cells that are differentiated only by their size, as we notice in the parietal and temporal lobe as well. The only thing that distinguishes the frontal cortex is the restricted variation of the size of the pyramidal cells due to the lack of large and giant pyramidal cells. My studies allow me to admit that toward the feet of the frontal gyri appear the large pyramidal cells covering the background of the precentral sulcus. Moreover, I managed to prove that the frontal cortex contains more cells per square millimeter compared with the Rolandic and the temporal regions. I could not deduce from this fact that the absolute number of cells would be greater in the frontal region compared to the Rolandic or the temporal regions, because the latter have a very high density".

At about the same time, Campbell (1905), a pioneer of cortical cytoarchitectonic parcellation, compiled clinical, anatomical and physiological evidence as a guide to function (ffytche and Catani, 2005). However, it was Brodmann's (1909) opus magnum that changed the view of histological localization in the human cerebral cortex once and for all (cf. Garey, 2006). Brodmann (1913) also produced a subsequent study concentrating on the frontal cortex (Elston and Garey, 2004).

3.2. Myeloarchitectonics

Having studied preparations with the Weigert method (Fig. 6), Jakob (1906c) argued: "As far as frontal myeloarchitectonics is concerned I notice the same disposition of radiating fibers as in other regions... The so-called association layers are identical to the ones of the other lobes and the tangential layer is well developed. On the contrary, the supraradial layer stands out in showing remarkably fewer myelinated fibers... I am inclined to

see a structural inferiority, an idea that is reinforced by the following facts: a diminished total density and density of the various layers, a smaller average cell volume and a less developed supraradial layer." Regarding Flechsig's proposal of a parallel development of myelination pathways and intellect, Jakob wrote: "While the central tracts of the frontal lobe are not completed until several months after birth, Flechsig demonstrated that other regions of the brain develop in a similar fashion, for instance parts of the parietal and temporal lobes, the insula, and the so-called associative centers... Any chronological difference is not of much importance since a child has his frontal center perfectly myelinated before reaching six months of age. However, a newborn infant and one of six months are not easily differentiated with respect to their cognitive development" (Jakob, 1906a).

Myelination in humans continues well into the second decade of life (Yakovlev and Lecours, 1967). Structural magnetic imaging studies have shown gray matter changes in the frontal lobe from adolescence to adulthood (Sowell et al., 1999). In support of Flechsig's claim, the myelination of the frontal lobe has been repeatedly correlated with the development of higher cognitive functions, such as working memory (Nagy et al., 2004) and language (Pujol et al., 2006), while incomplete myelination has been blamed as the underpinning of weak decision-making skills in adults (Giedd, 2004).

Campbell's cytoarchitectonic data led to conclusions close to those of Jakob: "The structural development of the prefrontal cortex is exceedingly low. It presents an extreme of fibre poverty; all its fibre elements are of delicate calibre, and its association system is particularly deficient. Its cell representation is on a similar scale. The cortex is also shallow" (Campbell, 1905).

4. Pathophysiological studies

For the most part of the 19th century, the literature emphasized the role of the frontal lobe based on cases of damage that resulted in profound personality changes. Having studied human brains with frontal lobe tumors, injuries and degeneration, Jakob (1909) pointed to the rareness of 'pure cases'; he highlighted the characteristics that may render pathological specimens inappropriate for drawing secure conclusions. He emphasized that (a) the appearance of symptoms does not necessarily coincide with the onset of the disease; thus, progression may be difficult to determine; (b) tumors compress the brain parenchyma; (c) lesions of vascular origin lead to widespread degeneration; and (d) brain damage may cause inflammation or concussion which may affect the whole brain (Jakob, 1906c, 1909).

von Monakow (1904, 1910) underlined certain factors that had been overlooked by other investigators who studied lesions, i.e., the effects of inflammation, the lack of aseptic conditions during surgery, and the distant effects of local damage over time (Finger, 1994). Further caution has been expressed by Teuber (2009) about the contradictions found in the clinical literature: case studies may involve either massive lesions extending beyond the frontal lobe or small, unilateral, or asymmetric lesions with correspondingly small and easily compensated effects.

5.   Comparative studies

Jakob's phylogenetic studies, from the human brain to over 100 species of the Patagonian fauna (Jakob, 1912a, 1912b; Jakob and Onelli, 1913, Triarhou, 2010), provided him with the bases for formulating the following ideas:

"The development of the frontal lobes increases from lower to higher mammals in a continuous and constant relation, whereas in other vertebrates there are no hemispheres with a cortex comparable to those of mammals. It is obvious that the region located in front of the cruciate sulcus (a structure homologous to the central sulcus) increases in size and in the number of gyri it possesses from the marsupial to the rodent, from the rabbit to the dog, from the dog to the monkey, and approaches the size and complexity of humans only in anthropomorphous apes...[1] Although the external morphology progresses from lower to higher scale in a constant manner, the same process does not occur in the internal structure... We see, then, that what is true about the process of comparative development in the frontal lobe is true in all the other lobes as well. Perhaps there are greater variations in one structure than another; but such variations are slight and it would be a highly difficult, if not impossible, venture to find a fundamental exception for the frontal lobe" (Jakob, 1906a).

Elsewhere, he wrote: "When the frontal lobes of the different mammals, of ape and man are compared, the concord of the fine cortical structure strikes our attention; it is hard to encounter well defined differences... I myself noticed that the radiating fibers of the frontal lobe in apes are of a smaller calibre in comparison to other regions, a fact that has already been mentioned for humans. As far as the pyramidal cells are concerned ape shows all the different human types... What distinguishes the human frontal lobe is only the number of large and giant pyramidal cells... If the frontal lobe were such a superior center that it would differentiate by its functions humans from animals, then we should have met more evident differences in histological structure. According to my studies, I am inclined to believe that the similarities between the frontal cortical regions of some higher animals (for example apes) and humans are greater than the differences. This fact comes to demonstrate that the problem of the superior human functions does not lie in their localization in this or that lobe, but in factors of another nature" (Jakob, 1906c).

From the beginning of the 20th century, the extraordinary human cognitive development has been attributed to the large size of the frontal lobes. Cytoarchitectonic studies show a very similar organization between human and macaque monkey prefrontal cortex (Petrides, 2005; Petrides et al., 2012, this issue). Magnetic resonance imaging studies (Semendeferi et al., 2002) show that the frontal cortex of humans and great apes occupies a similar proportion of the cortex of the cerebral hemispheres. Accordingly, the enlargement of the human brain has generally preserved the relationship

between its major lobes (Risberg, 2006). A relative increase of association cortex due to encephalization cannot lead to a regional expansion of the frontal association areas since all four cerebral lobes have both primary and association cortices; therefore, such an expansion should be common to all (Allen, 2009). For further discussion see also Petrides et al., (2012, this issue), Yeterian et al., (2012, this issue) and Thiebaut de Schotten et al., (2012, this issue).

6.   Experimental animal studies

Laboratories where experiments on animals were conducted have been one of the most vivid battlefields in the localizationist–antilocalizationist controversy. The experimental confirmation of motor cortex in dog brain by Fritsch and Hitzig (1870) was a landmark in the history of functional localization (Catani and Stuss, 2012, this issue). This tradition continued with new mosaïcists and holists. Jakob (1906b) points out: "Goltz, Ferrier, Hitzig and Bianchi observed that animals that had both frontal lobes removed present remarkable alterations in intellect and character, such that they become irritable and have an increased tendency to bite... These experimenters did not sufficiently prolong their observations, and neither were they able to exclude as an explanation the consecutive inflammation or infections caused by the operation. New experimental verification, performed with meticulous care by Munk, Grossglik, Horsley and Schafer (1888), did not absolutely verify any of the previous observations. They found that once the animals had passed the first moments of postoperative excitation, they all returned to their status quo".

The vulnerability of the first series of experiments was also highlighted by Jacobsen et al. (1936), who attributed it to (a) the lack of objective measures of the degree and nature of behavioral deficits and (b) the lack of the demonstration that lesions of equal extent in other cortical regions do not cause dementia of the same severity.

7.   Frontal lobe and higher cortical functions

According to Jakob's model, intelligence, memory and the like are needed for handling abstract concepts (Jakob, 1906c). Similarly, the view that psychical terms do not have localizable physiological correlates was expressed by Jackson and embraced by Freud, Goldstein, Pick, and Head (Meyer, 1974).

In 1906 Jakob wrote: "Consciousness is formed gradually as a result of the chaining of different cortical operations. It is impossible to view it as a localizable, special power separate from such processes. Consciousness is the manifestation of the synchronization of its components, since it is afflicted whenever any one of such components is afflicted. Intelligence is a quality par excellence that represents the rapid and safe function of the sensory, motor and associative apparatus. It cannot be localized, because it is a phenomenon inseparable from the overall cerebral process. Character is a mode of motor reactions congenitally imprinted on cortical elements. It intervenes in the transformation of the sensory and the motor functions and it is manifested in every action. Character is a quality, not a substantial power; therefore it could

---

[1] This was not a new observation: it is found, for instance, in the anatomy of Owen (1866–1868).

not be localizable. With the word 'memory' we designate an essential function that touches upon all the biological processes in the wider sense and especially upon the cortical processes. Will is the result of the inhibitory or productive influence that is exerted by gradually acquired associations on the inferior reflex actions via the motor centers. It has its origin in the association centers that cover the entire cortex. Only a determined voluntary act may be limited in a specific portion of the grand apparatus; nevertheless, for the production of such an act all the hemispheric regions intervene with greater or lesser intensity" (Jakob, 1906c).

Jakob's neurophilosophical writings became gradually refined and expanded in the course of his career. Well before neurophilosophy emerged as a formal scientific discipline, Jakob had written at least 14 philosophical works (López Pasquali, 1965), touching upon issues such as the relation between biogenesis and philosophy (Jakob, 1914) and the philosophical meaning of the human brain (Jakob, 1945; Théodoridou and Triarhou, 2010). Today, theories seek to elucidate the neural correlates of consciousness (cf. Crick & Koch, 1990). The so-called 'hard problem' lies in the consideration of consciousness as an 'emergent' property 'arising' from functional elements of the neurocognitive structure without attributing a dualistic character to it (Kouider, 2009). For example, according to Edelman and Tononi's model of a constantly shifting dynamic core (cf. Edelman, 1992; Tononi and Edelman, 1998), consciousness arises from the fast integration within a dynamic core of interacting elements. Other neurobiological theories, such as the global neuronal workspace (Dehaene et al., 1998; Dehaene and Naccache, 2001) highlight the interconnection between multiple cerebral modules that enables the broadcasting of information (Kouider, 2009). Whereas "proving the case for synchronization in the human brain" is still considered technically demanding (Zeman, 2001), Jakob conceived, in an impressive manner, the idea of synchronization of neuronal activity as the underlying mechanism of consciousness, more than a hundred years ago. Jakob's interpretation is consistent with the view that consciousness is to be correlated with a non-continuous event determined by synchronous activity in the thalamocortical system (Ribary et al., 1994). The transient synchronization of brain operations is considered to have the potential to construct unified and relatively stable neural states that underlie conscious states (Fingelkurts et al., 2005). The perception of volition seems to be generated in specific networks with the parallel activation of the global neuronal workspace (Hallett, 2007). The role of inheritance in behavior has been shown by selection and strain studies for animal behavior and by twin and adoption studies for human behavior (Plomin, 1990). Further evidence for the endogenous nature of traits derives from studies of behavior genetics, parent-child relations, personality structure, animal personality, and the longitudinal stability of individual differences (McCrae et al., 2000).

To conclude, Jakob tackled the 'terra incognita' of cognition with a multi-level approach in order to avoid bias. He was critical of oversimplifying localization explanations. Further, Jakob understood that it is essential to realize the limitations and misdirections involved in any attempt to decipher the brain—mind relationship. Being aware of such limitations, he searched for diverse clues, and largely relied on the anatomo-clinical approach. His concrete knowledge of neuroanatomy, coupled with his ingenuity, enabled him to produce knowledge that can be corroborated today via sophisticated tracing techniques. In a broad framework, studying Jakob's papers helps to correct and reconstruct an important episode in neurological history. Moreover, new English translations of such works will make them accessible by a wider audience. Given that the riddle of the human frontal lobe remains a central issue in modern neurobiology, Jakob's early views, a century later, may still provide meaningful clues.

Acknowledgments

Part of this work was presented at the 15th Annual Meeting of the International Society for the History of Neurosciences, Paris, France, June 15—19, 2010. The authors gratefully acknowledge the courtesy of the staff at the Ibero-Amerikanisches Institut and Staatsbibliothek Preussischer Kulturbesitz zu Berlin, the Library of Congress and the National Library of Medicine of the United States, as well as the anonymous reviewers for their constructive criticism.

REFERENCES

Alexander GE, DeLong MR, and Strick PL. Parallel organization of functionally segregated circuits linking basal ganglia and cortex. *Annual Review of Neuroscience*, 9: 357—381, 1986.
Allen JS. *The Lives of the Brain: Human Evolution and the Organ of Mind.* Cambridge, MA: Belknap Press of Harvard University Press, 2009.
Avillac M, Denève S, Olivier E, Pouget A, and Duhamel JR. Reference frames for representing visual and tactile locations in parietal cortex. *Nature Neuroscience*, 8(7): 941—949, 2005.
Baddeley AD. Exploring the central executive. *Quarterly Journal of Experimental Psychology*, 49A(1): 5—28, 1996.
Baddeley AD. Fractionating the central executive. In Stuss DT and Knight RT (Eds), *Principles of Frontal Lobe Function*. Oxford: Oxford University Press, 2002: 246—260.
Barutta J, Hodges J, Ibanez A, Gleichgerrcht E, and Manes F. Argentina's early contributions to the understanding of frontotemporal lobar degeneration. *Cortex*, 47(5): 621—627, 2011.
Broca P. Perte de la parole, ramollissement chronique et destruction partielle du lobe antérieur gauche du cerveau. *Bulletin de la Société d'Anthropologie (Paris)*, 2: 235—238, 1861.
Brodmann K. *Vergleichende Lokalisationslehre der Grosshirnrinde in ihren Prinzipien dargestellt auf Grund des Zellenbaues.* Leipzig: Barth, 1909.
Brodmann K. Neue Forschungsergebnisse der Grosshirnrindenanatomie mit besonderer Berücksichtigung anthropologischer Fragen. *Verhandlungen der Gesellschaft Deutscher Naturforscher und Ärzte*, 85: 200—240, 1913.
Campbell AW. *Histological Studies on the Localisation of Cerebral Function.* Cambridge: University Press, 1905.
Cappe C, Rouiller EM, and Barone P. Multisensory anatomical pathways. *Hearing Research*, 258(1/2): 28—36, 2009.
Catani M and Stuss DT. At the forefront of clinical neuroscience. *Cortex*, 48(1): 1—6, 2012.
Catani M, Dell'Acqua F, Vergani F, Malik F, Hodge H, Roy P, et al. Short frontal lobe connections of the human brain. *Cortex*, doi: 10.1016/j.cortex.2011.12.001.
Catani M and ffytche DH. The rises and falls of disconnection syndromes. *Brain*, 128(10): 2224—2239, 2005.

Chow TW and Cummings JL. Frontal subcortical circuits. In Miller BL and Cummings JL (Eds), *The Human Frontal Lobes: Functions and Disorders*. New York: Guilford Press, 1999: 25–43.

Crick F and Koch C. Towards a neurobiological theory of consciousness. *Seminars in Neuroscience*, 2: 263–275, 1990.

Cubillo A, Halari R, Smith A, Taylor E, and Rubia K. A review of fronto-striatal and fronto-cortical brain abnormalities in children and adults with Attention Deficit Hyperactivity Disorder (ADHD) and new evidence for dysfunction in adults with ADHD during motivation and attention. *Cortex*, doi:10.1016/j.cortex.2011.04.007.

Dehaene S, Kerszberg M, and Changeux JP. A neuronal model of a global workspace in effortful cognitive tasks. *Proceedings of the National Academy of Sciences of USA*, 95(24): 14529–14534, 1998.

Dehaene S and Naccache L. Towards a cognitive neuroscience of consciousness: Basic evidence and a workspace framework. *Cognition*, 79(1–2): 1–37, 2001.

Duncan J and Miller EK. Cognitive focus through adaptive neural coding in the primate prefrontal cortex. In Stuss DT and Knight RT (Eds), *Principles of Frontal Lobe Function*. Oxford: Oxford University Press, 2002: 278–291.

Duncan J and Owen AM. Common regions of the human frontal lobe recruited by diverse cognitive demands. *Trends in Neurosciences*, 23(10): 475–483, 2000.

Edelman GM. *Bright Air, Brilliant Fire: On the Matter of the Mind*. New York: Basic Books, 1992.

Elston GN and Garey L. *New Research Findings on the Anatomy of the Cerebral Cortex of Special Relevance to Anthropological Questions*. Australia: University of Queensland Press, 2004.

ffytche DH and Catani M. Beyond localization: From hodology to function. *Philosophical Transactions of the Royal Society of London (Biology)*, 360(1456): 767–779, 2005.

Fingelkurts AA, Fingelkurts AA, and Kahkonen S. Functional connectivity in the brain – Is it an elusive concept? *Neuroscience and Biobehavioral Reviews*, 28(8): 827–836, 2005.

Finger S. *Origins of Neuroscience: A History of Explorations Into Brain Function*. New York: Oxford University Press, 1994.

Finger S. *Minds Behind the Brain: A History of the Pioneers and Their Discoveries*. Oxford: Oxford University Press, 2000.

Flechsig P. *Meine myelogenetische Hirnlehre mit biographischer Einleitung*. Berlin: Julius Springer, 1927.

Fogassi L, Gallese V, Fadiga L, Luppino G, Matelli M, and Rizzolatti G. Coding of peripersonal space in inferior premotor cortex (area F4). *Journal of Neurophysiology*, 76(1): 141–157, 1996.

Fritsch GT and Hitzig E. On the electrical excitability of the cerebrum. In Von Bonin G (Ed), *Some Papers on the Cerebral Cortex*. Springfield, IL: Charles C. Thomas, 1870. p. 1960.

Fuster JM, Bodner M, and Kroger JK. Cross-modal and cross-temporal association in neurons of frontal cortex. *Nature*, 405(6784): 347–351, 2000.

Fuster J. The cognit: A network model of cortical representation. *International Journal of Psychophysiology*, 60(2): 125–132, 2006.

Garey LJ. *Brodmann's Localisation in the Cerebral Cortex*. New York: Springer, 2006.

Ghazanfar AA and Schroeder CE. Is neocortex essentially multisensory? *Trends in Cognitive Sciences*, 10(6): 278–285, 2006.

Giedd JN. Structural magnetic resonance imaging of the adolescent brain. *Annals of the New York Academy of Sciences*, 1021: 77–85, 2004.

Godefroy O, Cabaret M, Petit-Chenal V, Pruvo JP, and Rousseaux M. Control functions of the frontal lobes. Modularity of the central-supervisory system? *Cortex*, 35(1): 1–20, 1999.

Goldman-Rakic PS and Porrino LJ. The primate mediodorsal (md) nucleus and its projection to the frontal lobe. *Journal of Comparative Neurology*, 242(4): 535–560, 1985.

Goldman-Rakic P. Localization of function all over again. *NeuroImage*, 11(5): 451–457, 2000.

Graziano MSA, Reiss LA, and Gross CG. A neuronal representation of the location of nearby sounds. *Nature*, 397(6718): 428–430, 1999.

Graziano MSA, Yap GS, and Gross CG. Coding of visual space by premotor neurons. *Science*, 266(5187): 1054–1057, 1994.

Hallett M. Volitional control of movement: The physiology of free will. *Clinical Neurophysiology*, 118(6): 1179–1192, 2007.

Horsley V and Schafer EA. A record of experiments upon the functions of the cerebral cortex. *Philosophical Transactions of the Royal Society of London (Biology)*, 179: 1–45, 1888.

Jacobsen CF, Elder JH, and Haslerud GM. Studies of cerebral function in primates. *Comparative Psychology Monographs*, 13: 1–60, 1936.

Jakob C. *Atlas des gesunden und kranken Nervensystems nebst Grundriss der Anatomie, Pathologie und Therapie desselben*. München: J. F. Lehmann, 1895.

Jakob C. *An Atlas of the Normal and Pathological Nervous Systems. Together With a Sketch of the Anatomy, Pathology, and Therapy of the Same* (transl. by J. Collins). New York: William Wood & Co., 1896.

Jakob C. *Atlas des gesunden und kranken Nervensystems nebst Grundriss der Anatomie, Pathologie und Therapie desselben*. 2. Aufl. München: J. F. Lehmann, 1899.

Jakob C. *Atlas of the Nervous System Including an Epitome of the Anatomy, Pathology, and Treatment* (transl. by E. D. Fisher). 2nd edn. Philadelphia–London: W. B. Saunders & Co, 1901.

Jakob C. La leyenda de los lóbulos frontales cerebrales como centros supremos psíquicos del hombre. *Arquivos de Psiquiatría, Criminología y Ciencias Afines*, 5: 679–699, 1906a.

Jakob C. Nueva contribución á la fisio-patología de los lóbulos frontales. *La Semana Médica*, 13(50): 1325–1329, 1906b.

Jakob C. Estudios biológicos sobre los lóbulos frontales cerebrales. *La Semana Médica*, 13(52): 1375–1381, 1906c.

Jakob C. Sobre la sintomatología de las afecciones del lóbulo frontal. *La Semana Médica*, 14(43): 1285, 1907a.

Jakob C. Sobre apraxia. *La Semana Médica*, 14(44): 1344, 1907b.

Jakob C. La légende des lobes frontaux en tant que centres supérieurs du psychisme de l'homme (abstracted by F. Deleni). *Revue Neurologique*, 15: 595, 1907c.

Jakob C. Estudios anátomoclínicos sobre los lóbulos frontales del cerebro humano (Comunicación presentada al IV Congreso Médico Latinoamericano, Rio de Janeiro, 1–8 de agosto de 1909). *Argentina Médica*, 7(36): 463–472, 1909.

Jakob C. *Das Menschenhirn: Eine Studie über den Aufbau und die Bedeutung seiner grauen Kerne und Rinde*. München: J. F. Lehmann, 1911.

Jakob C. Über die Ubiquität der senso-motorischen Doppelfunktion der Hirnrinde als Grundlage einer neuen, biologischen Auffassung des corticalen Seelenorgans. *Journal für Psychologie und Neurologie (Leipzig)*, 19(1): 379–382, 1912a.

Jakob C. Ueber die Ubiquität der senso-motorischen Doppelfunktion der Hirnrinde als Grundlage einer neuen biologischen Auffassung des kortikalen Seelenorgans. *Münchener Medizinische Wochenschrift*, 59(9): 466–468, 1912b.

Jakob C. Los problemas biogenéticos en sus relaciones con la filosofía moderna. *Revista del Círculo Médico Argentino y Centro Estudiantes de Medicina (Buenos Aires)*, 14(150): 87–98, 1914.

Jakob C. *Folia neurobiológica Argentina, tomo III. El lóbulo frontal: Estudio monográfico anatomoclínico sobre base neurobiológica*. Buenos Aires: Aniceto López - López y Etchegoyen, 1943.

Jakob C. El cerebro humano: Su significación filosófica. Ensayo de un programa psico-bio-metafísico, después de 50 años de dedicación neurobiológica. *Revista Neurológica de Buenos Aires*, 10(2): 89–110, 1945.

Jakob C and Onelli C. *Atlas del cerebro de los mamíferos de la República Argentina: Estudios anatómicos, histológicos y biológicos comparados sobre la evolución de los hemisferios y de la corteza cerebral*. Buenos Aires: Guillermo Kraft, 1913.

Koechlin E, Ody C, and Kouneiher F. The architecture of cognitive control in the human prefrontal cortex. *Science*, 302(5648): 1181–1185, 2003.

Kouider S. Neurobiological theories of consciousness. In Banks W (Ed)Encyclopedia of Consciousness. Amsterdam: Elsevier, 2009: 87–100.

Krause M, Mahant N, Kotschet K, Fung VS, Vagg D, Wong CH, et al. Dysexecutive behaviour following deep brain lesions – A different type of disconnection syndrome. Cortex, 48(1): 96–117, 2012.

Langen M, Leemans A, Johnston P, Ecker C, Daly E, Murphy CM, et al. Fronto-striatal circuitry and inhibitory control in autism: Findings from diffusion tensor imaging tractography. Cortex, doi:10.1016/j.cortex.2011.05.018.

López Pasquali L. Christfried Jakob. Su obra neurológica, su pensamiento psicológico y filosófico. Buenos Aires: López Libreros Editores S.R.L, 1965.

Lorch MP. The merest Logomachy: The 1868 Norwich discussion of aphasia by Hughlings Jackson and Broca. Brain, 131(6): 1658–1670, 2008.

McCrae RR, Costa PT, Ostendorf F, Angleitner A, Hrebiććkovać M, Avia MD, Sanz J, Saćnchez-Bernardos ML, Kusdil ME, Woodfield R, Saunders PR, and Smith PB. Nature over nurture: Temperament, personality, and life span development. Journal of Personality and Social Psychology, 78(1): 173–186, 2000.

Meyer A. The frontal lobe syndrome, the aphasias and related conditions: A contribution to the history of cortical localization. Brain, 97(3): 565–600, 1974.

Mitchell BD and Cauller LJ. Corticocortical and thalamocortical projections to layer I of the frontal neocortex in rats. Brain Research, 921(1): 68–77, 2001.

von Monakow C. Über den gegenwärtigen Stand der Frage nach der Lokalisation im Grosshirn. Ergebnisse der Physiologie, 3(2): 100–122, 1904.

von Monakow C. Neue Gesichtspunkte in der Frage nach der Lokalisation im Grosshirn. Zeitschrift für Physiologie der Sinnesorgane, 54: 161–182, 1910.

Moyano BA. Christfried Jakob, 25/12/1866–6/5/1956. Acta Neuropsiquiátrica Argentina, 3: 109–123, 1957.

Nagy Z, Westerberg H, and Torkel K. Maturation of white matter is associated with the development of cognitive functions during childhood. Journal of Cognitive Neuroscience, 16(7): 1227–1233, 2004.

Orlando JC. Christofredo Jakob: su vida y obra. Buenos Aires: Editorial Mundi, 1966.

Owen R. On the Anatomy of Vertebrates. London: Longmans, Green, and Co, 1866–1868.

Petrides M, Tomaiuolo F, Yeterian EH, and Pandya DN. The prefrontal cortex: Comparative architectonic organization in the human and the macaque monkey brains. Cortex, 48(1): 45–56, 2012.

Petrides M. Lateral prefrontal cortex: Architectonic and functional organization. Philosophical Transactions of the Royal Society of London (Biology), 360(1456): 781–795, 2005.

Plomin R. The role of inheritance in behaviour. Science, 248(4952): 183–188, 1990.

Pujol J, Soriano-Mas C, Ortiz H, Sebastian-Galles N, Losilla JM, and Deus J. Myelination of language-related areas in the developing brain. Neurology, 66(3): 339–343, 2006.

Raichle ME. Foreword. In Stuss DT and Knight RT (Eds), Principles of Frontal Lobe Function. Oxford: Oxford University Press, 2002. vii-ix.

Ribary U, Llinács R, and Joliot M. Human oscillatory brain activity near 40Hz coexists with cognitive temporal binding. Proceedings of the National Academy of Sciences of USA, 91(24): 11748–11751, 1994.

Risberg J. Evolutionary aspects on the frontal lobes. In Risberg J and Grafman J (Eds), The Frontal Lobes: Development, Function, and Pathology. Cambridge: Cambridge University Press, 2006: 1–20.

Semendeferi K, Lu A, Schenker N, and Damasio H. Humans and great apes share a large frontal cortex. Nature Neuroscience, 5(3): 272–276, 2002.

Shallice T. Fractionation of the supervisory system. In Stuss DT and Knight RT (Eds), Principles of Frontal Lobe Function. Oxford: Oxford University Press, 2002: 261–277.

Shallice T and Burgess P. The domain of supervisory processes and temporal organization of behaviour [and discussion]. Philosophical Transactions of the Royal Society of London (Biology), 351(1346): 1405–1412, 1996.

Sowell ER, Thompson PM, Holmes CJ, Jernigan TL, and Toga AW. In vivo evidence for postadolescent brain maturation in frontal and striatal regions. Nature Neuroscience, 2(10): 859–860, 1999.

Stuss DT. Frontal lobes and attention: Processes and networks, fractionation and integration. Journal of the International Neuropsychological Society, 12(2): 261–271, 2006.

Stuss DT, Alexander MP, Floden D, Binns MA, Levine M, McIntosh AR, Rajah N, and Hevenor SJ. Fractionation and localization of distinct frontal lobe processes: Evidence from focal lesions in humans. In Stuss DT and Knight RT (Eds), Principles of Frontal Lobe Function. Oxford: Oxford University Press, 2002: 392–407.

Teuber HL. The riddle of frontal lobe function in man. Neuropsychology Review, 19(1): 25–46, 2009.

Théodoridou ZD and Triarhou LC. Christfried Jakob's 1921 theory of the gnoses and praxes as fundamental factors in cerebral cortical dynamics. Integrative Psychological and Behavioral Science, doi:10.1007/s12124-010-9145-4 (online first) 2010.

Thiebaut de Schotten M, Dell'Acqua F, Valabregue R, and Catani M. Monkey to human comparative anatomy of the frontal lobe association tracts. Cortex, 48(1): 81–95, 2012.

Tononi G and Edelman GM. Consciousness and complexity. Science, 282(5395): 1846–1851, 1998.

Triarhou LC. Revisiting Christfried Jakob's concept of the dual onto-phylogenetic origin and ubiquitous function of the cerebral cortex: A century of progress. Brain Structure and Function, 214(4): 319–338, 2010.

Triarhou LC and del Cerro M. Semicentennial tribute to the ingenious neurobiologist Christfried Jakob (1866–1956). 1. Works from Germany and the first Argentina period, 1891–1913. European Neurology, 56(3): 176–188, 2006a.

Triarhou LC and del Cerro M. Semicentennial tribute to the ingenious neurobiologist Christfried Jakob (1866–1956). 2. Publications from the second Argentina period, 1913–1949. European Neurology, 56(3): 189–198, 2006b.

Triarhou LC and del Cerro M. Pioneers in Neurology: Christfried Jakob (1866–1956). Journal of Neurology, 254(1): 124–125, 2007.

Tsapkini K, Vivas AB, and Triarhou LC. 'Does Broca's area exist?' Christofredo Jakob's 1906 response to Pierre Marie's holistic stance. Brain and Language, 105(3): 211–219, 2008.

Wallace MT, Meredith MA, and Stein BE. Integration of multiple sensory modalities in cat cortex. Experimental Brain Research, 91(3): 484–488, 1992.

Yakovlev PI and Lecours AR. The myelogenetic cycles of regional maturation of the brain. In Minkowski A (Ed), Regional Development of the Brain in Early Life. London: Blackwell, 1967: 3–70.

Yeterian EH, Pandya DN, Tomaiuolo F, and Petrides M. The cortical connectivity of the prefrontal cortex in the monkey brain. Cortex, 48(1): 57–80, 2012.

Zeman A. Consciousness. Brain, 124(7): 1263–1289, 2001.

Zülch KJ. Pyramidal and parapyramidal motor systems in man. In Zülch KJ, Creutzfeldt O, and Galbraith GC (Eds), Cerebral localization: An Otfrid Foerster symposium. Berlin-Heidelberg: Springer, 1975: 32–47.

Integr Psych Behav (2011) 45:247–262
DOI 10.1007/s12124-010-9145-4

COMMENTARY

Christfried Jakob's 1921 Theory of the Gnoses and Praxes as Fundamental Factors in Cerebral Cortical Dynamics

Zoë D. Théodoridou · Lazaros C. Triarhou

Published online: 13 October 2010
© Springer Science+Business Media, LLC 2010

Abstract This study aims at reviving an important contribution by the pioneer neurobiologist and neurophilosopher Christfried Jakob (1866–1956) to the understanding of higher cortical functions. Jakob studied cortical dynamics at multiple levels by comparing gnoses and praxes and their corresponding pathological states, i.e. the agnosias and the apraxias. We herein provide a complete English translation of Jakob's original Spanish article dating to 1921, and further consider some key points under the scope of the neuropsychological knowledge available then, and the research evidence available 90 years later.

Keywords Christfried Jakob · Cerebral cortex · History of neuroscience · Gnosis · Praxis · Agnosia · Apraxia

Introduction

The progress in aphasiology effected since the second half of the 19th century has enhanced research on the cortical localization of mental functions. The description and the study of multiple forms of agnosia and apraxia have contributed to that end.

The term 'agnosia' was introduced by Sigmund Freud (1856–1939) in his monograph *On aphasia* (1891) to denote functional disturbances between the concept of an object and the concept of the word corresponding to it (Macmillan 2004; Goldberg 2005, p. 103). The first use of the term 'apraxia' is attributed to the

L. C. Triarhou (✉)
University of Macedonia, 156 Egnatia Ave., Rm. Z-312, 54006 Thessaloniki, Greece
e-mail: triarhou@uom.gr

Z. D. Théodoridou · L. C. Triarhou
Economo-Koskinas Wing for Integrative and Evolutionary Neuroscience, University of Macedonia, Thessaloniki, Greece

🍷 Springer

German psychiatrist Hugo Liepmann (1863–1925), who defined motor apraxia and distinguished it from agnosia and sensory apraxia in his 1900 classic *Das Krankheitsbild der Apraxie* (Devinsky and D'Esposito 2004, p. 236; Etcharry-Bouyx and Ceccaldi 2007, p. 36). The term had been used in 1871 in a similar sense by Heymann Steinthal (1823–1899) (Cockburn 2008, p. 210; Liepmann 1988, p. 3).

The pioneer neurobiologist-neurophilosopher Christfried Jakob (1866–1956) (Triarhou and del Cerro 2006a, b, 2007) studied agnosias and apraxias both clinically and anatomically in order to shed light on cortical dynamics from a structural, functional and evolutionary viewpoint (Jakob 1921). The present study aims at reviving a particular contribution of Jakob on the representation and production of higher gnosic-praxic functions.

Jakob was born in 1866 in Bavaria, Germany (Moyano 1957). He studied medicine at the University of Erlangen and graduated in 1890 (López Pasquali 1965; Orlando 1966; Triarhou and del Cerro 2006a). His doctoral dissertation dealt with aortitis syphilitica and was supervised by Albert von Zenker (1825–1898) (Triarhou and del Cerro 2007). In the early 1890s Jakob worked as assistant at the Erlangen Medical Clinic headed by Adolph von Strümpell (1853–1925), and also practised privately in Bamberg (Moyano 1957). Having attained worldwide renown through his early brain atlases, Jakob moved to Argentina in 1899 to head the Laboratory of the Psychiatric and Neurological Clinic of the Hospicio de Las Mercedes, affiliated with the University of Buenos Aires (López Pasquali 1965; Orlando 1966). His first name became 'castillianized' to Christofredo when he was naturalized as an Argentinian citizen (Orlando 1966). In 1922 Jakob occupied the chair of biology of the nervous system in the Department of Educational Sciences of the University of La Plata (Triarhou and del Cerro 2006b) and the following year he published his textbook of neurobiology (Jakob 1923). Overall, Jakob authored 30 books and 250 articles covering developmental, evolutionary, anatomical, pathological and philosophical topics in neurobiology (Triarhou and del Cerro 2006a).

Several of Jakob's papers have philosophical ramifications, with Kantian influences often conspicuous. The consideration of gnoses into the axes of space and time—the a priori conditions of our internal intuition (Kant 1999)—becomes apparent. Jakob wrote 14 philosophical works (López Pasquali 1965). Some of the issues he tackled include the relation between biogenesis and philosophy (Jakob 1914) and the philosophical meaning of the human brain (1945a). He authored a monograph entitled 'Biophilosophical Documents' (Jakob 1946), the fifth in the *Folia neurobiológica Argentina* series. In that sense, Jakob can be considered as one of the earliest neurophilosophers.

In a path of enquiry spanning over half a century, the frontal lobe occupied Jakob's thought constantly, forming a major motive that led to the formulation of a dynamic theory on cerebral cortical function (Pedace 1949). Within such a framework, gnoses play a key role as the preparatory acts, and praxes as the productive acts, of all psychogenetic processes (Jakob 1941). According to Jakob, psychogenesis (<Gk. *psyche*=soul and *genesis*=origin) refers to a structuralistic developmental process taking place in the human cortex and leading to the formation of abstract thought (Jakob 1941). The term became widely used in psychiatry (Freud 1955; Jung 1960) and it was adopted by Jean Piaget (1896–

1980) to denote the formation of knowledge (Piaget 1972, p. 19), an explanation not far from Jakob's. Figure 1 presents an outline of Jakob's evolutionary components of psychogenesis.

This work carries special weight because it underscores the emergence of Jakob's foremost theories regarding (a) the biological basis of memory and (b) the phylogeny of the kineses. Both of these theories were published in 1935 (Jakob 1935a, b) and are considered essential in understanding his psychobiological thought (Moyano 1957).

According to Jakob's postulate (Jakob 1935b), phylogeny occurs in two phases. The first, 'plasmodynamic' phase entails elementary biological phenomena such as tropism and pulsatility. The second ('neurodynamic') phase is divided into three stages: a phylogenetically older *archikinetic* stage, where reflex actions emerge; a *paleokinetic* stage characterized by instinctive reactions; and a *neokinetic* stage, which elaborates conscious motor responses. The *neokineses* include three kinds of complex neurocognitive processes: (a) *gnoses*, which secure the conscious orientation in one's environment, (b) *praxes*, which underlie active individual intervention and (c) *symbolisms*, which subserve the communication by means of abstract ideas. Each one of these three stages corresponds to different levels of organization in the vertebrate CNS. The archikinetic stage corresponds to the archineuronal, the paleokinetic to the paleoneuronal, and the neokinetic to the neoneuronal.

The original presented article, written in Spanish, was published in 1921 in the Peruvian journal *La Crónica Médica* (Fig. 2). It highlights the 'middle period' of Jakob's work on cortical dynamics. When it was published, the 55-year-old Jakob was at the prime of his neurobiological thought (Triarhou and del Cerro 2006b). An earlier version had been presented a couple of years earlier to the 'Argentinian Medical Circle and Medical Student Center' in Buenos Aires (Jakob 1919; López Pasquali 1965; Orlando 1966).

We provide an English translation of Jakob's full 1921 paper, followed by a discussion of some key points, considered under the scope of the neuropsychological knowledge available at the time, and the experimental evidence available today.

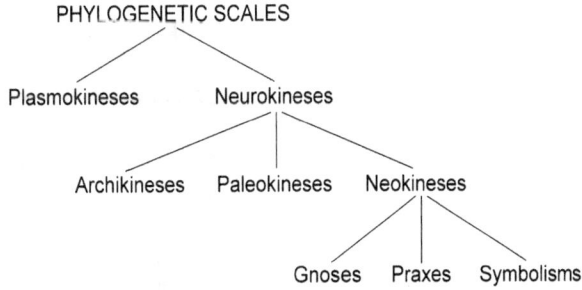

Fig. 1 According to Jakob's evolutionary postulate, *archikineses* represent hereditary reflex actions of psychism, *paleokineses* correspond to instinctive reactions, and *neokineses*—which represent the phylogenetically most recent type of the kineses—form the core of human consciousness (Jakob 1935b; c.f., Szirko 1995; Triarhou 2008; Triarhou and del Cerro 2006b)

Fig. 2 Frontispiece of Jakob's original 1921 article on gnoses and praxes (Jakob 1921)

LA CRONICA MEDICA

Año XXXVIII   LIMA - PERU - 1921   Sanmartí y Cía. Impresores.

La teoría actual de las gnosias y praxias como factores fundamentales en el dinamismo de la corteza cerebral

Por el Dr. CHRISTOFREDO JAKOB

Entre los múltiples problemas que ocupan la Fisiología moderna, no hay indudablemente ninguno más importante y complejo y que al mismo tiempo más interese al hombre como biotipo *sui géneris*, que el del dinamismo cerebro-cortical y su relación con las facultades mentales.

The Theory of the Gnoses and Praxes as Fundamental Factors in Cerebral Cortical Dynamics (1921)

by Christfried Jakob

Among the numerous problems that occupy modern physiology, there is undoubtedly no more important and complex a problem—and at the same time more interesting to man as a biotype sui generis—than that of cerebral cortical dynamics and its relation to the mental faculties.

A scientific concept on the psychophysiological conditions of cortical functions becomes equally necessary in psychology, in physiology, and in the neuropsychiatric clinic. With reasonable satisfaction, we may state that modern psychology owes its knowledge on the matter to medicine and its fosterers, both in the neurological clinic and laboratory, and in neurophysiological experimentation. On the other hand, psychology has only known to confuse and to entangle problems—which are sufficiently difficult on their own—with an imprecise terminology.

I present a short summary of the history of our relevant knowledge; it becomes clear how psychological concepts gradually disappear, and become replaced with clinical-physiological, and finally biological terms. Thus, concepts gradually turn from theoretical and fictitious into natural, capable of being subsequently subjected to critical scientific study.

Leaving aside the old theory of the 'animal spirits', a remnant of the theory of 'animism' sustained by the old philosophy and its localization (pneumatic ventricular theory), we find in principle scientific concepts only in the 17th and 18th centuries, with Bartholin and Willis and their schools localizing mental processes to the cerebral matter for the first time; they had only been precisely delineated by Gall and his disciples, in the cerebral gyri, i.e. in the internal organization. But what these authors localized were still extremely complex functions; sensitivity, will, memory, imagination or the various mental and moral qualities that distinguish humans. A localization of such abstract concepts, virtues and talents, was evidently psychological, but had nothing psychological about it.

Springer

It is to the clinic that we owe the rise of a new era: localization and language studies (Broca, Wernicke, Déjerine) [and] motor and sensory functional studies (Jackson, Hitzig, Munck, Goltz, etc.) rejected the psychological qualities and localized sensory-motor physiological functions of different modality and localization to projection areas for the first time. The shadow of old psychology continued to exert its influence, because in all those theories a zone (albeit smaller and smaller) was tacitly sustained for higher mental functions.

Such a view was presented concretely by Flechsig, based on embryological data, hastily interpreted as the frontal and temporoparietal 'association areas', like an 'ecclesiastic reserve'[1] for the elaboration of consciousness, ideation, abstraction, etc.

The normal and pathological anatomy rejected such theories and safely established that the so-called association areas receive important contingents of afferent pathways (of projection); projection areas have the same number of association pathways as well.

Thus, the functional dualism of both categories of areas could not be sustained anatomically. Besides, it has been impossible to find sufficient characteristic dispositions in regional cortical structure considered as projection and association, as I was able to show in my studies on cortical histoarchitecture.

One of the most difficult points in localization theories was the question of where does one localize the respective associative functions. A doubt persisted as to whether they are executed at the same time and place as perceptive processes. It is worth considering how such functions are distributed in sensory-motor projection areas, or even in areas considered as association; we can assume the existence of one or some memory areas or rather a mnemonic apparatus for each projection area. In the latter case, it would be necessary to interrupt the unity of association areas and to structure them into other such mnemonic areas, equalling in number the projection areas.

This last concept was mainly elaborated by Ramón y Cajal. In his cortical theory, Cajal established a mosaïc of projection areas surrounded by other mnemonic (association of first order), further surrounded by second and third order, committed to concrete and abstract ideative productions. Moreover, it is interesting to note the cortical functional unity established by Flourens's physiology, which has progressively resulted in much smaller dynamic centers, always correlated with each other. However, the resulting dualist criterion of the influence of psychology on different localization from the physiological phase on one hand, and from the mental on the other, was in essence still conserved. And it is with such inventory, more or less selectively fixed, that current psychology is handled. For the sake of curiosity, I mention here the recent creation—psychological rather than physiological—the occurrence of psycho-hormones,[2] a real resurrection of 'animal spirits' in the endocrine domain.

---

[1] According to the Peace of Augsburg (1555), a treaty between Charles V, Holy Roman Emperor, and the Lurtheran princes, German lay people could freely and unconditionally select their religion. In this respect Catholics recognized Lutheranism. Still, there was a clause consisting in the divestment of all the goods including the territory of an ecclesiastic in case one embraced Protestantism [translators' note].

[2] The author might conceivably be referring to von Monakow's contemporary concept of *horme*, i.e. 'a hypothetical, self-actualizing force that brought individual processes together into a moral and functional whole' (c.f., Finger 1994, p. 58) [translators' note].

 Springer

While the above theories were being elaborated, the clinic began to bring new, fertile concepts to psychophysiolology, as it had done in the past. These are the pathological phenomena described as *agnosias and apraxias*; their clinical-anatomical analysis has shed new light on normal cortical dynamics. Thus, the so-called *astereognosis* or *stereoagnosia* that appears in certain cases of injuries to the parietal lobe has been known for a long time through the work of Wernicke, Déjerine, Horsley, Starr, and others. *Astereognosis* or *stereoagnosia* consists in an inability of the patient to recognize only by tact and grasp, objects that are given to him to hold, despite the fact that tactile sensitivity is not substantially altered. The patient feels that he takes something in his hand, but he cannot remember if this object is long, short, round, smooth, etc., with his eyes closed. Visual, auditory, olfactory, gustative agnosias, etc. have been described in an analogous form. In such cases, the respective sense, though injured, is not abolished. Thus, the patient cannot integrate sensory information—normally gotten by a certain number of isolated perceptions of distance, color, forms, intensity, etc.,—which characterizes an object one has seen, heard, tasted, etc. One then normally arrives at a state of 'apperceptive condensation and associative correlation' for the analogous impressions that allow one to finally construct 'the notion of the object', namely its *complete gnosis*.

It is evident that such agnosias result from injuries of complex cortical dynamics of momentary perceptions with previously fixed associations. Thus, *gnosis*, i.e. the positive process, consists in the synthetic condensation of a previous experience with an analogous current situation. In brief, it results from an intricate game of sequential cortical elaborations.

Tactile (haptognoses), thermal, tactile-muscular (stereognoses), visual, auditory, olfactory gnoses, etc. work then with isolated, experienced, correlated and repeated senses. They distribute and organize them in order as the securing of orientation in space and time demand it. Therefore, they stabilize one's experience of the external environment.

'Gnosic (or cognitive) processes' are not naturally the result of a special cortical power of gnosis. Indeed, that would bring us back to the old error, i.e. the theory of projection and association areas. In that theory, memory, consciousness, will and intellect were thought of as substantiated powers. On the contrary, it is only modalities here that accompany all cortical neurobiological processes to a greater or smaller intensity and extent.

The gnosic mechanism, like all nervous processes, is made up of a trilogy of elements: receptors, assimilators and combined effectors.

The receptive factors represent all sensory systems which, directed by the posterior half of the basal and dorsal thalamus, radiate towards the posterior half of the hemispheres. Thus, they include the entire cortex behind the central sulcus of Rolando: parietal, temporal and occipital (thalamo-parietal, temporal and occipital radiations).

The assimilator elements are formed by short and long inter- and intra-cortical pathways, and the association of these regions.

The effector elements represent the motor apparatus of attention, which, from auditory, visual, olfactory, tactile centers etc., stimulates the movements of attention of the ear and its accessory apparatus, of the eyes and their motor apparatus, of the nose and the related respiratory movements; for the tactile and muscular regions the

 Springer

effector apparatus represents the same pyramidal tract with its motor impulse on the limbs and body. The exact boundaries among different cortical gnosic centers cannot be drawn as yet; that is a question for future clinical study.

The gnosic dynamics in its turn rests fundamentally on the congruence of the analogous sensory-affective reactions. Similar percipient situations produce equal central reactions regarding the corresponding location, association and attention. Thus, they raise essentially equal affective states. Gnosis, then, is elaborated on the basis of the parallelism of outer and inner experiences through the matching of an identical perceptive situation with an analogous affective tone.

The intensity of the affection (I call it interest) during the elaboration of a gnosis of a new object gradually loses its initial tension and finally maintains a very reduced value in numerous gnosic acts that are then called automatic. Still, attention can always return to its initial value.

In my opinion, the fundamental fact is that all parieto-occipito-temporal cortical zones contribute in the elaboration of gnoses both in animals and in humans. Thus, we can crystallize the localization of the gnoses as represented in the posterior half of the cerebral hemispheres. Nevertheless, I insist that gnosis consists in the elaboration and condensation of sensory-motor acts, and not only sensory, as the old theory of the 'association and projection areas' claimed.

Specifically and in a detailed manner, there may plausibly be different gnoses with different localizations: labial, lingual, digital and tactile-gnoses localized in the posterior central gyrus; thermo-gnoses localized in the superior parietal gyrus; oral, manual and ocular stereo-gnoses localized in the supramarginal and angular gyri; visual-gnoses of form, color and perspective localized in the entire occipital lobe and the angular gyrus; auditory-gnoses of noises, sounds, rhythm and melodies localized in the posterior two-thirds of the temporal lobe; olfactory-gnoses localized in the hippocampus; and gustatory-gnoses possibly localized in the temporal pole. In sympathetic areas (cingulate gyrus) respective processes for visceral-gnoses etc. (condition of the bladder, of the stomach, endo-gnoses) will occur, as well (Fig. 3).

In this way, all these regions are unlashed into projection and association areas simultaneously. I absolutely reject the possibility of localizing perceptive and associative process, which combined characterize gnosis, into different zones for each function.

For example, when we have acquired the gnosis of a pencil, the momentary visual perception of a pencil evokes the acquired partial gnoses of form, color, surface, weight, hardness etc. The cortical constitutive elements reside in the nuclear complexes of visuoretinal, visuomotor, tactile-motor etc. areas, connected to each other in the certain combination that has been elaborated during childhood, the gnosic notion of the pencil.

Then, we do not need special mnemonic centers, because what distinguishes the mnemonic image and the immediate sensory perception is only the degree of affective tension, lower in the former case and greater in the latter. Nevertheless, they are the same elements combined in the same form that with their dynamics produce the image or the sensation.

This can be proven experimentally in animals and in humans, when partial resections of cortical segments only lead to transient gnosic alterations. This means that the neighboring zones, according to the old 'associative theory', gradually

 Springer

Fig. 3 'Golden section' of
the human cerebral hemisphere
(frontal praxic sectors I, II,
III; gnosic sectors I, II, III, IV,
V) according to Jakob (1943,
p. 37). Lateral **a** and midsagittal
b views

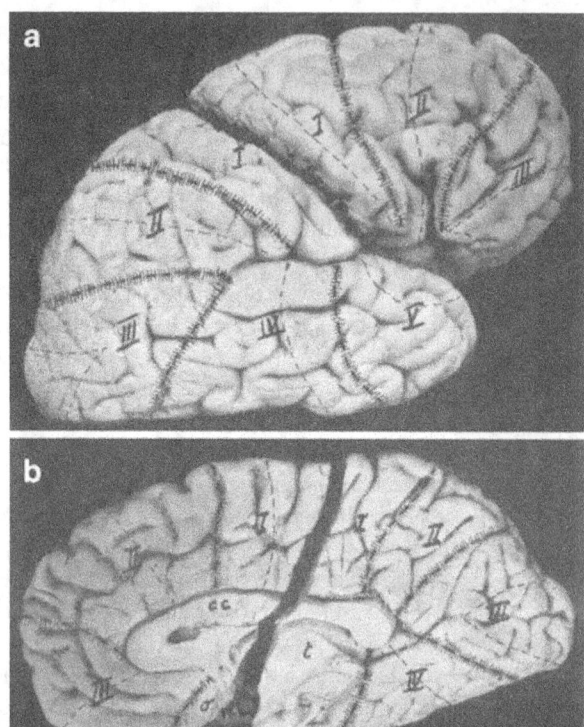

compensate for the defect. That would be impossible a priori; they would only have the ability to receive and store gnosic functions in equal form, exactly like the area previously destined to carry out such work.

In an analogous form, the study of the apraxias has contributed to our understanding of cortical function; their analysis offers a major importance to the concept of the gnoses, as a 'cortical faculty' sui generis.

Apraxia has been studied by Heilbronner, Liepmann, Pick, von Monakow, and others; I have also contributed to this field with several works. It consists in the oblivion of an act or a series of necessary acts, previously learned, to make any intentional movement without a real paralysis of the respective muscle. Thus, an apraxic does not know how to take the pencil, to wind the clock or to smoke etc., because he has forgotten the series of the necessary movements, coordinated to this end.

The complete analogy with the concept of gnosis can be immediately observed: the background of memory defects, objects or qualities, movements or their coordination.

Like in the gnoses, different forms of apraxias can be distinguished, e.g. manual, digital, oral, labial, limb, etc. There is always a certain sequence of associated movements that the patient has forgotten. Frontal ataxia belongs to this group, i.e. the inability to maintain balance in cases of lesions in the superior frontal gyrus (frontal astasia-abasia without paralysis). Furthermore, the motor aphasia of Broca type following injury of the inferior frontal gyrus, whereby the patient, without having paralysis of his articulator apparatus, does not find the necessary movements

to produce the previously well-known word. The case of agraphia seems to belong to the apraxias as well.

If we now move on from the pathological process to its corresponding normal function, we must name as praxis the cortical process that associates in different combination the series of motor acts during the long learning process in infancy and childhood, to guarantee a determined movement of the limbs, the tongue, the lip, and the trunk or the entire body.

Thus, praxes are the walking, jumping and dancing regarding the lower limbs (localization in the superior frontal gyrus); the use of fork and knife, writing, and every technical-manual work for the hands (superior and mainly middle frontal gyrus); mastication, imitation, language, singing and whistling for mouth and tongue (middle frontal gyrus). The mechanism of praxis is formed by the apparatus of three systems, receptors, assimilators and effectors, just like in the gnoses. Its receiving systems formed by indirect pathways of cutaneous sense muscle (kinesthetic) crossing the cerebellum project via the red nucleus to the anterior thalamus; from there, the anterior, frontal and central radiations of the thalamus penetrate towards the cortical areas in front of the central sulcus of Rolando. It is virtually undoubted that the seat of the elaboration of praxes is the whole anterior half of the cerebral hemispheres, i.e. the anterior Rolandic and entire frontal lobe with its related associative systems (I shall not discuss the major dominion of the left hemisphere for both higher gnoses and praxes).

Their effector pathways are represented by the fronto-hypothalamic and pontine systems on the one hand, destined to strengthen the muscular coordination en bloc. Especially the pyramidal and the operculo-bulbar tract serve as effector systems as well. These turn out to be at the disposition of the two great cortical powers: gnoses and praxes become mainly discharged by means of the pyramidal tract. Hence results the great importance of such motor pathways among other secondary pathways.

Praxes become gradually automatic as well. The affective tone that initially accompanies and stimulates their acquisition finally occurs with a lesser effort. We see, then, that the gnosic and praxic cortical automatism cannot be explained, as it has been claimed by physiology and psychology [the famous polygon of Grasset (1912) shown in Fig. 4], through a mechanism of specific location, but through dynamics different from its constitutive elements.

Without entering into histophysiological details of such a process and the relative participation of cortical elements [cf. my previous study on cortical neurobiology (Jakob 1913)], I establish in summary that, with the theory of the gnoses and the praxes, clinical neurological studies have a new potential in analyzing pathological, nervous and mental phenomena. In physiology, the fact that we have once and for all closed the books on psychophysiological dualism is equally important, replacing the old concept of sensory areas with those of gnosic functions. Thus, we establish a real and reasonable physiological meaning for the puzzling frontal lobe.

The complete mental process results from the assimilating, energetic gnosic-praxic condensation, and the 'idea of the pencil' is equal to the correlation of the gnosic and praxic dynamics of that object elaborated in the cortex; this is a totally new fact in psychology, which has been unaware of the importance of praxic factors in ideative elaboration.

Springer

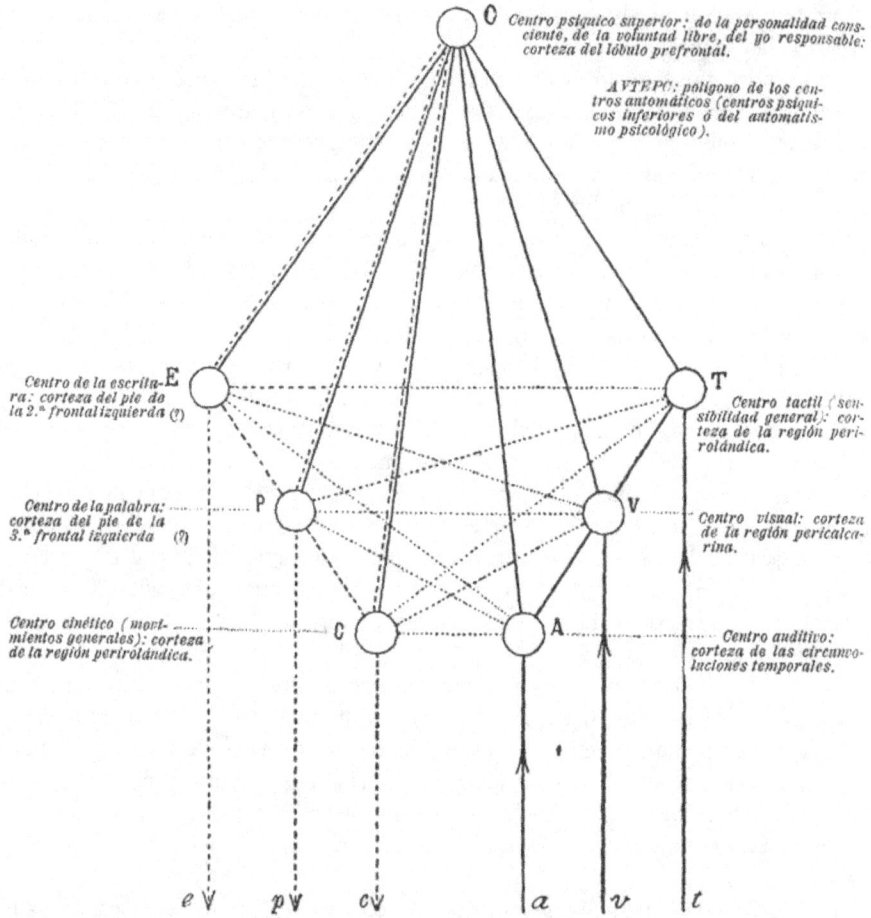

Fig. 4 General scheme of the higher anatomical centers (inferior and superior 'psychical' centers) according to Grasset (1912, p. 68). Abbreviations: *O*, superior psychical center of conscious personality, free will and responsible ego (prefrontal cortex); *AVTEPC*, polygon of automatic centers (inferior psychical centers) or of psychological automatism; *A*, auditory center (cortex of temporal gyri); *V*, visual center (cortex of pericalcarine region); *T*, tactile center or general sensitivity (cortex of perirolandic region); *C*, kinetic center or general movements (cortex of perirolandic region); *P*, language center (cortex of foot of left inferior frontal gyrus); *E*, writing center (cortex of foot of left middle frontal gyrus)

Gnoses and praxes are then neither sensory nor motor, but concomitantly sensory-motor processes and their a priori connection with the functions of the pyramidal tract give us the possibility of satisfactorily explaining the passage of gnosis and praxis to the definitive voluntary movement; a passage that the old physiology, and psychology to a lesser extent, had never been able to explain.

Mental functions cannot have for that reason localizations determined in 'associative areas' of first or third order. Rather, their characteristics reside in the transcortical dynamics that reunite isolated sensory-motor acts. Therefore, the only and true localizable elements of the process are physiological; existing primarily in the gnoses and praxes, they then create the mental phenomenon, the integrative fusion of both dynamics.

 Springer

The intervention of language and its explanation by identical gnosic and praxic processes completes the objective and abstract mental elaboration. I reserve myself for such a study on another occasion, which will give me the opportunity to further deepen into what I have previously established on cortical dynamics, ideas also useful in the physiogenesis of that supreme function of humans.

Regarding the phylogeny of praxic gnosic centers, we are also led to a deeper biological concept by establishing that gnosic centers are much older, and presently exist in all animal species with cerebral cortical matter. On the contrary, praxic cortical dynamics developed much later and appear more extended in higher mammals and especially in primates. This in turn means that the intensity of mental elaboration rests essentially with the praxic components that complete the gnosic cortical product. Such a clue teaches us about human psychology in an incomparable way, with productive (praxic) mentality predominating over the merely representative (gnosic) mentality.

Discussion

The idea of studying pathology to shed light on normal brain function was not new when Jakob (1921) worked on agnosias and apraxias. His ingenuity rests on the combination of different approaches in formulating an integrated theory of cortical dynamics.

From an evolutionary perspective, Jakob conveyed the idea, still valid today, that productive mentality derives from the frontal lobes. In contemporary terms, 'Homo sapiens, knowing man, is issued from Homo habilis, handy man' (de Duve 2002, p. 192). Jakob pointed out that this brain region evolved and expanded in a unique way in primates: 'The great development of the frontal lobes is typical of the brain of primates and in no way an exclusively human characteristic' (Jakob 1943, p. 89). Jakob's extensive studies on human brains and over 100 species of the Patagonian fauna helped him propose a theory of cortical phylogeny (Jakob 1912a, b; Triarhou 2010). The fact that humans and the great apes share a large frontal cortex is backed by modern research. The possibility of a parallel functional reorganization of this region may account for the special cognitive abilities that distinguish primates from other species (Semendeferi et al. 2002). The evolutionarily older gnosic centers are thought to reside in the postcentral 'microdynamics' (Capizzano 2006).

Concerning the ontogeny of gnoses and praxes, Jakob placed their development in infancy and childhood. Thus, one herein encounters a striking similarity between Piagetian and 'Jakobian' concepts. The term 'assimilation' was introduced by Piaget (1952, p. 6) to describe 'structuring through incorporation of external reality into forms due to the subject's activity'. Nonetheless, an early use of the term appears in Jakob's 1921 article, being subsequently refined (Jakob 1935a, 1945a, b), to imply the process of changing of qualities, modalities and relations through which the individual incorporates the external and internal world of objects, processes and situations.

Jakob recorded in detail the various types of gnoses, including tactile, thermal, tactile-muscular (stereognoses), visual, auditory and olfactory, each one being further classified into subtypes. For example, labial, lingual and digital gnoses fall into the

 Springer

category of tactile gnoses. Accordingly, Jakob held the view that gnosic processes are accompanied by modalities, a concept close to the modern interpretation of agnosia, which is considered a modality-specific inability to access semantic knowledge of an object or any other stimulus which cannot be attributed to an impairment of basic perceptual processes (Greene 2005).

With regard to localization theory, Jakob defended the existence of Broca's area from an anatomical-clinical standpoint (Jakob 1906; Tsapkini et al. 2008). He argued that every gnosic and praxic mechanism comprises localizable elements, such as receptors, assimilators and effectors (Jakob 1921); however, he considered the strict localization of mental function or dysfunction as misleading. By attributing apraxias to the disturbance of transcortical dynamics, Jakob highlighted the role of cortical communication. He paralleled neurocognitive functions to electrical current: 'it is only possible to localize the source' (Jakob 1941). Carl Wernicke (1848–1905) opposed the localization of higher functions to specific regions as well, stressing the importance of association areas (Catani and ffytche 2005) and claimed that apraxia results from the separation of brain regions (Finger 1994). Wernicke's line of thought influenced Heinrich Lissauer (1861–1891), an assistant at the Breslau Psychiatric Clinic (Shallice and Jackson 1988). In his 1890 paper, Lissauer subdivided visual agnosia into two subtypes, 'apperceptive' and 'associative' (Lissauer 1890; Lissauer and Jackson 1988). Such a distinction is considered to be the most influential in the history of research in agnosia (Shallice and Jackson 1988). Apperceptive agnosia is accompanied by impaired object recognition due to deficits in perceptual processing, whereas in association agnosia the primary deficit lies in difficulties in accessing the relevant knowledge about objects from memory (Eysenck 2004, p. 251). Jakob seems to agree with Lissauer's work when he refers to a dual premise for the construction of a complete gnosis: 'apperceptive condensation and mnemonic correlation'. Associationist models produced disconnectionist accounts of disorders of higher functions. Liepmann's apraxia model and Déjerine's pure alexia description fall into this tradition, which was revived with Geschwind's neo-associationism. Geschwind (1965a, b) attributed higher function deficits to disconnections that result either from white matter lesions or lesions of association areas, whereas, more recently, Catani and ffytche (2005) updated that model into a hodotopic framework.

Jakob (1921) argued that 'gnoses and praxes are neither sensory nor motor, but concomitantly sensory-motor processes'. Similarly, the idea of 'occasionally fluid boundaries' between agnosia and apraxia has been developed by several authors (Lange 1988, p. 176). In his renowned work 'Matter and memory' (Bergson 1896), the French philosopher Henri Bergson (1859–1941) argued that it is impossible to define where perception ends and movement begins (Blumen and Blumen 2002). Contemporary researchers seem to agree with this view; limb apraxias are considered higher-order disorders of sensory-motor integration (Leiguarda and Marsden 2000). Since apraxia is viewed as a type of motor agnosia, Jakob (1921) aptly notes that 'it is impossible for the patient to integrate sensory information'.

An interesting point in Jakob's work concerns the localization of sensory and associative functions in relation to cortical plasticity. Jakob clearly rejected the separation of the cerebral cortex into independent projection and association areas (Jakob 1912a, b; Triarhou 2010). Specifically, he claimed that there are no special associative centers apart from sensory areas, where a stimulus is both perceived and revived, thus arguing against Cajal's hypothesis of 'mnemonic centers' (Azmitia 2007), namely, a three-order system of neural networks that subserve associative functions.

Jakob explained the compensatory functions of the cerebral cortex by arguing that a functional take-over is only possible whenever brain regions show a certain equipotentiality as far as the elaboration of modality-specific stimuli is concerned. The experimental data gathered from the advent of sophisticated imaging methods lend credence to Jakob's reasoning: for example, Grafman (2000) attributes primary and secondary functional assignments to cortical regions; secondary functional assignments are inhibited until the normally responsible area suffers a damage that renders necessary the activation of the backup region. Neuroimaging techniques further shed light to cases of cross-modal plasticity (Fujii et al. 2009; Sadato 2005). In view of those considerations, one may understand how Jakob's theorizing ability compensated for the technical limitations in his times.

Jakob rejected mechanistic concepts and adopted a dynamic approach in explaining cortical function; he argued for an active exchange between external environment and the adaptive brain. Such a dynamic approach first appeared in his 1918 paper 'From the mechanism to the dynamics of the mind: A critical historical study of organic psychology' (Jakob 1918). Although dynamic concepts prevailed in physics at that time, it took a while for such ideas to be applied to brain theory. York (2009) contends that an era's broader historical, political and cultural framework is reflected in scientific trends. Theoretical dynamic approaches became popular in many fields after the 1940s. To our knowledge, the first reference to neurodynamics is attributed to Trigant Burrow (1943). Influenced by computer science, modern theories use the metaphor of cognition as a dynamic system sustained on spatiotemporal topology (Ibañez and Cosmelli 2008). Wiener's (1948) critical work in cybernetics opened up new vistas (François 1999). The Chilean neurobiologist Francisco Varela (1946–2001), a herald of modern brain dynamics and cybernetics, argued against 'brain-bound neural events' that constitute the mind (Rudrauf et al. 2003); he supported the view that 'consciousness depends crucially on the manner in which brain dynamics are embedded in the somatic and environmental context of an animal's life' (Thompson and Varela 2001). Such trends are compatible with Jakob's views: López Pasquali (1965) underlines that Jakob's work seems to anticipate cybernetics in certain aspects.

Scientists today highlight the importance of the study of praxes in relation to (a) the localization of function, (b) hemispheric potential and (c) the ability of the brain to compensate for injury (Goldmann-Gross and Grossman 2008). Jakob addressed these problems and formulated an integrative theory on the function of gnosio-praxic systems. Some of his views may share commonalities with Wernicke and Lissauer. A point worth emphasizing is the multi-level approach that Jakob adopted, by combining anatomo-functional and phylo-ontogenetic data.

Acknowledgments The authors gratefully acknowledge the anonymous reviewers for their constructive criticism which led to an improved manuscript, and the courtesy of the staff at the National Library of Medicine of the United States, the Bibliotheek van de Universiteit van Amsterdam, and the Ibero-Amerikanisches Institut Preussischer Kulturbesitz zu Berlin.

 Springer

References

Azmitia, E. C. (2007). Cajal and brain plasticity: insights relevant to emerging concepts of mind. *Brain Research Reviews, 55*, 395–405.

Bergson, H. (1896). *Matière et memoire: essai sur la relation du corps à l'esprit.* Paris: Félix Alcan.

Blumen, S. C., & Blumen, N. (2002). From the philosophy auditorium to the neurophysiology laboratory and back: from Bergson to Damasio. *The Israel Medical Association Journal, 4*, 163–165.

Burrow, T. (1943). The neurodynamics of behavior. A phylobiological foreword. *Philosophy of Science, 10*, 271–288.

Capizzano, A. (2006). Actualidad del pensamiento de Cristofredo Jakob. *Revista del Hospital Italiano de Buenos Aires, 26*, 71–73.

Catani, M., & ffytche, D. H. (2005). The rises and falls of disconnection syndromes. *Brain, 128*, 2224–2239.

Cockburn, J. (2008). Stroke. In B. Woods & L. Clare (Eds.), *Handbook of the clinical psychology of ageing* (pp. 201–218). West Sussex: John Wiley and Sons Ltd.

de Duve, C. (2002). *Life evolving: Molecules, mind, and meaning.* New York: Oxford University Press.

Devinsky, O., & D'Esposito, M. (2004). *Neurology of cognitive and behavioral disorders.* New York: Oxford University Press.

Etcharry-Bouyx, F., & Ceccaldi, M. (2007). Gestural apraxia. In O. Godefroy & J. Bogousslavsky (Eds.), *The behavioral and cognitive neurology of stroke* (pp. 36–52). Cambridge: Cambridge University Press.

Eysenck, M. W. (2004). *Psychology: An international perspective.* Hove: Psychology.

Finger, S. (1994). *Origins of neuroscience: A history of explorations into brain function.* New York: Oxford University Press.

François, C. (1999). Systemics and cybernetics in a historical perspective. *Systems Research and Behavioral Science, 16*, 203–219.

Freud, S. (1891). *Zur Auffassung der Aphasien—Eine kritische Studie.* Leipzig–Wien: Franz Deuticke.

Freud, S. (1955). The psychogenesis of a case of homosexuality in a woman (Translation by J. Strachey, A. Freud, A. Strachey and A. Tyson of *Über die Psychogenese eines Falles von weiblicher Homosexualität* [1920]). In: J. Strachey (Ed.), *The Standard Edition of the Complete Psychological Works of Sigmund Freud, Volume XVIII (1920–1922): Beyond the Pleasure Principle, Group Psychology and Other Works* (pp. 145–172). London: The Hogarth Press and the Institute of Psycho-Analysis.

Fujii, T., Tanabe, H. C., Kochiyama, T., & Sadato, N. (2009). An investigation of cross-modal plasticity of effective connectivity in the blind by dynamic causal modeling of functional MRI data. *Neuroscience Research, 65*, 175–186.

Geschwind, N. (1965a). Disconnexion syndromes in animals and man. Part I. *Brain, 88*, 237–294.

Geschwind, N. (1965b). Disconnexion syndromes in animals and man. Part II. *Brain, 88*, 585–644.

Goldberg, E. (2005). *The wisdom paradox.* New York: Gotham Books.

Goldmann-Gross, R. G., & Grossman, M. (2008). Update on apraxia. *Current Neurology and Neuroscience Reports, 8*, 490–496.

Grafman, J. (2000). Evidence for forms of neuroplasticity. *Journal of Communication Disorders, 33*, 345–356.

Grasset, J. (1912). *Tratado de fisiopatología clínica. III. Neurobiología, ontogenia y filogenia, herencia.* Barcelona: Salvat y Compañía.

Greene, J. D. W. (2005). Apraxia, agnosias, and higher visual function abnormalities. *Journal of Neurology, Neurosurgery and Psychiatry, 76*(Suppl 5), 25–34.

Ibañez, A., & Cosmelli, D. (2008). Moving beyond computational cognitivism: Understanding intentionality, intersubjectivity and ecology of mind. *Integrative Psychological & Behavioral Science, 42*, 129–136.

Jakob, C. (1906). Existe ó no un centro de Broca? *La Semana Médica (Buenos Aires), 13*, 677–678.

Jakob, C. (1912a). Über die Ubiquität der senso-motorischen Doppelfunktion der Hirnrinde als Grundlage einer neuen, biologischen Auffassung des corticalen Seelenorgans. *Journal für Psychologie und Neurologie (Leipzig), 19*, 379–382.

Jakob, C. (1912b). Über die Ubiquität der senso-motorischen Doppelfunktion der Hirnrinde als Grundlage einer neuen biologischen Auffassung des kortikalen Seelenorgans. *Münchener Medizinische Wochenschrift, 59*, 466–468.

Jakob, C. (1913). La psicología orgánica y su relación con la biología cortical. *Archivos de Psiquiatría. Criminología y Ciencias Afines (Buenos Aires), 12*, 680–698.

 Springer

Jakob, C. (1914). Los problemas biogenéticos en sus relaciones con la filosofía moderna. *Revista del Círculo Médico Argentino y Centro Estudiantes de Medicina (Buenos Aires), 14*, 87–98.

Jakob, C. (1918). Del mecanismo al dinamismo del pensamiento: Estudio histórico-crítico de psicología orgánica. *Anales de la Facultad de Derecho y Ciencias Sociales de la Universidad de Buenos Aires, 18*, 195–238.

Jakob, C. (1919). La teoría actual de las gnosias y praxias como factores fundamentales en el dinamismo cortical. *Revista del Círculo Médico Argentino y Centro Estudiantes de Medicina (Buenos Aires), 19*, 1266–1275.

Jakob, C. (1921). La teoría actual de las "gnosias y praxias" como factores fundamentales en el dinamismo de la corteza cerebral. *La Crónica Médica, 38*, 17–24.

Jakob, C. (1923). *Elementos de neurobiología*. La Plata: Biblioteca Humanidades.

Jakob, C. (1935a). Sobre las bases orgánicas de la memoria. *Revista de Criminología, Psiquiatría y Medicina Legal (Buenos Aires), 127*, 84–114.

Jakob, C. (1935b). La filogenia de las kinesias: sobre su organización y dinamismo evolutivo. *Anales del Instituto de Psicología de la Facultad de Filosofía y Letras de la Universidad de Buenos Aires, 1*, 109–127.

Jakob, C. (1941). La función psicogenética de la corteza cerebral y su posible localización (Aspectos de la ontopsicogénesis humana). *Anales del Instituto de Psicología de la Facultad de Filosofía y Letras de la Universidad de Buenos Aires, 3*, 63–80.

Jakob, C. (1943). *Folia neurobiológica Argentina, tomo III. El lóbulo frontal: Estudio monográfico anatomoclínico sobre base neurobiológica*. Buenos Aires: Aniceto López–López y Etchegoyen.

Jakob, C. (1945a). El cerebro humano: su significación filosófica. *Revista Neurológica de Buenos Aires, 10*, 89–110.

Jakob, C. (1945b). Sobre el origen de la conciencia: investigaciones neurobiológicas sobre la dinámica cortical en relación con su sectorización conmemorativa. In E. Mouchet (Ed.), *Temas actuales de psicología normal y patológica, publicado bajo el patrocinio de la Sociedad de Psicología de Buenos Aires* (pp. 345–381). Buenos Aires: Editorial Médico-Quirúrgica/Talleres 'The Standard'.

Jakob, C. (1946). *Folia neurobiológica Argentina, tomo V. Documenta biofilosófica: Folleto I: Biología y filosofía A. Aspectos de sus divergencias y concomitancias; B. Ensayo de psicogenia orgánica*. Buenos Aires: López & Etchegoyen.

Jung, C. G. (1960). *The psychogenesis of mental disease* (Translation by R. F. C. Hull of *Zur Psychogenese der Geisteskrankheiten* [1906]). *Volume 3 of the Collected Works—Bollingen Series XX*. New York: Pantheon Books.

Kant, I. (1999). *Critique of pure reason* (Translation by P. Guyer & A. W. Wood of *Kritik der reinen Vernunft* [1781]) (1999th ed.). Cambridge: Cambridge University Press.

Lange, J. (1988). Agnosia and apraxia (Translation by G. Dean, E. Perecman & J. W. Brown of *Agnosie und Apraxie* [1936]). In J. W. Brown (Ed.), *Agnosia and apraxia: Selected papers of Liepmann, Lange, and Pötzl* (pp. 43–226). Hillsdale: Lawrence Associates.

Leiguarda, R. C., & Marsden, C. D. (2000). Limb apraxias: higher-order disorders of sensorimotor integration. *Brain, 123*, 860–879.

Liepmann, H. (1988). Apraxia (Translation by G. Dean & E. Franzen of *Apraxie* [1920]). In J. W. Brown (Ed.), *Agnosia and apraxia: Selected papers of Liepmann, Lange, and Pötzl* (pp. 3–39). Hillsdale: Lawrence Erlbaum Associates.

Lissauer, H. (1890). Ein Fall von Seelenblindheit nebst einem Beitrage zur Theorie derselben. *Archiv für Psychiatrie und Nervenkrankheiten, 21*, 222–270.

Lissauer, H., & Jackson, M. (1988). A case of visual agnosia with a contribution to theory. *Cognitive Neuropsychology, 5*, 157–192.

López Pasquali, L. (1965). *Christfried Jakob—Su obra neurológica, su pensamiento psicológico y filosófico*. Buenos Aires: López Libreros Editores S.R.L.

Macmillan, M. (2004). "I could see, and yet, mon, I could na' see": William MacEwen, the agnosias, and brain surgery. *Brain and Cognition, 56*, 63–76.

Moyano, B. A. (1957). Christfried Jakob, 25/12/1866–6/5/1956. *Acta Neuropsiquiátrica Argentina, 3*, 109–123.

Orlando, J. C. (1966). *Christofredo Jakob—Su vida y obra*. Buenos Aires: Editorial Mundi.

Pedace, E. A. (1949). Contribución de la escuela neurobiológica Argentina del Prof. Chr. Jakob en el estudio del lóbulo frontal. *Archivos de Neurocirugía, 6*, 464–466.

Piaget, J. (1952). *The origins of intelligence in children* (Translation by M. Cook of *La naissance de l'intelligence chez l'enfant, 2ème édition* [1948]). New York: International Universities Press.

 Springer

Piaget, J. (1972). *The principles of genetic epistemology* (Translation by W. Mays of *Introduction à l'épistémologie génétique* [1950]). London: Routledge and Kegan Paul.

Rudrauf, D., Lutz, A., Cosmelli, D., Lachaux, J.-P., & Le Van Quyen, M. (2003). From autopoiesis to neurophenomenology: Francisco Varela's exploration of the biophysics of being. *Biological Research, 36*, 27–66.

Sadato, N. (2005). How the blind 'see' Braille: lessons from functional magnetic resonance imaging. *The Neuroscientist, 11*, 577–582.

Semendeferi, K., Lu, A., Schenker, N., & Damasio, H. (2002). Humans and great apes share a large frontal cortex. *Nature Neuroscience, 5*, 272–276.

Shallice, T., & Jackson, M. (1988). Lissauer on agnosia. *Cognitive Neuropsychology, 5*, 153–156.

Szirko, M. (1995). A la antropología ganglionar desde la kinesiología: un fallido ensayo de extrapolar lo orgánico. *Electroneurobiología, 2*, 101–191.

Thompson, E., & Varela, F. J. (2001). Radical embodiment: neural dynamics and consciousness. *Trends in Cognitive Sciences, 5*, 418–425.

Triarhou, L. C. (2008). Centenary of Christfried Jakob's discovery of the visceral brain: an unheeded precedence in affective neuroscience. *Neuroscience and Biobehavioral Reviews, 32*, 984–1000.

Triarhou, L. C. (2010). Revisiting Christfried Jakob's concept of the dual onto-phylogenetic origin and ubiquitous function of the cerebral cortex: a century of progress. *Brain Structure and Function, 214*, 319–338.

Triarhou, L. C., & del Cerro, M. (2006a). Semicentennial tribute to the ingenious neurobiologist Christfried Jakob (1866–1956). 1. Works from Germany and the first Argentina period, 1891–1913. *European Neurology, 56*, 176–188.

Triarhou, L. C., & del Cerro, M. (2006b). Semicentennial tribute to the ingenious neurobiologist Christfried Jakob (1866–1956). 2. Publications from the second Argentina period, 1913–1949. *European Neurology, 56*, 189–198.

Triarhou, L. C., & del Cerro, M. (2007). Pioneers in neurology: Christfried Jakob (1866–1956). *Journal of Neurology, 254*, 124–125.

Tsapkini, K., Vivas, A. B., & Triarhou, L. C. (2008). 'Does Broca's area exist?' Christofredo Jakob's 1906 response to Pierre Marie's holistic stance. *Brain and Language, 105*, 211–219.

Wiener, N. (1948). *Cybernetics or control and communication in the animal an the machine.* Paris: Hermann.

York, G. K., III. (2009). Localization of language function in the twentieth century. *Journal of the History of the Neurosciences, 18*, 283–290.

Zoë D. Théodoridou holds BA and MA degrees in Educational Policy from the University of Macedonia, Thessaloniki, Greece, where she is currently pursuing a doctorate in neuroeducation. She works as a special education teacher at the Second Elementary School in Chalastra, Greece.

Lazaros C. Triarhou is Professor of Neuroscience at the University of Macedonia, Thessaloniki, Greece. He obtained his MD from Aristotelian University (Greece), MSc from the University of Rochester, New York, and PhD from Indiana University. His research interests are centered on the evolution of ideas in neurobiology, mainly after the laboratory revolution.

🖄 Springer

Brain and Cognition 78 (2012) 179–188

Contents lists available at SciVerse ScienceDirect

Brain and Cognition

journal homepage: www.elsevier.com/locate/b&c

Theoretical Integration

Christfried Jakob's late views (1930–1949) on the psychogenetic function of the cerebral cortex and its localization: Culmination of the neurophilosophical thought of a keen brain observer

Zoë D. Théodoridou, Lazaros C. Triarhou *

Economo-Koskinas Wing for Integrative and Evolutionary Neuroscience, Department of Educational and Social Policy, University of Macedonia, 54006 Thessaloniki, Greece

ARTICLE INFO

Article history:
Accepted 14 November 2011
Available online 30 January 2012

Keywords:
Neurophilosophy
Christfried Jakob
Consciousness
History of neuroscience

ABSTRACT

This article follows the culmination of the scientific thought of the neurobiologist Christfried Jakob (1866–1956) during the later part of his career, based on publications from 1930 to 1949, when he was between 64 and 83 years of age. Jakob emphasized the necessity of bridging philosophy to the biological sciences, neurobiology in particular. Thus, we consider him as one of the early protagonists in the emergence of neurophilosophy in the 20th century. The topics that occupied his mind were the foundations for a future philosophy of the brain, and the 'neurobiogenetic', 'neurodynamic', and 'neuropsychogenetic' problems in relation to how consciousness emerges. Jakob's views have many elements in common with great thinkers of philosophy and psychology, including Immanuel Kant, William James, Edmund Husserl, Henri Bergson, Jean Piaget and Willard Quine. A common denominator can also be discerned between Jakob's dynamic approach and certain aspects of cybernetics and neurophenomenology. Jakob propounded the interdisciplinarity of sciences as an indispensable tool for ultimately solving the enigma of consciousness.

© 2011 Elsevier Inc. All rights reserved.

1. Introduction

With the progress effected in the brain sciences over the past 20 years, traditional philosophical questions have been steered into new directions (Churchland, 2008). Thus, the field of consciousness studies has been opened up to a growing body of biologists, neuroscientists, psychologists and philosophers (Blackmore, 2005). The investigation of philosophical theories in relation to neuroscientific hypotheses falls within the 'modern' domain of neurophilosophy (Northoff, 2004). Formalized by Churchland (1986), the term 'neurophilosophy' denotes the interdisciplinary attempt at unifying cognitive neurobiology. In the years following its foundation, neurophilosophy has grown exponentially. Its main theses have centered around: (a) psychological and neuroscientific theories, as well as intertheoretical relationships; (b) the opposition to the autonomy of either psychology or functionalism alone; and (c) a trend of rendering the cognitive neurosciences accessible to and comprehensible by a broader audience (Bickle, 2009, p. 3).

On the other hand, philosophy of neuroscience has gradually become a distinguishable field reflecting "an inquiry into foundational (especially epistemic and metaphysical) questions that apply to neuroscience" (Bechtel, 2001, p. 7). Such questions can be approached either descriptively, i.e. by depicting how neuroscience proceeds, or normatively, i.e. by implying how neuroscience should proceed (Bickle, Mandik, & Landreth, 2010).

The fluidity of the boundaries between neurophilosophy and the philosophy of neuroscience has led Brook and Mandik (2007) and Bickle (2009) to use the term 'Philosophy and Neuroscience', which entails ongoing transdisciplinary interactions, accommodating both endeavors. That is the definition we adopt in the present article.

Not until recently have philosophers started paying close attention to the data provided by the neurosciences. A few exceptions prior to the 1980s include the work of Nagel (1971), von Eckardt-Klein (1975), and Dennett (1978) as pointed out by Brook and Mandik (2007). The establishment and dissemination of reductionistic approaches in the 20th century, prompted to a great extent by Jacques Loeb (1859–1924) and Ivan P. Pavlov (1849–1936), relegated consciousness studies to philosophy, mysticism or 'soft' science, thus diminishing the influence of more integrated contemporary approaches, such as those of Sherrington and Lashey (Greenspan & Baars, 2005). As behaviorists were reacting to the earlier introspection—as exemplified e.g. in the thought of Wundt, James and Freud—with a desire for objectivity, they devised animal experiments, resolutely leaving the human mind out of the picture;

* Corresponding author. Address: University of Macedonia, Egnatia 156, Bldg. Z-312, 54006 Thessaloniki, Greece. Fax: +30 2310891388.
E-mail addresses: zoitheo@uom.gr (Z.D. Théodoridou), triarhou@uom.gr (L.C. Triarhou).

0278-2626/$ - see front matter © 2011 Elsevier Inc. All rights reserved.
doi:10.1016/j.bandc.2011.11.005

only around the middle of the 20th century did psychological theorizing swing back to studying the mind in the realm of cognitive psychology (Ochs, 2004, p. 356).

About a century ago, Christfried Jakob (1866–1956), a neurobiologist with an extraordinary scope of interests, put consciousness under scientific scrutiny, arguing that philosophy should be linked to the biological sciences (Triarhou & del Cerro, 2006b).

The son of Godofredo and Babette (née Körber) Jakob, Christfried Jakob was born on December 25, 1866 in Bavaria. His father, a cultivated teacher, recognized and encouraged Christfried's inclination in the natural sciences (Orlando, 1966). Jakob graduated in medicine in 1890 from the University of Erlangen with a prize of 1000 DEM, offered to the most distinguished student (Moyano, 1957). He next carried out his doctorate under the supervision of Friedrich Albert von Zenker (1825–1898), studying aortitis syphilitica (Triarhou & del Cerro, 2007). In the early 1890s he worked as an assistant to Adolf von Strümpell at the Erlangen Medical Clinic and privately practised medicine in Bamberg (Orlando, 1966). Jakob made a name for himself through his first brain atlas (Jakob, 1895; Jakob, 1899), which was translated into several languages (for details, see Triarhou & del Cerro, 2006a).

In 1899, Jakob accepted an offer from Domingo Cabred (1859–1929), the Argentinian Professor of Psychiatry, to direct the Laboratory of the Psychiatric and Neurological Clinic of the Hospital of Mercedes at the National University of Buenos Aires (Orlando, 1966). He went to Argentina having signed a three-year contract. One determining factor in his decision to leave Europe was the prospect of having 300 brains available for pathological study annually (López Pasquali, 1965). At that time, the Argentine population as well as the country's economy grew fervently as a result of immigration and a decreasing mortality (Véganzonès & Winograd, 1997), as Argentina was emerging as one of the ten richest countries in the world. By 1910, Jakob had produced critical works in anatomy, neurology, psychopathology and anthropology (Triarhou & del Cerro, 2007).

When his Argentinian contract expired in 1910, Jakob returned to Germany, where he promoted his original idea on the ubiquity of the dual sensory-motor function of the cerebral cortex (Jakob, 1911, 1912a, 1912b; Triarhou, 2010b). The works of his 'early' period (1890–1912) mostly centered around neuroanatomy, reflecting Jakob's training in the 'German school' (López Pasquali, 1965). His anatomical thinking during that early period has been presented elsewhere (Théodoridou & Triarhou, 2012).

The 'middle' period of Jakob's work (1913–1935) began with his permanent move to Argentina, where he assumed clinical, research and teaching duties. He was appointed Chief of the Neuropathological Institute at the National Psychiatric Hospital for Women in the Federal Capital, and Professor and Director of the Institute of Biology at the Faculty of Philosophy and Letters of the National University of La Plata. In 1922, Jakob was appointed Professor of Neurobiology at the Faculty of Humanities and Educational Sciences of the National University of La Plata. From 1921 to 1933, he held a joint appointment as Professor of Pathological Anatomy at the School of Medical Sciences of La Plata (Triarhou & del Cerro, 2006b).

The development of Jakob's 'dynamic approach' emerged in a 1918 article entitled 'From the mechanism to the dynamics of the mind: A critical historical study of organic psychology'. In his 1919–1921 work on gnoses and praxes as fundamental factors in cerebral cortical dynamics (Jakob, 1919, 1921) he further built on his original psychobiological ideas. Two subsequent studies (Jakob, 1935a, 1935b) have been considered by his colleague and biographer Moyano (1957) as vital constituents of Jakob's psychobiological theorizing; they were titled, 'On the biological bases of memory' (Jakob, 1935b) and 'The phylogeny of the organization and the evolutionary dynamics of the kineses' (Jakob, 1935a). An

account of the 'middle' period of Jakob's thought has been published as well (Théodoridou & Triarhou, 2011).

In the 'late' phase of his life, Jakob's thought became more synthetic. He blended his philosophical background with the clinical and research experience, maintaining that the utmost problem of science and philosophy converges in cerebral function (Théodoridou & Triarhou, in press). He suggested a scientific psychology and a corpus of philosophy (López Pasquali, 1965).

Jakob retired in 1945 (Orlando, 1966). However, he kept his formal appointment in Buenos Aires as Chairman of Pathological Anatomy and continued to work in his laboratory at the National Psychiatric Hospital for Women until 1954; he died in Buenos Aires in 1956 at the age of 90 (Triarhou & del Cerro, 2006a, 2006b, 2007).

In all, Jakob authored 30 monographs and 200 papers that cover developmental, evolutionary, anatomical, pathological and philosophical themes in neurobiology (Triarhou & del Cerro, 2006a). He is viewed as the father of Argentinian neuroscience (Pedace, 1949) and one of the great thinkers of the 20th century. His scientific caliber was such that von Economo and Koskinas (1925) express the view that future research on the cortex would have to be based on the fundamental works of three investigators: Theodor Kaes (1852–1913), Santiago Ramón y Cajal (1852–1934) and Christfried Jakob, further considering Jakob's ideas on cortical phylo-ontogeny as 'ingenious' (Triarhou & del Cerro, 2006a).

The present study examines the culmination of his neurophilosophical thought during the 'late' period of his life and career. Jakob's involvement with philosophy was neither superficial nor based on improvisations (López Pasquali, 1965). Quite the contrary. We consider Jakob as an 'early neurophilosopher' for the following reasons.

Firstly, he was most likely one of the first academics to formally teach neurobiology in a School of Education, at the National University of La Plata, Argentina. Thus, he introduced fundamentals of neuroeducation decades before that discipline was formalized in the current era. Fischer et al. (2007) have defined the fervently growing field of mind, brain and education as "the quest for the integration of disciplines that investigate human learning and development bringing together education, biology, and cognitive science".

Secondly, Jakob's philosophical background is evident throughout his work. It peaked during his late years. Jakob published 20 articles in philosophy and neuroscience, besides philosophical papers on Kant and Descartes (Jakob, 1926, 1937, 1938).

Thirdly, Jakob (1943a) proposed the term 'psychophilosophy' in an imaginative conference of philosophical discussions among a small number of interlocutors ('Conferencia magistral de introducción a la psicofilosofía') that recalls the Socratic dialogs. Jakob (1943a), being the professor alongside six alumni, deals with issues such as the existence and the perception of God by the human mind, the nature of philosophy and science, the scientific basis of psychology, the nature of ideas, and the existence of a priori conditions of our internal intuition.

In the following sections we highlight some of Jakob's most important ideas on philosophy and neuroscience.

2. The diffusion of neurobiological knowledge into philosophy

In the first part of his 'Biophilosophical Documents' ('Documenta Biofilosófica') (Fig. 1), Jakob (1946) presented the following arguments for the necessity of diffusing common and divergent aspects of neurobiology into philosophy:

(A) Issues concerning life, from general aspects of evolution to heredity and the diversity of the human species, form a justified base for an objective, rational and scientific development of the philosophical orientations.

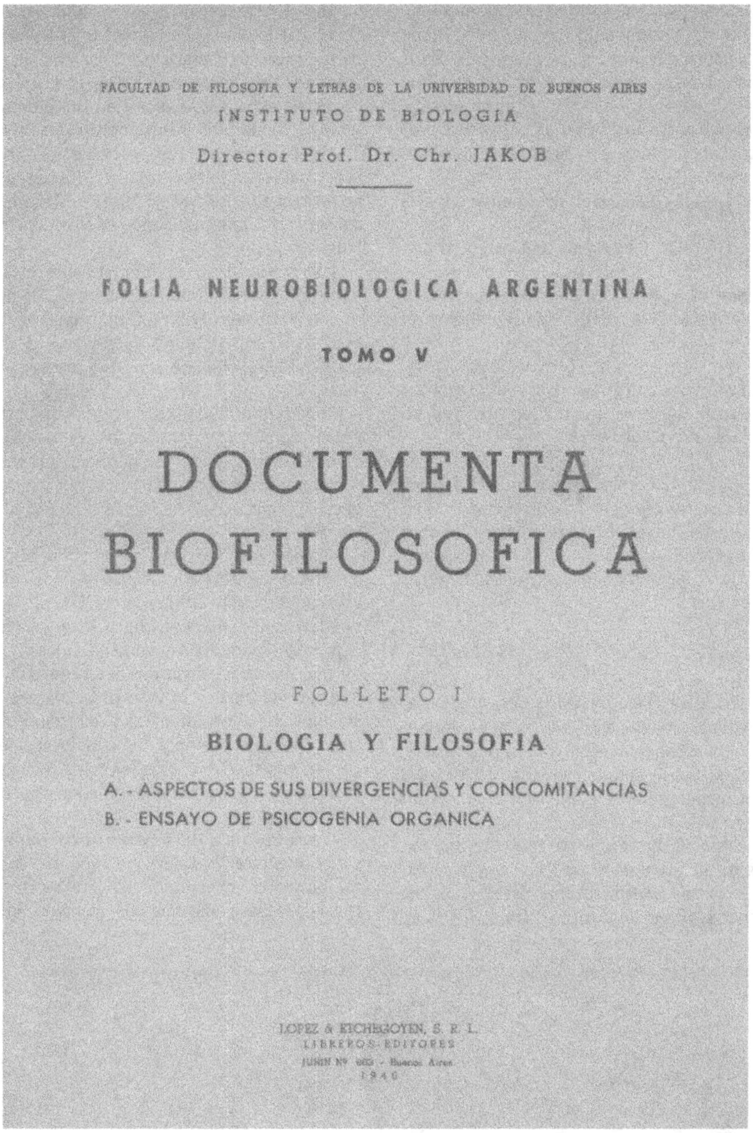

Fig. 1. Frontispiece of Jakob's monograph *Biophilosophical Documents*, being volume 5 of the *Folia Neurobiológica Argentina* series (Jakob, 1946).

(B) The scientific field of neurobiology that studies nervous structure and function (see e.g. Jakob, 1906b) under normal or pathological conditions, both evolutionarily and developmentally, is indispensable for psychology and its related sciences (cf. also Jakob, 1913).

(C) The knowledge of the morphophysiological evolution of the human brain in correlation with psychogenetic maturation, as well as brain alterations and their sequelae on memory, behavior, language and other abstract processes form the natural foundation of a conscious learning science.

(D) The creation and the preservation of higher cognitive functions (intellect, volition and emotions), instincts and reflexes depend on our cerebral organization.

Jakob (1946) argued that philosophical reasoning consists in elaborations stemming from a germ cell. Therefore, neurobiology should provide the organic basis of epistemology, logic, phenomenology, axiology, ethics, esthetics and metaphysics.

In that respect, Jakob (1945a) maintained that sciences dealing with the empirically accessible reality and philosophy which examines the possibilities that arise beyond experience need a philosopher who would above all master the former in order to treat the latter with composure.

3. Foundations for a future philosophy of the brain

Jakob (1945b) considered himself "a groundworker of a biocentric epistemology" ('*Como preliminares de una gnoseología biocéntri-*

ca...'). He described the theoretical background and main points that a future philosophy of the brain would have to treat. Along this line, he suggested that such a future discipline should consist of a synthesis of proven as universally valid neuro-psycho-dynamic theories concerning: (a) a universal, central organization; (b) heredity; and (c) the evocation and transformation of physical processes into psychological phenomena by means of neurohistological and physiological processes (Jakob, 1945b).

In this endeavor, Jakob (1945b) considered the following issues of primary importance:

(A) The laws that govern cerebral phylogeny and ontogeny and their stages.
(B) The microoganization of neuroblasts and the dynamics of their functional derivatives in normal and pathological conditions.
(C) The polyenergetic transformation at cosmological, biological, neurological and psychological levels in the integrative creation of the external (objective) and the internal (subjective) environment (see also Jakob, 1920).

Moreover, Jakob (1945b) argued that a philosophy of brain should shed light on the 'neurobiogenetic', 'neurodynamic', and 'neuropsychogenetic' problems. The first of these problems implies the various stages of the biological development of the nervous system, which is beyond the scope of the present article. The other two problems are detailed next.

4. The neurodynamic problem

In his neurodynamic postulate, (Jakob, 1921, 1935a; Théodoridou & Triarhou, 2011) explores psychogenesis from an evolutionary perspective (Fig. 2), with phylogeny occurring in two phases. The first or 'plasmodynamic' phase entails elementary biological phenomena such as tropism and pulsatility. The second or 'neurodynamic' phase is divided into three stages that correspond to different levels of organization in the vertebrate CNS: the 'archikinetic' stage corresponds to the archineuronal level; the 'paleokinetic' stage corresponds to the paleoneuronal level; and the 'neokinetic' stage corresponds to the neoneuronal level.

The phylogenetically older, archikinetic stage entails reflex actions that lead to the emergence of 'archipsychism'. The paleokinetic stage is characterized by the appearance of instinctive reactions that constitute 'paleopsychism'. The neokinetic stage is responsible for the elaboration of conscious responses, in other words it underpins 'neopsychism'. Neopsychism comprises three kinds of neurocognitive processes: (a) *gnoses*, which secure the conscious orientation in one's environment; (b) *praxes*, which underlie active individual intervention; and (c) *symbolisms*, which subserve the communication of abstract ideas by means of human language.

Jakob held that the notion of space results from the direction of a movement, the notion of time from its duration and, finally, the notion of causality from its intensity. From the angle of psychogenesis, space and time are deeply rooted psychogenetically in the gnosic sphere, whereas causality rests in the praxic sphere (Jakob, 1945b).

Topographically, gnoses are mostly positioned in retrorolandic regions, precisely in parieto–occipito–temporal regions. However, for the construction of the notion of a concrete object the dynamic collaboration of all cortical sectors (Fig. 3) is indispensable. Besides, gnoses maintain their sensory-motor character. Jakob (1921) wrote:

"Gnoses and praxes are then neither sensory nor motor, but concomitantly sensory-motor processes and their a priori connection with the functions of the pyramidal tract give us the possibility of satisfactorily explaining the passage of gnosis and praxis to the definitive voluntary movement...Mental functions cannot have for that reason localizations determined in such-and-such associative areas of the first or third order. Rather, their characteristics reside in transcortical dynamics that reunite isolated sensory-motor acts. Therefore, the only and true localizable elements of the process are physiological, existing primarily in gnoses and praxes; they then create the psychic phenomenon, the integrative fusion of both dynamics".

Gnoses express the reconstructive thinking and they are necessarily orientated in space and time. The praxic zone is apparent in the brains of lower mammals. Still, it crescents in primates with the appearance of functions of active intervention individually

Fig. 2. *Left:* The edentate 'pichiciego pampeano' *(Chlamyphorus truncatus)*, the pink fairy of the armadillo species also known as *ratoncito cascarudo*, shown at a ½ scale (Jakob, 1943, p. 12). *Right:* Manual praxis in a chimpanzee (Jakob, 1943b, p. 22; photo by Clemente Onelli).

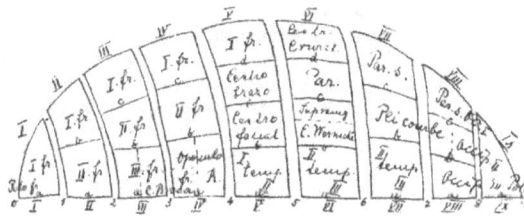

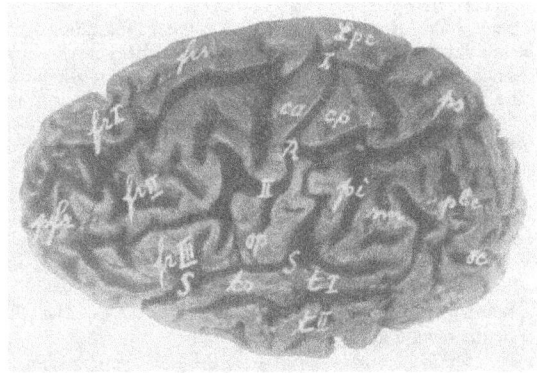

Fig. 3. *Upper:* Jakob's (1906b, p. 363) neuroanatomical procedure for studying hemispheric 'sectors' in the human cerebrum, consisted in defining a line, which he took as a basis, with the brain situated in its normal position, and dividing it through a system of coordinates dictated by the plan of cerebral morphology itself. *Lower:* A human brain with two atypical interruptions of the Rolandic area (x, x), fully encompassing the trigyral type of the hemispheric convexity in the Rolandic region (Jakob, 1941, p. 77). See also Fig. 3 in Théodoridou and Triarhou (2011) for lateral and midsagittal views of the human cerebral hemispheres and the 'golden section' concept with its praxic and gnosic sectors (Jakob, 1943, p. 37).

acquired by means of motor responses (for example, walking and language).

Jakob maintained that a higher mental activity that enables the emergence of the neopsychic sphere is reached only in primates, and especially humans, due to symbolic language, whereby the sphere of inner (visceral) feelings becomes integrated with environmental gnosio-praxic experiences (Triarhou, 2010a). He associated the emergence of conscious processes with the appearance of "the organ of consciousness, the cerebral cortex that allows the individual elaboration of the essential condition of memory" (Jakob, 1945b). Jakob (1945b) argued that the first commemorative sector and therefore creator of something conscious is the Ammonic cortex, termed 'paleocortex'. He attributed the commemorative ability of the Ammonic cortex to the fact that it houses two layers of cortical elements, one receiving stimuli (dentate area) and one effector (Ammonic area), that form reciprocal sets of fixation, residual of elaborated experiences (Jakob, 1945b).

By arguing that only in the human brain the commemorative cortical superiority enables the elevation of comparative thought to abstract reasoning, Jakob highlighted memory as a basic component or prerequisite for conscious processes. In fact, he claimed that the sense of continuity that is given by memory underpins the emergence of the conscious self. Jakob (1941, 1946) tackled the development of such processes in his neuropsychogenetic postulate.

5. The neuropsychogenetic problem

Jakob delved into the neuropsychogenetic problem in his 1941 article entitled: 'The psychogenetic function of the cerebral cortex

and its possible localization: aspects of human ontopsychogenesis'. He used much of the text in the second part of his 'Biophilosophical Documents' (Jakob, 1946) under the title 'An essay on organic psychogenesis'.

Jakob (1941, 1946) theorized that two great worlds, 'like battlefields', create our cortical organ during its ontogeny forming an a priori unit: the external ('*ambiental*') and the internal ('*introyental*') milieu. The external, environmental factors act as stimulating material and the internal as hereditary germ capital, the maturation of which gives birth to the adequate central organic assimilation system. The external and internal domains were considered to enable the elaboration of personal and conscious experiences through a process of "internal-environmental frontalization", whereas the "commemorative accumulation" of such experiences was thought to allow an individual to plan and execute future actions (Barutta, Hodges, Ibáñez, Gleichgerrcht, & Manes, 2010). The endogenous sector is represented on the medial facies of the mammalian cerebral hemispheres, including the cingulate gyrus, and is charged with vegetative-autonomic functions, whereas the exogenous is represented on the convexity of the cerebral hemispheres and serves somatic functions (Triarhou, 2008). Nonetheless, Jakob (1906a, 1906–1908) rejected strict localization, arguing that it is impossible to view consciousness as a localizable, special power separate from processes chaining cortical operations.

The external and the internal milieux differentiate and complement each other via two essential psychogenetic acts, 'somatization', i.e. a course of action that leads to the formation of the position toward the external milieu, and 'sympathization', i.e. the course of action that leads to the formation of the position toward the internal milieu. A somatic act consists in a process of acceptance or rejection accompanied by the corresponding affective intonation. For example, when the infant encounters an obstacle (stimulus), unity becomes divorced: this 'object' will be 'environmental', and the organ that hits against the obstacle with all its neuromuscular organization will be 'internal'. On the other hand, a sympathetic act corresponds to a process of emotional intonation of pleasure or pain. For example, the infant satisfies its hunger by sucking. Milk, along with mother and chest, will be environmental; the tranquilization of the visceral needs, along with all of the glandulo–musculo–neural apparatus, belongs to the internal milieu.

Jakob (1941, 1946) suggested that psychogenesis is effected in three developmental stages—an 'infantile', a 'juvenile', and a 'mature' stage—and eventually leads to the construction of the external milieu and the creation of the somatic ego ("*La psicogenia realiza, en general, en su fase evolutiva, las siguientes tres etapas: una infantil, otra juvenil y uno última de maduración*").

In the first or *infantile* stage, there is a primitive perception of constellations of objects and processes. This stage elaborates the elementary knowledge of experience via cortical macro–microdynamic successions and associations. ("*La primera etapa representa la fijación de situaciones enteras que en sucesión macro–microdinámica cortical y colaboración asociativa primitiva gnoseo-práxica, elaboran las nociones elementales de la experiencia y en donde objeto y proceso están completamente fusionados formando un solo complejo conjunto..., una «situación completa»—constelación.*")

Therefore, the elemental reflex formation is created through a process of cortical 'synergy'. Memory provides accumulated material transferred via cortical elements organized in macro- and micro-dynamic systems (Fig. 4) for the emergence of orientation and intervention processes, i.e. gnoses and praxes. For this to happen, constellations must be transformed into differentiated objects and processes, forming concrete phenomena of the experienced world through a process of comparison and identification in the second or *juvenile* stage. ("*Enlazando y comparando tales «complejos» se separan poco a poco los elementos estacionarios de la situación de los movidos, transformándolos en objeto y proceso y creándose así,*"

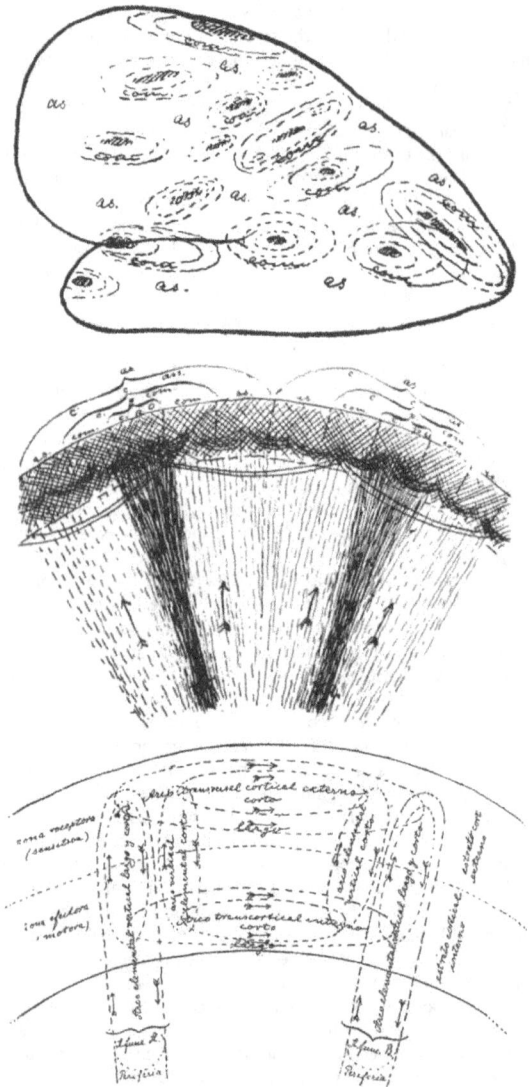

Fig. 4. *Upper:* A schematic drawing of perception areas by Jakob (1906–1908, p. 301), outlining the annular commemorative and the association centers. *Middle:* Schematic drawing by Jakob (1906–1908, p. 304) of two adjacent centers of perception, *v.o.* and *v.a.*, representing the central afferent pathways; within their confines, the other centripetal and centrifugal fibers of the commemorative (*c. com.*) and the association center (*c. acc.*); *c.p.o.* and *c.p.a.*, perception (centrofocal) center. *Lower:* Outline of the major cortical streams (elementary long and short vertical arc, external or internal cross-arc, long and short) according to Jakob (1913, p. 694). See also Figs. 5 and 6 in Triarhou and del Cerro (2006b) for Jakob's depiction of neocortical histotopography with its microdynamic organization and macrodynamic events (Jakob, 1945a, pp. 99–100).

por encima de situaciones análogas o diferentes, las ideas concretas del mundo objetivo experimentado.")

The third or *mature* stage reflects human consciousness, or according to Jakob's terminology 'neopsychism'. Within this phase objects and processes are organized in the dimensions of space, time and causality by means of complex neurocognitive processes, i.e. gnoses and praxes. The human mind then becomes capable of generating abstract ideas through the symbolic code of language,

a human ability that Jakob termed 'symbolism'. ("*En la tercera fase psicogenética de maduración mental, se procede a la seriación de objetos y procesos, creándose por la subsumpción simbolizante del lenguaje de los fenómenos concretos, la ideación abstracta. Situaciones complejas aisladas, nociones concretas totalizadas y, finalmente, seriadas en ideas abstractas simbolizadas son, entonces, sucesivamente los productos psicogeneticos de la labor gnoseo-práxica-cortical.*")

Jakob (1945b) described a cortical apparatus organized in such a way that accumulates and guards its traces in the form of cortical microdynamisms, linking them in a continuous and therefore conscious current.

Jakob (1906a) defined consciousness as the manifestation of the synchronization of its components. He argued that consciousness does not just emerge, but it is gradually formed as a result of cortical elaborations, attributing an adaptive character to the brain. ("*La conciencia se forma en el niño poco á poco como resultante del encadanamiento de las diferentes operaciones corticales...Ella es la manifestación den sincronismo de sus componentes...*")

He thought that there is a circular, reactive process between the object and the subject. The dynamics of consciousness—stemming from his views on cerebral cortical dynamics (Fig. 4)—consist in the simultaneous evocation of somatic reactions orientated to the external environment and sympathetic reactions orientated to the internal environment. Their synthesis links the external with the internal world in a constant adaptation.

6. Discussion

Jakob's understanding of evolutionary anatomy and biological mechanisms led him into viewing the cerebral cortex as a historical product of the external environment and at the same time as the human organ of active adaptation (López Pasquali, 1965). In a similar way, Ochs (2004) argues that the accumulated historical experiences have allowed the evolution of human social groups and subsequently the emergence of civilization.

Jakob supported the idea of a constant dynamic exchange between the internal and the external milieu, sensation and motion, perception and action (Théodoridou & Triarhou, in press). He developed such views especially throughout his 'middle' and 'late' periods. In his 1921 article he described explicitly his dynamic approach that highlights a circular flow, wherein the brain gets informed, updated and finally orientated in its environment in order to actively intervene in it. Such an approach anticipated, in certain aspects, the field of cybernetics (cf., Wiener, 1948), which, in turn, is considered as one of the critical antecedents of contemporary cognitive science (Gardner, 1985).

In defining 'systemics and cybernetics' we follow François (1999): "a metalanguage of concepts and models for transdisciplinarian use still evolving within a slow process of accretion through inclusion and interconnection of many notions, which came and are still coming from very different disciplines". Some common points between Jakob's ideas and the theories of cybernetics are discussed next.

Jakob and Copello (1948) wrote: "Life in general and the human organism in particular receive stimuli for their reactive neuronal phylogeny and ontogeny from two sources: an endogenous, generic, inner source that gives rise to the vegetative-sympathetic sphere, and an exogenous, individually orientated one that gives rise to the environmental-somatic sphere. They both create the neurodynamic nature and the personal consciousness of their carrier in a continuous reciprocal amalgamation. Neurobiology demonstrates that the same structurally bipartite and functionally integrated neural plan is applied as much on amoebae as on men. Even in protozoa there is a mutual contact of the organism with the external milieu and an internal regulatory mechanism that secures the preservation of the organism". Furthermore, within the tripartite model that he presented in his neurodynamic postulate, Jakob conceived 'psychism' as "the neurobiophylactic [neural life-protecting] complex of neuroenergetic reception, assimilation and reaction, which regulates the organism's vital necessities against variable factors in the external and internal milieu" (Jakob, 1939, p. 8). In addition, López Pasquali (1965) underlines the fact that Jakob's studies on assemblies of circuits might have anticipated the concept of autoregulation in cybernetics.

Bernard (1878, 1974) had formulated his ideas on the internal environment to unify the explanations concerning the fundamental physiologies of the body under the general principle of the preservation of stability (Gross, 1998). Bernard's momentous pronouncements, including his final account of the conception of an internal environment ("le milieu intérieur"), were gathered and published posthumously in the first volume of the 'Lectures on the Phenomena of Life Common to Animals and Plants' (Olmsted & Olmsted, 1952). At the time, the general concepts of 'living system' and 'regulation' were latent (François, 1999).

Jakob's hypothesis on the integrated function of perception and action may have parallels in diverse fields. The concept that higher processes enter at the most elementary stage of sensation was introduced by Kant; perception is then far from a simple construct following on passively received sensory reception (Ochs, 2004). von Uexküll (1934) described a functional cycle of perceptual and motor field, considered as an early account of Biocybernetics. Within his theory, perceptual and effector fields together form a closed unit, a systematic whole, the Umwelt (von Uexküll, 1934). von Weizsäcker (1950) attempted to represent the unit of perception and movement in the theoretical basis of Gestalt psychology introducing the concept of Gestaltkreis (Théodoridou & Triarhou, in press).

The idea of a 'perception–action cycle' flourished within the confines of ecological psychology. Gibson (1986) saw perception in dynamic terms and emphasized the importance of sensory feedback from movement (Hurley, 2001). Arbib (1981) put the concept into the framework of computational neuroscience, whereas Fuster (2006) is credited with the designation of the perception–action cycle in the cerebral cortex. According to the latter's theorizing (Fuster, 2006), the upper stages of the biocybernetic cycle compose the perception–action cycle, where sensory information is analyzed in the context of existing perceptual 'cognits', i.e. basic units of memory or knowledge comprised of distributed, interactive, and overlapping networks of neurons, and processed in the context of existing executive 'cognits'.

In a biocybernetic framework, Maturana and Varela (1980, 1987) developed the concepts of 'autopoiesis' and 'operational closure'. Autopoiesis, a multi-connected concept significant for problems of cognition but also for the self-reproduction of living systems is associated with the concepts of self-closure, self-reference and self-production (François, 1999).

Varela introduced neurophenomenology arguing against 'brain-bound neural events' that constitute the mind (Rudrauf, Lutz, Cosmelli, Lachaux, & Le Van Quyen, 2003); he supported the view that "consciousness depends crucially on the manner in which brain dynamics are embedded in the somatic and environmental context of an animal's life" (Thompson & Varela, 2001). Varela's conception of mind and ultimately of experience is concerned with the constraints exerted by the specific phenomenology of our concrete coping upon our internal dynamics as autonomous systems, and reciprocally, the effects of the latter upon the former, in a circular framework (Rudrauf et al., 2003). In this sense, one could argue that Jakob's views herald neurophenomenology (cf., Varela, 1996).

As far as the neuropsychogenetic problem is concerned, Jakob (1941) perceived psychogenesis (<Gk. psyche = soul and genesis = origin) as a dynamic process leading to the formation of abstract thought. Piaget shared a common view relating 'psychogenesis' to cognitive development, maintaining the literal meaning of the word contrary to its wide and popular psychiatric use (Freud, 1920; Jung, 1960).

Further similarities are found between the mechanisms and laws that rule Jakob's stages of organic psychogenesis and concepts encountered in Piaget's formulations. In particular, in his attempt to explain how the forms of intellectual activity are constructed at the sensory-motor level and subsequently how the world is constructed in the child's mind, Piaget (1952, 1954) conceived and described the functions of assimilation and accommodation that proceed from a state of chaotic undifferentiation to a state of differentiation with correlative coordination: "At first the universe consists in mobile and plastic perceptual images centered about personal activity...The external world, therefore, begins by being confused with the sensations of a self unaware of itself, before the two factors become detached from one another and are organized correlatively" (Piaget, 1954).

In Jakob's theoretical framework, the conscious self arises while one elaborates interacting internal (sympathetic) and external (somatic) experiences; external experiences issue from the external milieu, the notion of which is created when the child realizes that his/her body is separate from the objects found in his/her environment (López Pasquali, 1965). Jakob (1941, 1946) argued that the first notion of something external, the divorce between the self and the world, comes with the satisfaction of hunger, an internal need. In the first stage of the construction of the external milieu "object and process are completely fused to form a single, joint complex consisting of blurred, moving or variable elements" (Jakob, 1941, 1946). Elsewhere (Jakob & Copello, 1948), he argued that neither in an evolutionarily primitive nor in a developmentally infantile stage do humans discriminate the 'inner' from the 'outer' being subjected to a 'genuine monism'. Such a monism turns into the dualism of the two milieux, internal and external, only by means of experience.

To Piaget (1954) "assimilation ceases merely to incorporate things in personal activity and establishes, through the progress of that activity, an increasingly tight web of coordinations among the schemata that define it and consequently among the objects that such schemata are applied to. From this time on, the universe is built up into an aggregate of permanent objects connected by causal relations that are independent of the subject and are placed in objective space and time."

Jakob (1941, 1946) described the second psychogenetic stage of the somatic ego as a process that enables the separation and differentiation of the stable and unaltered elements of a situation from the blurred, moving, or variable elements through the connection and comparison of the 'complexes' of the first stage. The complexes are thus transformed into objects and processes, and they create concrete ideas of the experienced objective world through a process of identification or differentiation within the juvenile phase. This process seems to share a common element with Quine's (1960) 'similarity standard', i.e. the ability to relate things to the world as similar to or different from one another, according to their properties and the state of our perceptual scheme at the time.

The third psychogenetic phase of mental maturation elaborates the "sequencing of objects and processes". Eventually, "isolated complex situations, integrated concrete notions, and series of symbolized abstract ideas successively comprise the psychogenetic products of the gnosio-praxico-cortical work" (Jakob, 1941, 1946). According to Jakob (1941, 1946), our psycho-dynamic creation moves forward to three correlated dimensions: the spatial, the temporal, and the causal.

Maintaining that inherited dynamics are organized in the dimensions of space, time and causality, Jakob (1921, 1943a,

1943b, 1945b)—like Piaget—adopted Kant's (1781) a priori conditions. However, this aprioristic conservative principle is counterbalanced by the aposterioristic flexible principle that rises due to the openness of the brain to the stimuli of the external world (Capizzano, 2006).

Living an ocean apart, Jakob and Piaget left their marks of interdisciplinarity in biopsychology and philosophy, and formulated convergent propositions. For instance (Szirko, 1999), they adopted a similar approach to the study of the procedure through which extramental organic regulations prolong themselves into certain cognitive processes (Jakob, 1906–1908, 1922, 1948; Piaget, 1967, 1976). The two scientists were born and died almost a quarter-century apart. We do not know whether they ever communicated or exchanged ideas directly. An extra degree of difficulty stems from Jakob's rather informal style of citing references—like many authors in his era, and a common trend in scientific writing at the time. Thus, he refers e.g. to ideas conceived by Kant or Aristotle, without quoting specific sources. Piaget published his first relevant work in the 1920s (Fondation Jean Piaget, 2011), and Jakob was fluent in French (see for example Jakob, 1905, 1907). Some of the statements made by Jakob early on could clearly be independent of Piaget's writing. On the other hand, might some later statements and terminology, in the 1940s and 1950s, have been conceivably influenced by reading Piaget? That interpretation, and the question whether Jakob borrowed ideas from Piaget or vice versa, remain open to future historical research.

Jakob argued that the conscious self is born through the binding of the external and internal spheres and it becomes manifest by the synchronization of its components. Whereas "proving the case for synchronization in the human brain" is still considered technically demanding (Zeman, 2001), Jakob conceived the idea of synchronization of neuronal activity as the underlying mechanism of consciousness more than a hundred years ago (Théodoridou & Triarhou, 2012).

Jakob seems to have conceived ideas that were much ahead of his time; he anticipated the emergence of critical aspects in the incessant attempt to elucidate the neural correlates of consciousness. For example, the idea that consciousness crucially depends on memory was expressed by Crick and Koch (1990) in their theory of consciousness, where attentional mechanisms render possible the firing of neurons in a coherent semi-oscillatory way, so that a global unity would be imposed on the brain with the subsequent activation of working memory. The role of memory in consciousness was also stressed by Bergson (1934), who claimed that the continuous growth of memory equals consciousness.

Two landmark works particularly stand out with regard to the psychological and philosophical study of the experience of time (Andersen & Grush, 2009). These are William James' 'Principles of Psychology' (James, 1952), first published in 1890, and Edmund Husserl's (1928) papers on the 'Phenomenology of Inner Time Consciousness', compiled by Heidegger. They both convey the idea that the contents grasped by consciousness are built upon duration and therefore they are temporally solid (Andersen & Grush, 2009). In line with James' 'stream of consciousness' the 'Cartesian theater model' assumes that there is a locus of synthesis in the brain where experience enters consciousness (Dennett, 1991).

On the other hand, the 'multiple drafts' model (Dennett, 1991), which appeared as an alternative to the dualistic 'Cartesian theater model', holds that neural events that discriminate various perceptual contents are distributed in both space and time in the brain. However, none of these temporal properties is thought to determine subjective order, since there is no single, constitutive 'stream of consciousness' but rather a parallel stream of conflicting and continuously revised contents (cf. Dennett & Kinsbourne, 1992).

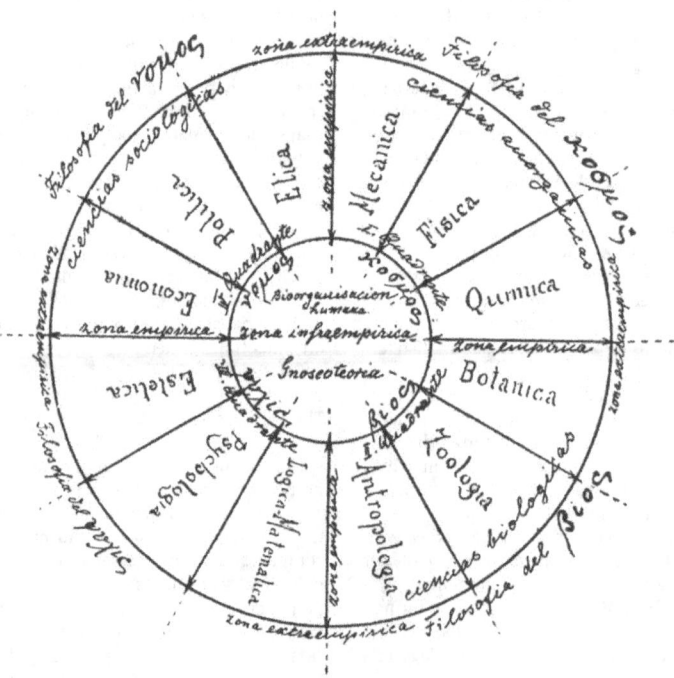

Fig. 5. Schematic synopsis of the zone of experience with the four quadrants of the exact sciences and their infra- and extra-empirical philosophical projections according to Jakob (1920, p. 30). See also Fig. 4 in Triarhou and del Cerro (2006b), based on Jakob (1945a, p. 90).

It has been further suggested that the function of time in human consciousness primarily resides in systems that maintain synchrony and allow the 'binding' between the internal environment as it is shaped by the brain's chemicals and the influence of the external environment (Dawson, 2004). Thus, time is considered as an organizing parameter to the binding problem (Dawson, 2004). The binding problem, i.e. the generation of the unity of conscious experience, may have made its first appearance in Kant's 'Critique of Pure Reason' where the "principle of transcendental unity of apperception" describes the synthesis of the "knowledge of the manifold" (Mashour, 2004).

The richness and variety of mechanisms by which animals and humans, including infants, can represent the dimensions of space, time and number to integrate diverse sensory elements for the creation of conscious experience is complex and suggests the existence of evolutionary processes and neural mechanisms by which Kantian intuitions might universally arise (Dehaene & Brannon, 2010). The evolution of the concept of time carries special weight due to its role in promoting the biological ability of the species to survive and adapt to environmental demands. From a developmental viewpoint, its importance is further reflected in the consequences of temporal disorganization on consciousness as they are observed in psychopathological conditions (Broome, 2005), aging and drug use (Dawson, 2004).

In spite of remarkable advances in neuroscience and the diffusion of biological evidence into philosophy in the modern field of neurophilosophy, the so-called 'hard problem of consciousness', i.e. the problem of explaining why and how cerebral elaborations give rise to conscious phenomenal experience (cf. Chalmers, 1995) remains a puzzle for scientists, philosophers and lay people to date. Jakob's (1945b) stance towards such a puzzle is summarized in the following paragraph: "The origin of errors and misconceptions in the attempt to explain consciousness lies in the excessive though inevitable specialization of the diverse scientific fields that are getting involved in this attempt. Given that the external and the internal spheres, the environment and the brain or macrocosmos and microcosmos, create a cycle, the answers can be found only in a synthetic approach. Therefore, physics, histology, physiology, psychology and philosophy should form a new arena of scientific exploration" (Fig. 5).

The aim of this article has been to shed light on some early key works that fall within the domain of 'Philosophy and Neuroscience', largely overlooked. The introduction of philosophical ideas into neuroscience can be considered as some of the most critical and pioneering among Christfried Jakob's contributions. We further attempted to draw parallels and to provide evidence of convergence between the ideas of Jakob and other great thinkers of the 20th century; a definite documentation of precedence remains open to future research. In any case, Jakob's synthesis may be meaningful in the formulation of new theories of cortical function.

Disclosure statement

The authors declare that there is no conflict of interest regarding the work presented in the manuscript.

Acknowledgments

These results were included in the dissertation submitted by Z.D.T. to the University of Macedonia in partial fulfillment of the requirements for the Doctorate in Educational and Social Policy. The authors gratefully acknowledge the anonymous reviewers for their constructive criticism, and the courtesy of the staff at the Ibero-Amerikanisches Institut Preussischer Kulturbesitz zu Berlin, the British Library, the Library of Congress and the National Library of Medicine of the United States.

References

Andersen, H. K., & Grush, R. (2009). A brief history of time-consciousness: Historical precursors to James and Husserl. *Journal of the History of Philosophy, 47*, 277–307.

Arbib, M. A. (1981). Perceptual structures and distributed motor control. In V. B. Brooks (Ed.). *Handbook of physiology; Nervous system* (Vol. II, pp. 1448–1480). Bethesda, MD: American Physiological Society.

Barutta, J., Hodges, J., Ibáñez, A., Gleichgerrcht, E., & Manes, F. (2010). Argentina's early contributions to the understanding of frontotemporal lobar degeneration. *Cortex, 47*, 621–627.

Bechtel, W. (2001). *Philosophy and the neurosciences: A reader*. Malden, MA: Blackwell.

Bergson, H. (1934). *La pensée et le mouvant: Essais et conférences*. Paris: Félix Alcan.

Bernard, C. (1878). *Leçons sur les phénomènes de la vie communs aux animaux et aux végétaux, tome premier* (pp. 113–114). Paris: J.-B. Baillière et fils.

Bernard, C. (1974). *Lectures on the phenomena common to animals and plants* [1878] (translated by H.E. Hoff, R. Guillemin, L. Guillemin). Springfield, IL: Charles C. Thomas.

Bickle, J. (2009). *The Oxford handbook of philosophy and neuroscience*. Oxford: Oxford University Press.

Bickle, J., Mandik, P., & Landreth, A. (2010). The philosophy of neuroscience. In E. N. Zalta (Ed.), *The Stanford encyclopedia of philosophy*. <http://plato.stanford.edu/archives/sum2010/entries/neuroscience> Accessed 24.06.11.

Blackmore, S. J. (2005). *Consciousness: A very short introduction*. Oxford: Oxford University Press.

Brook, A., & Mandik, P. (2007). The philosophy and neuroscience movement. *Analyse und Kritik, 26*, 382–397.

Broome, M. R. (2005). Suffering and eternal recurrence of the same: The neuroscience, psychopathology, and philosophy of time. *Philosophy, Psychiatry, & Psychology, 12*, 187–194.

Capizzano, A. (2006). Actualidad del pensamiento de Cristofredo Jakob. *Revista del Hospital Italiano de Buenos Aires, 26*, 71–73.

Chalmers, D. J. (1995). The puzzle of conscious experience. *Scientific American, 273*, 80–86.

Churchland, P. S. (1986). *Neurophilosophy: Toward a unified science of the mind-brain*. Cambridge, MA: MIT Press.

Churchland, P. S. (2008). The impact of neuroscience on philosophy. *Neuron, 60*, 409–411.

Crick, F., & Koch, C. (1990). Towards a neurobiological theory of consciousness. *Seminars in the Neurosciences, 2*, 263–275.

Dawson, K. A. (2004). Temporal organization of the brain: Neurocognitive mechanisms and clinical implications. *Brain and Cognition, 54*, 75–94.

Dehaene, S., & Brannon, E. M. (2010). Space, time, and number: A Kantian research program. *Trends in Cognitive Sciences, 14*, 517–519.

Dennett, D. C. (1978). Why you can't make a computer that feels pain. *Synthese, 38*, 415–449.

Dennett, D. C. (1991). *Consciousness explained*. Boston: Little, Brown and Co.

Dennett, D. C., & Kinsbourne, M. (1992). Time and the observer: The where and when of consciousness in the brain. *Behavioural and Brain Sciences, 15*, 183–247.

Fischer, K. W., Daniel, D. B., Immordino-Yang, M. H., Stern, E., Battro, A., & Koizumi, H. (2007). Why mind, brain, and education? Why now? *Mind, Brain and Education, 1*, 1–2.

Fondation Jean Piaget (2011). *Bibliographie*. <http://www.fondationjeanpiaget.ch/fjp/site/bibliographie> Accessed 10.11.11.

François, C. (1999). Systemics and cybernetics in a historical perspective. *Systems Research and Behavioral Science, 16*, 203–219.

Freud, S. (1920). The psychogenesis of a case of homosexuality in a woman. *Standard Edition, 18*, 145–172.

Fuster, J. (2006). The cognit: A network model of cortical representation. *International Journal of Psychophysiology, 60*, 125–132.

Gardner, H. (1985). *The mind's new science: A history of the cognitive revolution*. New York: Basic Books.

Gibson, J. J. (1986). *The ecological approach to visual perception*. Hillsdale, NJ: Lawrence Erlbaum.

Greenspan, R. J., & Baars, B. J. (2005). Consciousness eclipsed: Jacques Loeb, Ivan P. Pavlov, and the rise of reductionistic biology after 1900. *Consciousness and Cognition, 14*, 219–230.

Gross, C. G. (1998). Claude Bernard and the constancy of the internal environment. *Neuroscientist, 4*, 380–385.

Hurley, S. (2001). Perception and action: alternative views. *Synthese, 129*, 3–40.

Husserl, E. (1928). *Vorlesungen zur Phänomenologie des inneren Zeitbewusstseins* [1893–1917] (herausgegeben von Martin Heidegger). Halle: Max Niemeyer.

Jakob, C. (1895). *Atlas des gesunden und kranken Nervensystems nebst Grundriss der Anatomie, Pathologie und Therapie desselben*. München: J.F. Lehmann.

Jakob, C. (1899). *Atlas des gesunden und kranken Nervensystems nebst Grundriss der Anatomie, Pathologie und Therapie desselben* (2nd ed.). München: J.F. Lehmann.

Jakob, C. (1905). Contribution à l'étude de la morphologie des cerveaux des Indiens. *Revista del Museo de La Plata, 12*, 59–72.

Jakob, C. (1906a). Estudios biológicos sobre los lóbulos frontales cerebrales. *La Semana Médica (Buenos Aires), 13*, 1375–1381.

Jakob, C. (1906b). Estudio anátomo-topográfico acerca de las relaciones entre los hemisferios cerebrales y el cráneo. Revista de la Sociedad Médica Argentina (Buenos Aires), 14, 353–378.

Jakob, C. (1906–1908). Localización del alma y de la inteligencia, pt. I–IX. El Libro–Órgano de la Asociación Nacional del Profesorado (Buenos Aires), 1, 151–159, 281–291, 433–445, 553–567; 2, 3–16, 171–186, 293–308, 537–552, 695–710.

Jakob, C. (1907). Problèmes actuels de l'embryologie humaine. Revue de la Clinique Obstétricale et Gynécologique (Buenos Aires), 2, 19–32, 105–120.

Jakob, C. (1911). Das Menschenhirn: Eine Studie über den Aufbau und die Bedeutung seiner grauen Kerne und Rinde. München: J.F. Lehmann.

Jakob, C. (1912a). Über die Ubiquität der senso-motorischen Doppelfunktion der Hirnrinde als Grundlage einer neuen, biologischen Auffassung des corticalen Seelenorgans. Journal für Psychologie und Neurologie (Leipzig), 19, 379–382.

Jakob, C. (1912b). Ueber die Ubiquität der senso-motorischen Doppelfunktion der Hirnrinde als Grundlage einer neuen biologischen Auffassung des kortikalen Seelenorgans. Münchener Medizinische Wochenschrift, 59, 466–468.

Jakob, C. (1913). La psicología orgánica y su relación con la biología cortical. Archivos de Psiquiatría, Criminología y Ciencias Afines (Buenos Aires), 12, 680–698.

Jakob, C. (1919). La teoría actual de las gnosias y praxias como factores fundamentales en el dinamismo cortical. Revista del Círculo Médico Argentino y Centro de Estudiantes de Medicina (Buenos Aires), 19, 1266–1275.

Jakob, C. (1920). Filosofía de la naturaleza: Un curso de conferencias dictadas en la Facultad de Filosofía y Letras en 1920, Cátedra de Biología. Revista del Jardín Zoológico de Buenos Aires [Época II], 16, 28–55.

Jakob, C. (1921). La teoría actual de las gnosias y praxias como factores fundamentales en el dinamismo de la corteza cerebral. Crónica Médica (Lima), 38, 17–24.

Jakob, C. (1922). Del tropismo a la teoría general de la relatividad. Revista Humanidades (La Plata), 3, 45–58.

Jakob, C. (1926). El espíritu de la música en la filosofía pre y postkantiana. Revista Humanidades (La Plata), 13, 119–132.

Jakob, C. (1935a). La filogenia de las kinesias: Sobre su organización y dinamismo evolutivo. Anales del Instituto de Psicología de la Facultad de Filosofía y Letras de la Universidad de Buenos Aires, 1, 109–127.

Jakob, C. (1935b). Sobre las bases orgánicas de la memoria. Revista de Criminología, Psiquiatría y Medicina Legal, 127, 84–114.

Jakob, C. (1937). Descartes en la biología. In L. J. Gerrero (Ed.), Descartes–Homenaje en el tercer centenario del "Discurso del método", tomo I (pp. 57–66). Buenos Aires: Instituto de Filosofía de la Facultad de Filosofía y Letras.

Jakob, C. (1938). La psicología de Descartes a través de tres siglos. Anales del Instituto de Psicología de la Facultad de Filosofía y Letras de la Universidad de Buenos Aires, 2, 297–327.

Jakob, C. (1939). El neoencéfalo: Su organización y dinamismo. La Plata – Buenos Aires: Universidad Nacional de La Plata, Facultad de Humanidades y Ciencias de la Educación – Imprenta López.

Jakob, C. (1941). La función psicogenética de la corteza cerebral y su posible localización: Aspectos de la ontopsicogénesis humana. Anales del Instituto de Psicología de la Facultad de Filosofía y Letras de la Universidad de Buenos Aires, 3, 63–80.

Jakob, C. (1943a). Folia Neurobiológica Argentina, Tomo II. El pichiciego (Chlamydophorus truncatus): Estudios neurobiológicos de un mamífero misterioso de la Argentina. Buenos Aires: Aniceto López.

Jakob, C. (1943b). Folia Neurobiológica Argentina, Tomo III. El lóbulo frontal: Un estudio monográfico anatomoclínico sobre base neurobiológica. Buenos Aires: Aniceto López.

Jakob, C. (1945a). El cerebro humano: Su significación filosófica. Revista Neurológica de Buenos Aires, 10, 89–110.

Jakob, C. (1945b). Sobre el origen de la conciencia: Investigaciones neurobiológicas sobre la dinámica cerebral en relación con su sectorización conmemorativa. In: Mouchet, E. (Ed.), Temas actuales de psicología normal y patológica, publicados bajo el patrocinio de la Sociedad de Psicología de Buenos Aires (pp. 345–381). Buenos Aires: Editorial Médico-Quirúrgica/Talleres 'The Standard'.

Jakob, C. (1946). Folia Neurobiológica Argentina, Tomo V. Documenta Biofilosófica, Folleto I. Biología y filosofía. (A) Aspectos de sus divergencias y concomitancias. (B) Ensayo de psicogenia orgánica. Buenos Aires: López y Etchegoyen.

Jakob, C. (1948). La psicointegración introyento-ambiental orgánica y sus problemas para la neuropsiquiatría y psicología, primera parte: su filogenia constructiva. Revista Neurológica de Buenos Aires, 13, 115–141.

Jakob, C., & Copello, A. R. (1948). La psicointegración introyento-ambiental orgánica y sus problemas para la neuropsiquiatría y psicología. Revista Neurológica de Buenos Aires, 13, 63–79.

James, W. (1952). The principles of psychology [1890]. Chicago: University of Chicago Press – Encyclopaedia Britannica.

Jung, C. G. (1960). The psychogenesis of mental disease. New York: Pantheon Books.

Kant, I. (1781). Critik der reiner Vernunft. Riga: Johann Friedrich Hartknoch.

López Pasquali, L. (1965). Christfried Jakob: Su obra neurológica, su pensamiento psicológico y filosófico. Buenos Aires: López.

Mashour, G. A. (2004). The cognitive binding problem: From Kant to quantum neurodynamics. NeuroQuantology, 2, 29–38.

Maturana, H. R., & Varela, F. J. (1980). Autopoiesis and cognition. The realization of the living. Dordrecht: Reidel.

Maturana, H. R., & Varela, F. J. (1987). The tree of knowledge: The biological roots of human understanding. Boston, MA: New Science Library – Shambhala Publications.

Moyano, B. A. (1957). Christfried Jakob, 25/12/1866–6/5/1956. Acta Neuropsiquiátrica Argentina, 3, 109–123.

Nagel, T. (1971). Brain bisection and the unity of consciousness. Synthese, 22, 396–413.

Northoff, G. (2004). What is neurophilosophy? A methodological account. Zeitschrift für Allgemeine Wissenschaftstheorie, 35, 91–127.

Ochs, S. (2004). A history of nerve functions: From animal spirits to molecular mechanisms. Cambridge – New York: Cambridge University Press.

Olmsted, J. M. D., & Olmsted, E. H. (1952). Claude Bernard and the experimental method in medicine. New York: Henry Schuman.

Orlando, J. C. (1966). Christofredo Jakob: Su vida y obra. Buenos Aires: Editorial Mundi.

Pedace, E. A. (1949). Contribución de la escuela neurobiológica Argentina del Prof. Chr. Jakob en el estudio del lóbulo frontal. Archivos de Neurocirugía, 6, 464–466.

Piaget, J. (1952). The origins of intelligence in children. New York: International Universities Press.

Piaget, J. (1954). The construction of reality in the child. New York: Basic Books.

Piaget, J. (1967). Biologie et connaissance. Essai sur les relations entre les régulations organiques et les processus cognitifs. Paris: Gallimard.

Piaget, J. (1976). Le comportement, moteur de l'évolution. Paris: Gallimard.

Quine, W. (1960). Word and object. Cambridge, MA: MIT Press.

Rudrauf, D., Lutz, A., Cosmelli, D., Lachaux, J.-P., & Le Van Quyen, M. (2003). From autopoiesis to neurophenomenology: Francisco Varela's exploration of the biophysics of being. Biological Research, 36, 27–66.

Szirko, M. (1999). Constructivism is not pantopoiesis. Karl Jaspers Forum. <http://www.kjf.ca/15-C32SZ.htm> Retrieved 08.11.11.

Théodoridou, Z. D., & Triarhou, L. C. (in press). Evolution of Christfried Jakob's views on the frontal lobe, 1899–1949. In: Cavanna, A.E. (Ed.), Frontal lobe: Anatomy, function and injury. Hauppauge, NY: Nova Science Publishers.

Théodoridou, Z. D., & Triarhou, L. C. (2011). Christfried Jakob's 1921 theory of the gnoses and praxes as fundamental factors in cerebral cortical dynamics. Integrative Psychological and Behavioral Science, 45, 247–262.

Théodoridou, Z. D., & Triarhou, L. C. (2012). Challenging the supremacy of the frontal lobe: Early views (1906–1909) of Christfried Jakob on the human cerebral cortex. Cortex, 48, 15–25.

Thompson, E., & Varela, F. J. (2001). Radical embodiment: Neural dynamics and consciousness. Trends in Cognitive Sciences, 5, 418–425.

Triarhou, L. C. (2008). Centenary of Christfried Jakob's discovery of the visceral brain: An unheeded precedence in affective neuroscience. Neuroscience and Biobehavioral Reviews, 32, 984–1000.

Triarhou, L. C. (2010a). Final publications of Christfried Jakob: On the frontal lobe and the limbic region. In C. E. Flynn & B. R. Callaghan (Eds.), Neuroanatomy research advances (pp. 165–169). Hauppauge, NY: Nova Science Publishers.

Triarhou, L. C. (2010b). Revisiting Christfried Jakob's concept of the dual onto-phylogenetic origin and ubiquitous function of the cerebral cortex: A century of progress. Brain Structure and Function, 214, 333–338.

Triarhou, L. C., & del Cerro, M. (2006a). Semicentennial tribute to the ingenious neurobiologist Christfried Jakob (1866–1956). 1. Works from Germany and the first Argentina period, 1891–1913. European Neurology, 56, 176–188.

Triarhou, L. C., & del Cerro, M. (2006b). Semicentennial tribute to the ingenious neurobiologist Christfried Jakob (1866–1956). 2. Publications from the second Argentina period, 1913–1949. European Neurology, 56, 189–198.

Triarhou, L. C., & del Cerro, M. (2007). Pioneers in neurology: Christfried Jakob (1866–1956). Journal of Neurology, 254, 124–125.

Varela, F. (1996). Neurophenomenology: A methodological remedy to the hard problem. Journal of Consciousness Studies, 3, 330–350.

Véganzonès, M.-A., & Winograd, C. (1997). Argentina in the 20th century: An account of long-awaited growth. Paris: Development Centre of the Organisation for Economic Co-operation and Development.

von Eckardt-Klein, B. (1975). Some consequences of knowing everything (essential) there is to know about one's mental states. Review of Metaphysics, 29, 3–18.

von Economo, C., & Koskinas, G. N. (1925). Die Cytoarchitektonik der Hirnrinde des erwachsenen Menschen. Wien: Springer.

von Uexküll, J. (1934). A stroll through the worlds of animals and men. In C. Schiller (Ed.), Instinctive behavior (pp. 5–80). New York: International Universities Press.

von Weizsäcker, V. (1950). Der Gestaltkreis. Stuttgart: Thieme.

Wiener, N. (1948). Cybernetics or control and communication in the animal and the machine. Paris: Hermann.

Zeman, A. (2001). Consciousness. Brain, 124, 1263–1289.

Cavanna, A.E. (ed.) *Frontal Lobe: Anatomy, Function and Injury*
© 2012 Nova Science Publishers, Hauppauge, New York

Evolution of Christfried Jakob's views on the frontal lobe, 1890–1949

Zoe D. Theodoridou and Lazaros C. Triarhou

Department of Educational and Social Policy, University of Macedonia,
Thessaloniki, Greece

Abstract

We review the evolution of the ideas of Christfried Jakob (1866–1956) on the cerebral cortex, with special emphasis on the frontal lobe. For more than five decades, Jakob studied the frontal lobe from the macroscopic to the microscopic level, its function and structure, development, evolution, and pathology. He developed his views on frontal lobe function based chiefly on anatomical works during his 'early' period of the 1890s through the 1910s. In the 1920s, he formulated his psychobiological thought, and in the 1930s and 1940s synthetic neurobiological and neurophilosophical ideas. Arguing that the human cerebral cortex carries within its long natural and social history, he suggested that natural demands have created nervous structures as dictated, at the same time, by the need for cortical specialization and communication. Thus, Jakob attributed a 'humanizing' element to the frontal lobe that lies principally in its praxic character.

Introduction

We track the evolution of the ideas of Christfried Jakob (1866–1956) on the cerebral cortex, particularly on the frontal lobe. Jakob was a German-born neurobiologist, who spent most of his life in Argentina, where he established one of the most important neuropathological laboratories in all of South America. He is considered to be the father of Argentinian neurosciences (Pedace, 1949) and one of the great thinkers of the 20th century. Among his works, the frontal lobe occupied a central part for over five decades (Moyano, 1957).

We have thus divided those works into three 'periods'. During an 'early' period (1890s–1910s), Jakob mostly carried out anatomical works. In his 'middle period' (1920s) he formulated his psychobiological thought. In the 'late' period (1930s–1940s), he developed synthetic neurobiological and neurophilosophical concepts.

Jakob completed his medical studies at the University of Erlangen in 1890. In the early 1890s he worked as an assistant to Adolf von Strümpell at the Erlangen Medical Clinic and privately practised medicine in Bamberg (Orlando, 1995). He published his first book in 1895, an atlas of the normal and pathological anatomy of the nervous system (for details, see Triarhou & del Cerro, 2006a). In 1897, Jakob published an atlas of methods of clinical investigation (an epitome of internal medicine), which was translated into French and English (Triarhou & del Cerro, 2006a).

In 1899, Jakob went to Argentina to direct the Laboratory of the Psychiatric and Neurological Clinic of the Hospicio de Las Mercedes at the National University of Buenos Aires. One of the elements that was crucial in his decision to leave Europe was the prospect of having available 300 brains yearly for pathoanatomical study (López Pasquali, 1965). For a dozen years, Jakob produced works in anatomy, neurology, psychopathology and anthropology (Triarhou & del Cerro, 2007). Then, with his Argentinian contract having ended, he returned to Germany to further his knowledge. At that time, he completed two landmark works on phylogeny, putting forth his original idea on the ubiquity of the sensory-motor dual function of the cerebral cortex (Jakob, 1911; Jakob, 1912a; Jakob, 1912b; Triarhou, 2010a). The works of that 'early' period are mostly centered around the anatomy of the nervous system, reflecting Jakob's German scientific training (López Pasquali, 1965).

The beginning of the 'middle' period of Jakob's work is signalled by his permanent move to Argentina in 1913 (López Pasquali, 1965). At that time, he assumed a triple role consisting of clinical, research and teaching duties. He was appointed Chief of the Neuropathological Institute at the Hospicio Nacional de Alienadas (Mental Asylum for Women) in the Federal Capital, and Professor and Director of the Institute of Biology at the Faculty of Philosophy and Letters of the National University of La Plata. In 1922, Jakob was named Professor of Neurobiology at the Faculty of Humanities and Educational Sciences of the Universidad Nacional de La Plata (Triarhou & del Cerro, 2006b). From 1921 to 1933, he held a joint appointment as Professor of Pathological Anatomy at the School of Medical Sciences of La Plata (Triarhou & del Cerro, 2006b).

A key work from that period is his 1918 article 'From the mechanism to the dynamics of the mind: A critical historical study of organic psychology', in which Jakob pursued his 'dynamic approach'. During that time, Jakob wrote two works which are considered by his biographer and colleague Braulio Moyano (1957) as vital constituents of his psychobiological theorizing: his article 'On he biological bases of memory' (Jakob, 1935b) and his 'Theory on the phylogeny of the kineses' (Jakob, 1935a). However, the emergence of his psychobiological theories becomes obvious in his 1919–1921 work on gnoses and praxes as fundamental factors in cerebral cortical dynamics (Jakob, 1921). Those ideas formed the core for Jakob's future theories (Jakob, 1935a, 1935b).

Gradually, Jakob's thought became more synthetic. He coupled his philosophical background with clinical and research experience. Maintaining that the utmost problem of science and philosophy converges in cerebral function, Jakob studied the brain from the macroscopic to the microscopic level, considered its functional aspects, and pursued its development, evolution, and pathology. He suggested a scientific psychology and a corpus of philosophy (López Pasquali, 1965), in works such as 'The psychogenetic function of the cerebral cortex and its possible localization' (Jakob, 1941), 'The philosophical meaning of the human brain' (Jakob, 1945a), 'The origin of consciousness' (Jakob, 1945b), and 'Common and diverge aspects between biology and philosophy' (Jakob, 1946a) (reviewed in Theodoridou & Triarhou, 2011b, 2011c).

The 'Early' Period (1891–1912)

Between 1906 and 1909 Jakob published eight papers (Jakob, 1906d, 1906f, 1906e; 1906a, 1906b, 1907b, 1907a; 1909), which address biological, anatomo-clinical, pathophysiological and psycholinguistic aspects of the frontal lobe, as well as a series of articles on localization under the title 'Localization of the soul and of intelligence' (Jakob, 1906e). Jakob (1906d) cast doubt on the 'supremacy' previously attributed to the frontal lobe. He argued that "the question concerning superior human functions cannot be answered pointing out their localization in one or another brain lobe but, instead, taking into account issues of another kind" (Barutta, Hodges, Ibañez, Gleichgerrcht, & Manes, 2010).

Jakob's arguments witness a sceptical stance toward strict localization. He highlighted putative historical reasons—from classical Greek philosophy—that might explain the importance attached by various authors to the frontal lobe (Théodoridou & Triarhou, 2011a). In particular, Jakob (1906e) considered the 'Olympian forehead' that is artistically depicted in the sculptures of Zeus as the symbol of 'humanization'.

Having studied preparations with the Weigert method in a series of what are considered as classical contributions, Jakob (1906b) rejected the superiority of frontal myeloarchitectonics by pinpointing at a diminished total density and density of the various layers, a smaller average cell volume and a less developed supraradial layer of the frontal lobe. Campbell (1905) expressed similar views to Jakob regarding the comparatively moderate structural development of the prefrontal cortex, in terms of the low fiber numbers and their delicate nature, as well as the absence of an association system. Regarding Flechsig's proposal of a parallel development of myelination pathways and intellect, Jakob (1906d) remained sceptical.

In regard to the arguments drawn from the clinical literature and placing emphasis on the relation between frontal lobe damage and profound personality changes, Jakob highlighted the rarity of 'pure cases' (1909) in neuropathology. Having studied human brains with frontal lobe tumors, injuries and degeneration, he underlined that (a) the appearance of symptoms does not necessarily coincide with the onset of the disease, and thus, progression may be difficult to determine; (b) tumors compress the brain parenchyma; (c) lesions of vascular origin lead to widespread degeneration; and (d) brain damage may cause either inflammation or concussion that may affect the entire brain (Jakob, 1906b, 1909). Contradictions still exist in the clinical literature that make researchers cautious; case studies may involve either massive lesions extending beyond the frontal lobe or small, unilateral, or asymmetric lesions with correspondingly small and easily compensated effects (Teuber, 2009).

Jakob was fascinated by phylogenetics and viewed it as a means for getting answers about higher human functions, by differentiating between different attributes of the human species and the corresponding evolutionary correlates (Barutta et al., 2010). Jakob's phylogenetic studies on the human brain and over 100 species of the Patagonian fauna (Jakob 1912a, 1912b; Jakob & Onelli, 1913; Triarhou 2010b) provided him with material for formulating original ideas. Jakob (1906d) observed that the evolution of the frontal lobe proceeds from lower to higher mammals in a continual and constant fashion, whereas some other vertebrates do not possess hemispheres with a cortex comparable to that of mammals. Highlighting the similarities between the frontal regions of humans and higher mammals he maintained that productive mentality derives from the frontal lobes. Since the beginnings of the 20th century, human cognitive development was attributed to the large size of the frontal lobes. Modern cytoarchitectonic studies show a very similar organization between human and macaque monkey prefrontal cortex (Petrides, 2005). Moreover, magnetic resonance imaging studies (Semendeferi, Lu, Schenker, & Damasio, 2002) show that the frontal cortex of humans and the great apes occupies a similar proportion of the cortex of the cerebral hemispheres. Accordingly, the enlargement of the human brain has generally preserved the relative proportion of its major lobes (Risberg, 2006).

In the 19th century attempts were made in laboratories where experiments on animals were conducted to show the superiority of the frontal lobe. The experimental confirmation of a motor cortex in the dog brain by Fritsch and Hitzig (1870) was a landmark in the history of functional localization. This tradition continued with new mosaicists and holists. Jakob (1906f) described caveats in such methods, which he considered inappropriate for reaching conclusions about higher human brain functions.

Always considering morphology in a functional context (Tsapkini et al., 2008), Jakob claimed that the elucidation of the anatomical connections of the frontal lobe would decipher its functions (Théodoridou & Triarhou, 2011a). In studying the structure of the frontal lobe, he did not notice any substantial differences from the remaining lobes of the cerebral hemispheres as far as the categories of fibers are concerned, i.e. afferent and efferent projection fibers, association, and commissural fibers (Jakob, 1906f). Emphasizing the importance of studying connections, he anticipated the hodological trend (cf. Catani and ffytche, 2005; ffytche and Catani, 2005). Jakob's writings on cortical connectivity are further attuned to recent theories of frontal systems and neural networks, such as Alexander, de Long and Strick's (1986) concept of parallel but segregated frontal-subcortical circuits, which has been further put into a clinical framework by Chow and Cummings (1999).

According to Jakob (1911), the various centripetal pathways course into all cortical sectors; thus, the cortex has a perceptive activity over its entire extent (Triarhou, 2010b). Based on his anatomical observations, Jakob (1906b) viewed the major part of the frontal lobe as a central station with multiplier and combinatorial characteristics, constantly receiving stimuli from all the motility organs via multiple pathways. He described the sensory-muscular pathways which arrive at the frontal lobe via the cerebellum, the red nucleus and the thalamus, concluding that numerous muscular sensory inputs enter the frontal lobe (Jakob, 1906b).

Based on anatomical and electrophysiological observations, Cappe, Rouiler, & Barone (2009) argued that a connectivity network that includes cortical and thalamocortical pathways as well as the diversity of interactions observed across the thalamus, the cortical sensory or associative areas is involved in multisensory interplay. Thus, most areas in the parietal, temporal, or frontal regions of primates are thought to have connection patterns that relate them to more than one sensory modalities (Cappe, Rouiler, & Barone, 2009).

Jakob's positions apparently antedate some of the modern views on the function of the anterior parts of the human brain, which consider the prefrontal cortex as a locus of synthesis of the outputs of various neuronal systems in providing the basis for the orchestration of complex behavior (Duncan & Miller, 2002). Frontal and prefrontal regions have been linked to visual, auditory and somatosensory inputs (Fogassi et al., 1996; Graziano, Yap, & Gross, 1994; Graziano, Reiss, & Gross, 1999; Wallace, Meredith, M.A., & Stein, 1992). Sensory, mnemonic and response signals that a single neuron displays provide strong evidence that prefrontal neurons behave as sensorimotor integrators (Goldman-Rakic, 2000). Thus, mounting evidence shows that much if not all of the neocortex is involved in multisensory integration (Ghazanfar & Schroeder, 2006). Moreover, the role of the frontal lobe in integrating information from multiple brain areas supports its crucial involvement in learning, comprehension and reasoning (Baddeley, 2002). According to Fuster (2006), actions related to human behavior, reasoning, and language are organized by means of interactions between prefrontal and posterior networks at the top of the 'perception-action cycle'.

The 'Middle' Period (1913–1935)

From 1913 to 1935 Jakob's dynamic approach was gradually refined. He viewed the human brain not as an isolated organ, but as an organ that regulates the internal as well as the external environment (López Pasquali, 1965). In his 1918 paper 'From the mechanism to the dynamics of the mind: A critical historical study of organic psychology' Jakob rejected mechanistic concepts and presented a dynamic approach in explaining cortical function. He argued for an active exchange between the external environment and the adaptive brain (Jakob, 1918).

Jakob incorporated his dynamic views into a phylo-ontogenetic theory of cortical function. He kept moulding his framework (Jakob, 1921; Jakob, 1935a; Jakob, 1946) to include multiple aspects of brain research. According to Jakob's evolutionary postulate (Jakob 1935a; Théodoridou & Triarhou, 2011d; Triarhou & del Cerro, 2006b), phylogeny occurs in two phases. In the first, 'plasmodynamic' phase—or plasmopsychism—elementary biological phenomena such as tropism and pulsatility emerge. The second phase, called 'neurodynamic' and corresponding to 'neuropsychism', is divided into three stages: a phylogenetically older 'archikinetic' stage, where reflex actions emerge; a 'paleokinetic' stage that entails instinctive reactions; and a 'neokinetic' stage, which elaborates conscious motor reactions. 'Neokineses' consist of three kinds of higher neurocognitive processes: (a) 'gnoses', which secure the conscious orientation in one's environment, (b) 'praxes', which underpin active individual intervention and (c) 'symbolisms', which subserve the communication of abstract ideas by means of language, art, etc. Each of these stages corresponds to different levels of organization in the vertebrate C.N.S. The archikinetic stage corresponds to the archineuronal, the paleokinetic to the paleoneuronal, and the neokinetic to the neoneuronal.

Between 1919 and 1921, Jakob presented his theory on gnoses and praxes as fundamental factors in cerebral cortical dynamics (Jakob, 1919, 1921; Théodoridou & Triarhou, 2011d). Gnoses play a key role as the preparatory acts, and praxes as the productive acts, of all psychogenetic processes (Jakob, 1941). For the identification of objects perceived by the senses and the subsequent orientation in the world the integration of the specific features of each object as they are registered in the brain is imperative (López Pasquali, 1965).

Jakob (1921) describes the gnosio-praxic dynamism as follows:

"Through the integration of the sensory information that one normally gets by a certain number of isolated perceptions of distance, color, form, intensity, etc., which characterizes an object one has seen, heard, tasted, etc. normally arrives at a state of 'apperceptive condensation and associative correlation' for the analogous impressions that finally allow the construction of 'the notion of the object', namely its complete gnosis. Thus, gnosis consists in the synthetic condensation of a previous experience with an analogous current situation on the basis of the parallelism of the external and the internal milieu... Gnoses do not result from a special cortical power but from an intricate game of sequential cortical elaborations. Therefore, a gnosis or a sensory perception is not an illusion of the world but a correct approximate representation, proven in practice... Gnoses distribute and organize experience as the securing of orientation in space and time demand it...

All parieto-occipito-temporal cortical zones contribute in the elaboration of gnoses both in animals and in humans. Thus gnoses are represented in the posterior half of the cerebral hemispheres. Nevertheless, gnoses consist in the elaboration and condensation of sensory-motor acts...

Praxis is the cortical process that associates in different combination the series of motor acts during the long learning process in infancy and childhood, to guarantee a determined movement of the limbs, the tongue, the lip, and the trunk or the entire body... The seat of the elaboration of praxes is the whole anterior half of the cerebral hemispheres that is, the anterior Rolandic and the entire frontal lobe with its related associative systems."

Jakob claimed that psychisms are processes of growing complexity (Barutta et al., 2010) and that the human cerebral cortex reflects its long natural and social history (López Pasquali, 1965); the natural demands have created nervous structures in the human brain as they are dictated by the need for cortical specialization and communication at the same time. Based on comparative anatomical studies (Jakob, 1912; Triarhou, 2010b), Jakob pointed out that the frontal lobe evolved and expanded in a unique way in primates. He wrote (Jakob, 1943, p. 89): "The great development of the frontal lobe is typical of the brain of primates". The evolutionarily older gnosic centers are thought to reside in postcentral 'microdynamics', whereas frontal regions are primarily responsible for praxic processes (Capizzano, 2006).

On the other hand, Jakob considered the strict localization of mental function or dysfunction as misleading. He argued that every gnosic and praxic mechanism comprises localizable elements, such as receptors, assimilators and effectors, but gnosio-praxic dynamics per se are transcortical (Jakob, 1921). By attributing apraxias to the disturbance of transcortical dynamics, Jakob (1921) highlighted the role of cortical communication. Wernicke also opposed the localization of higher functions to specific regions, stressing the importance of association areas (Catani & ffytche, 2005) and claimed that apraxia results from the separation of brain regions (Finger, 1994). Associationist models produced disconnectionist accounts of disorders of higher functions. Liepmann's apraxia model and Déjerine's pure alexia description fall into this tradition, which was revived with Geschwind's neo-associationism. Geschwind (1965a, 1965b) attributed higher function deficits to disconnections that result either from white matter lesions or lesions of association areas, whereas, more recently, Catani and ffytche (2005) updated that model into a hodotopic framework.

Jakob (1921) argued that "gnoses and praxes are neither sensory nor motor, but concomitantly sensory-motor processes" (Jakob, 1911; Jakob, 1912a; Jakob, 1912b; Triarhou, 2010b). Jakob supported the idea of a constant dynamic exchange between the internal and the external milieu, sensation and motion, perception and action.

In his famous book 'Matter and memory' (1896), the French philosopher Henri Bergson (1859–1941) argued that it is impossible to define where perception ends and movement begins (Blumen & Blumen, 2002). The idea of a perception-action cycle has been expressed in various frameworks: in theoretical biology and biosemiotics, by Jakob Johann von Uexküll (1934) to denote perceptual and effector fields that together form a closed unit, the 'Umwelt'; in biocybernetics, by Maturana and Varela (1980; 1987) with the concepts of 'autopoiesis' and 'operational closure'. The German physician and physiologist Viktor Freiherr von Weizsäcker (1950) attempted to represent the unit of perception and movement in a theoretical basis introducing the concept of 'Gestaltkreis', an elaboration of Gestalt psychology. Within ecological psychology, Gibson (1986) saw perception in dynamic terms and emphasized the importance of sensory feedback from movement (Hurley, 2001). Perception and action were thought to be interdependent creating a continuous circle of causes and effects of action (Gibson, 1986). Arbib (1981) put the concept into the framework of computational neuroscience. However, it is Fuster (2006) who is credited with the designation of the perception-action cycle in the cerebral cortex. Therein, the upper stages of the biocybernetic cycle constitute the perception-action cycle where, the sensory information is analyzed in the context of existing perceptual cognits and processed in the context of existing executive cognits.

Although dynamic concepts prevailed in physics in the early part of the 20th century, it took some time for such concepts to be applied to brain theory. Dynamic approaches became popular in many fields after the 1940s confirming the idea that scientific trends reflect an era's broader historical, political and cultural framework (York, 2009). To our knowledge, the first reference to neurodynamics prior to Jakob was made by Trigant Burrow (1943). Wiener's (1948) critical work in cybernetics opened up new vistas (François, 1999). Francisco Varela, a pioneer of modern brain dynamics and cybernetics (cf. Rudrauf, Lutz, Cosmelli, Lachaux, & Le Van Quyen, 2003) supported the view that "consciousness depends crucially on the manner in which brain dynamics are embedded in the somatic and environmental context of an animal's life" (Thompson & Varela, 2001). Thus, influenced by computer science, modern theories use the metaphor of cognition as a dynamic system sustained on spatiotemporal topology (Ibañez & Cosmelli, 2008). Such trends are close to Jakob's views: López Pasquali (1965) emphasizes the fact that certain aspects of Jakob's work such as his studies on the assemblies of circuits seem to anticipate concepts of cybernetics, e.g. autoregulation and feedback.

The 'Late' Period (1935–1949)

Jakob's later years reflect the integration of his thought through the clinical, research and educational experiences, in conjunction with his background in philosophy. Viewing his own self as "a groundworker of a biocentric epistemology", Jakob (1945b) strived to set the foundations of a new interdisciplinary field that would diffuse neurobiological evidence into philosophy. Recognizing the 'humanizing' role of the frontal lobe, Jakob (1943) described the meaning of its dynamics for science and philosophy in a monograph and left that work "rather as a plan for future research and not as an essay with solutions" (Jakob, 1943).

In particular, as far as biology is concerned, he related the progressive perfection that derives from frontal dynamics with the accumulated commemorative function. In neurology, frontal dynamics are thought to be expressed through the evolution from the brutal commands of the generic and irresistible, hereditary and universally obligatory instincts to an elevating liberation that results from an individually orientated intervention in the sphere of consciously caused aims (Jakob, 1943). In psychology, the substitution of the media of communication, conveying exclusively affective information by other, conveying intellectual, was rendered possible by means of the concrete gnosio-praxic experience represented by abstract symbols in language (Jakob, 1943). In sociology, frontal dynamics are related to the possibility of extension and intensification in time and space of the individual and the collective productivity that affects the economic, the cultural and the political spheres in order to progressively become more 'human' (Jakob, 1943). In education, the development of a social intellect that will reinforce individual inclinations and will put emphasis on the active engagement of the student in the formation of concrete knowledge issues from the importance of frontal lobe functions (Jakob, 1943). In neuropsychiatry, it leads to the emergence of objective neuro-psycho-analytical processes for the elaboration of the factors that derive from complex neuropsychological symptoms, reactions and phenomena (Jakob, 1943).

The systematic application of anatomo-clinical methods to study physiological phenomena behind normal and pathological cortical localization and communication leads to the replacement of vague verbal constructions by histo-physio-pathological concrete concepts enabling, thus, the transition from a mechanic to a dynamic phase in neuropsychiatry (Jakob, 1943). Our biopsychism has created unique intellectual, aesthetic and ethical values that enable the explanation of the nature, mental liberty and understanding of the demands and the limits of our mind, thus shedding light onto reality and representation (Jakob, 1943). Praxic (motor) dynamics lead the course of life with the individual and the collective aims; the gnosic intellect, on the other hand, expresses the capability of orientating in one's environment (Jakob, 1943).

Jakob (1943) further proposed that the frontal lobe underpins at the neuronal level the phylogenetic transition from 'blind' desires dictated by impulses to conscious aims planned and executed in order to bring a result. In his late period he revised and updated his phylogenetic postulate in the clinical framework of Pick disease (frontotemporal lobar degeneration). Jakob (1946b) made the supposition that Pick disease represents a model for progressive disintegration of a hierarchical cognitive system (Barutta et al., 2010).

In this revised view, Jakob (1946b) explained that by the Aristotelian concept of 'psyche' he refers to the integrative dynamics issued from sensory-motor regulations. The 'phylopsyche' carries brain activities inherited from phylogenetically older species and is comprised by the 'archipsyche', which contributes to the reflex functions and the 'paleopsyche', which contributes to instinctual functions. The most recent phylogenetic acquisition and typical of the human brain, the 'ontopsyche', is responsible for the elaboration of the individual brain activity mediated by individual experience processes. Ontopsyche or neopsyche, is further divided it into the trophopsyche, an internal milieu regulator whose activity is carried out by the limbic system; the somatopsyche, an external milieu regulator whose activity is performed by the suprasylvian gyri; and the logopsyche, mediator of our symbolization of the world, through the perisylvian gyrus (Barutta et al., 2010). According to Jakob (1946b), Pick disease results in a 'diaschisis', a disruption between the paleopsyche and the neopsyche leading to intellectual and affective disorders. Such a disruption implies the dissociation between the internal and environmental aspects, the subjective and objective world, respectively (Barutta et al., 2010). Therefore, the creation of the most important human ideals is affected because— even though they become realized into the sphere of the intellect—they are rooted in the sphere of the emotions.

Jakob's last publication on the frontal lobe (co-authored with his pupil Eduardo A. Pedace and dated 1949) is entitled 'The task of the frontal lobe in connection with a synthetic quantification of its constitutive elements' (see Triarhou, 2010a). Jakob thought that the frontal lobe reflects the latest acquisitions in the ascending neurophylogeny (Barutta et al., 2010). The fact that the frontal lobe represents 25% of the human brain, i.e. about 350–370 g of cerebral mass, solely considered from a quantitative standpoint, must alone confirm the higher task of their cortical functions. In the frontal lobe, Jakob (1949) recognized the centers of experiential accumulation resulting from personal intervention, progressively elaborated for the elemental and highest human skills, stimulated by the corresponding affective manifestations.

Jakob (1949) viewed the frontal cortex as a locus of interaction between afferent and efferent pathways, the system of transformation of specific stimulations and reactions (endogenous-exogenous frontalization) and with it the final accumulation of its elaborations (frontalized commemorative function). Both areas, in close gnosio-praxic collaboration, execute the conscious activation of human mentality in its creative labor from the concrete to the abstract in an intimate synthesis between their endogenous and exogenous domains, i.e. from their affectivity and intellectuality, reciprocally (Jakob, 1949).

Conclusion

The fact that the human frontal cortex covers about 30% of the total cortical surface has prompted clinicians and basic scientists to hope that unravelling frontal lobe function might eventually explain human behavior (Raichle, 2002). As the riddle of the frontal lobes remains central in modern neurobiology, Jakob's views are still meaningful.

Current theories on the functional localization of cognitive processes in the frontal lobe range from fractionated approaches to central concepts, with concomitant attempts to reconcile contrasting views (Théodoridou & Triarhou, 2010a). The common element in fractionated approaches (cf. Koechlin, Ody, & Kouneiher, 2003; Shallice, 2002; Shallice & Burgess, 1996; Stuss et al., 2002) is the view that there is no unitary frontal lobe process. The anterior part of the brain rather subserves multiple distinct control processes that underpin executive functions (Godefroy, Cabaret, Petit-Chenal, Pruvo, & Rousseaux, 1999). Modularity and fractionation may pertain even to higher human abilities (Baddeley, 1996; Stuss et al., 2002). A more central concept has been put forth by Duncan and Miller (2002) who reject the fixed functional specialization and highlight the adaptability of selected regions of the prefrontal cortex in order to complete a goal-directed activity. Finally, Stuss (2006) argues that the debate between fractionation and adaptability is a false debate and suggested that brain networks may be both locally segregated and functionally integrated. Evidence on the recruitment of the same frontal regions for different cognitive demands (Duncan & Owen, 2000) indicates that in spite of fractionation, frontal processes are applicable to many domain-specific modules; therefore, frontal processes are domain-general (Stuss, 2006).

Jakob used a multilevel approach in studying the frontal lobe in an attempt at being as unbiased as possible. Having understood the limitations and misdirections inherent in any effort to decipher brain-mind relationships, he remained critical of oversimplifying localization explanations (Théodoridou & Triarhou, 2011d) and looked for clues relying on anatomical-clinical correlations. Thus, Jakob's work over 50 years clearly shows a trend that has been discovered anew by modern researchers.

References

Alexander, G. E., DeLong, M. R., & Strick, P. L. (1986). Parallel organization of functionally segregated circuits linking basal ganglia and cortex. *Annual Review of Neuroscience, 9*, 357–381.

Arbib, M. A. (1981). Perceptual structures and distributed motor control. In: V.B. Brooks (Ed.), *Handbook of Physiology; Nervous System, vol. II.* (pp. 1448–1480). Bethesda: American Physiological Society.

Baddeley, A. D. (1996). Exploring the central executive. *Quarterly Journal of Experimental Psychology, 49A*, 5–28.

Baddeley, A. D. (2002). Fractionating the central executive. In: D. T. Stuss, & R. T. Knight (Eds.), *Principles of frontal lobe function* (pp. 246–260). Oxford: Oxford University Press.

Barutta, J., Hodges, J., Ibañez, A., Gleichgerrcht, E., & Manes, F. (2010). Argentina's early contributions to the understanding of the frontotemporal lobar degeneration. *Cortex, 47*, 621–627.

Bergson, H. (1896). *Matière et memoire: essai sur la relation du corps à l'esprit.* Paris: Félix Alcan.

Blumen, S. C., & Blumen, N. (2002). From the philosophy auditorium to the neurophysiology laboratory and back: From Bergson to Damasio. *The Israel Medical Association Journal, 4*, 163–165.

Burrow, T. (1943). The neurodynamics of behavior. A phylobiological foreword. *Philosophy of Science, 10*, 271–288.

Campbell, A.W. (1905). Histological studies on the localisation of cerebral function. Cambridge: University Press.

Cappe, C., Rouiller, E. M., & Barone, P. (2009). Multisensory anatomical pathways. *Hearing Research, 258*, 28–36.

Capizzano, A. (2006). Actualidad del pensamiento de Cristofredo Jakob. *Revista del Hospital Italiano de Buenos Aires, 26*, 71–73.

Catani, M., & ffytche, D.H. (2005). The rises and falls of disconnection syndromes. *Brain, 128*, 2224–2239.

Chow, T. W., & Cummings, J. L. (1999). Frontal subcortical circuits. In: B. L. Miller, & J. L. Cummings (Eds), *The human frontal lobes: Functions and disorders* (pp. 25–43). New York: Guilford Press.

Duncan, J., & Miller, E. K. (2002). Cognitive focus through adaptive neural coding in the primate prefrontal cortex. In: D. T. Stuss, & R. T. Knight (Eds.), *Principles of frontal lobe function* (pp. 278–291). Oxford: Oxford University Press.

Duncan, J. & Owen, A. M. (2000). Common regions of the human frontal lobe recruited by diverse cognitive demands. *Trends in Neurosciences, 23*, 475–483.

ffytche, D. H., & Catani, M. (2005). Beyond localization: From hodology to function. *Philosophical Transactions of the Royal Society of London. Series B, Biological Sciences, 360*, 767–779.

Fogassi, L., Gallese, V., Fadiga, L., Luppino, G., Matelli, M., & Rizzolatti, G. (1996). Coding of peripersonal space in inferior premotor cortex (area F4). *Journal of Neurophysiology, 76*, 141–157.

François, C. (1999). Systemics and cybernetics in a historical perspective. *Systems Research and Behavioral Science, 16*, 203–219.

Fritsch, G. T., & Hitzig, E. (1870). On the electrical excitability of the cerebrum. In: G. Von Bonin (Ed.), *Some Papers on the Cerebral Cortex* (pp. 73–96). Springfield IL: Charles C. Thomas (1960).

Fuster, J. (2006). The cognit: A network model of cortical representation. *International Journal of Psychophysiology, 60*, 125–132.

Geschwind, N. (1965a). Disconnexion syndromes in animals and man. I. *Brain : a Journal of Neurology, 88*, 237–294.

Geschwind, N. (1965b). Disconnexion syndromes in animals and man. II. *Brain : a Journal of Neurology, 88*, 585–644.

Ghazanfar, A. A., & Schroeder, C. E. (2006). Is neocortex essentially multisensory? *Trends in Cognitive Sciences, 10*, 278–285.

Gibson, J. J. (1986). The ecological approach to visual perception. Hillsdale, New Jersey: Lawrence Erlbaum.

Godefroy, O., Cabaret, M., Petit-Chenal, V., Pruvo, J. P., & Rousseaux, M. (1999). Control functions of the frontal lobes. Modularity of the central-supervisory system? *Cortex, 35*, 1–20.

Goldman-Rakic, P. (2000). Localization of function all over again. *NeuroImage, 11*, 451–457.

Graziano, M. S. A., Reiss, L. A., & Gross, C.G. (1999). A neuronal representation of the location of nearby sounds. *Nature, 397*, 428–430.

Graziano, M. S. A., Yap, G.S., & Gross, C. G. (1994). Coding of visual space by premotor neurons. *Science, 266*, 1054–1057.

Hurley, S. (2001). Perception and action: alternative views. *Synthese, 129*, 3–40.

Ibañez, A., & Cosmelli, D. (2008). Moving beyond computational cognitivism: Understanding intentionality, intersubjectivity and ecology of mind. *Integrative Psychological & Behavioral Science, 42*, 129–136.

Jakob, C. (1906a). Consideraciones anátomo-biológicas sobre los centros del lenguaje. *La Semana Médica (Buenos Aires), 13*, 733–737.

Jakob, C. (1906b). Estudios biológicos sobre los lóbulos frontales cerebrales. *La Semana Médica, 13*, 1375–1381.

Jakob, C. (1906c). Existe ó no un centro de Broca? *La Semana Médica (Buenos Aires), 13*, 677–678.

Jakob, C. (1906d). La leyenda de los lóbulos frontales cerebrales como centros supremos psíquicos del hombre. *Arquivos de Psiquiatría, Criminología y Ciencias Afines, 5*, 679–699.

Jakob, C. (1906e). *Localización del alma y de la inteligencia.* Buenos Aires: El libro.

Jakob, C. (1906f). Nueva contribución á la fisio-patología de los lóbulos frontales. *La Semana Médica, 13*, 1325–1329.

Jakob, C. (1907a). Sobre apraxia. *La Semana Médica, 14*, 1344.

Jakob, C. (1907b). Sobre la sintomatología de las afecciones del lóbulo frontal. *La Semana Médica, 14*, 1285.

Jakob, C. (1909). Estudios anátomoclínicos sobre los lóbulos frontales del cerebro humano (Comunicación presentada al IV Congreso Médico Latinoamericano, Rio de Janeiro, 1–8 de agosto de 1909). *Argentina Médica, 7*, 463–472.

Jakob, C. (1911). *Das Menschenhirn: Eine Studie über den Aufbau und die Bedeutung seiner grauen Kerne und Rinde.* J. F. Lehmann, München.

Jakob, C. (1912a). Über die Ubiquität der senso-motorischen Doppelfunktion der Hirnrinde als Grundlage einer neuen, biologischen Auffassung des corticalen Seelenorgans. *Journal für Psychologie und Neurologie (Leipzig), 19,* 379–382.

Jakob, C. (1912b) Ueber die Ubiquität der senso-motorischen Doppelfunktion der Hirnrinde als Grundlage einer neuen biologischen Auffassung des kortikalen Seelenorgans. *Münchener Medizinische Wochenschrift, 59,* 466–468.

Jakob, C. (1918). Del mecanismo al dinamismo del pensamiento: Estudio histórico-crítico de psicología orgánica. *Anales de la Facultad de Derecho y Ciencias Sociales de la Universidad de Buenos Aires, 18,* 195–238.

Jakob, C. (1919). La teoría actual de las gnosias y praxias como factores fundamentales en el dinamismo cortical. *Revista del Círculo Médico Argentino y Centro Estudiantes de Medicina (Buenos Aires), 19,* 1266–1275. Jakob, C. (1921). La teoría actual de las gnosias y praxias como factores fundamentales en el dinamismo de la corteza cerebral. *La Crónica Médica, 38,* 17–24.

Jakob, C. (1935a). La filogenia de las kinesias: Sobre su organización y dinamismo evolutivo. *Anales del Instituto de Psicología de la Facultad de Filosofía y Letras de la Universidad de Buenos Aires, 1,* 109–127.

Jakob, C. (1935b). Sobre las bases orgánicas de la memoria. *Revista de Criminología, Psiquiatría y Medicina Legal, 127,* 84–114.

Jakob, C. (1941). La función psicogenética de la corteza cerebral y su posible localización (Aspectos de la ontopsicogénesis humana. *Anales del Instituto de Psicología de la Facultad de Filosofía y Letras de la Universidad de Buenos Aires, 3,* 63–80.

Jakob, C. (1943). *Folia neurobiológica Argentina, tomo III. El lóbulo frontal: Estudio monográfico anatomoclínico sobre base neurobiológica.* Buenos Aires: Aniceto López-López y Etchegoyen.

Jakob, C. (1945a). El cerebro humano: su significación filosófica. *Revista Neurológica de Buenos Aires, 10,* 89–110.

Jakob, C. (1945b). Sobre el origen de la conciencia: investigaciones neurobiológicas sobre la dinámica cortical en relación con su sectorización conmemorativa. In: E. Mouchet (Ed.), *Temas actuales de psicología normal y patológica, publicados bajo el patrocinio de la Sociedad de Psicología de Buenos Aires* (pp. 345–381). Buenos Aires: Editorial Médico-Quirúrgica/Talleres 'The Standard'.

Jakob, C. (1946a). *Folia Neurobiológica Argentina, tomo V. Documenta Biofilosófica.* Buenos Aires: López & Etchegoyen.

Jakob, C. (1946b). La demencia progresiva: Un analisis neurobiologico de la enfermedad de Pick. *Revista Neurologica de Buenos Aires, 1,* 81-94.

Jakob, C., & Onelli, C. (1913). *Atlas del cerebro de los mamíferos de la República Argentina: estudios anatómicos, histológicos y biológicos comparados sobre la evolución de los hemisferios y de la corteza cerebral.* Buenos Aires: G. Kraft

Jakob, C., & Pedace, E. A. (1949). La misión del lóbulo frontal frente a una cuantificación sintética de sus elementos productores. *Archivos de Neurocirugía, 6,* 467–474.

Koechlin, E., Ody, C., & Kouneiher, F. (2003).The architecture of cognitive control in the human prefrontal cortex. *Science, 302,* 1181–1185.

López Pasquali, L. (1965). *Christfried Jakob. Su obra neurológica, su pensamiento psicológico y filosófico.* Buenos Aires: López.

Maturana, H. R., & Varela, F. J. (1980): *Autopoiesis and cognition. The realization of the living.* Dordrecht: Reidel.

Maturana, H. R., & Varela, F. J. (1987). *The tree of knowledge: The biological roots of human understanding.* Boston, MA, US: New Science Library/Shambhala Publications.

Moyano, B. A. (1957). Christfried Jakob, 25/12/1866–6/5/1956. *Acta Neuropsiquiátrica Argentina, 3,* 109–123.

Orlando, J. C. (1966). *Christofredo Jakob: Su vida y obra.* Buenos Aires: Editorial Mundi.

Pedace, E. A. (1949). Contribución de la escuela neurobiológica Argentina del Prof. Chr. Jakob en el estudio del lóbulo frontal. *Archivos de Neurocirugía, 6,* 464–466.

Petrides, M. (2005). Lateral prefrontal cortex: Architectonic and functional organization. *Philosophical Transactions of the Royal Society of London. Series B, Biological Sciences, 360,* 781–795.

Raichle, M. E. (2002). Foreword. In: D. T. Stuss, & R. T. Knight (Eds.), *Principles of frontal lobe function* (pp. vii–ix). Oxford: Oxford University Press.

Risberg, J. (2006). Evolutionary aspects on the frontal lobes. In: J. Risberg, & J. Grafman (Eds), *The frontal lobes: Development, function, and* pathology (pp. 1–20). Cambridge: Cambridge University Press.

Rudrauf, D., Lutz, A., Cosmelli, D., Lachaux, J.-P., & Le Van Quyen, M. (2003). From autopoiesis to neurophenomenology: Francisco Varela's exploration of the biophysics of being. *Biological Research, 36,* 27–66.

Semendeferi, K., Lu, A., Schenker, N., & Damasio, H. (2002). Humans and great apes share a large frontal cortex. *Nature Neuroscience, 5,* 272–276.

Shallice, T. (2002). Fractionation of the supervisory system. In: D. T. Stuss, & R. T. Knight (Eds.), *Principles of frontal lobe function* (pp. 261–277). Oxford: Oxford University Press.

Shallice, T., &Burgess, P. (1996). The domain of supervisory processes and temporal organization of behaviour [and discussion]. *Philosophical Transactions: Biological Sciences, 351,* 1405–1412.

Stuss, D. T. (2006). Frontal lobes and attention: Processes and networks, fractionation and integration. *Journal of the International Neuropsychological Society, 12,* 261–271.

Stuss, D. T., Alexander, M. P., Floden, D., Binns, M. A., Levine, M., McIntosh, A. R., Rajah, N., & Hevenor, S.J. (2002). Fractionation and localization of distinct frontal lobe processes: Evidence from focal lesions in humans. In: D. T. Stuss, & R. T. Knight (Eds.), *Principles of frontal lobe function* (pp. 392–407). Oxford: Oxford University Press.

Teuber, H. L. (2009). The riddle of frontal lobe function in man. *Neuropsychology Review, 19,* 25–46.

Théodoridou, Z. D., Triarhou L. C. (2011a). Challenging the supremacy of the frontal lobe: Early views (1906-1909) of Christfried Jakob on the human cerebral cortex. *Cortex.* doi:10.1016/j.cortex.2011.01.001.

Théodoridou, Z. D., Triarhou L. C. (2011b). Christfried Jakob's late views on cortical development, localization and neurophilosophy. *Neuroscience Letters* (in press).

Théodoridou, Z. D., Triarhou, L. C. (2011c). Christfried Jakob's late views (1930-1949) on the psychogenetic function of the cerebral cortex and its localization: Culmination of the neurophilosophical thought of a keen brain watcher. *Brain and Cognition* (in press).

Théodoridou, Z. D., Triarhou, L. C. (2011d). Christfried Jakob's 1921 theory of the gnoses and praxes as fundamental factors in cerebral cortical dynamics. *Integrative Psychological and Behavioral Science, 45*, 247–262.

Thompson, E., & Varela, F. J. (2001). Radical embodiment: Neural dynamics and consciousness. *Trends in Cognitive Sciences, 5*, 418–425.

Triarhou, L. C., & del Cerro, M. (2006a). Semicentennial tribute to the ingenious neurobiologist Christfried Jakob (1866–1956). 1. Works from Germany and the first Argentina period, 1891–1913. *European Neurology, 56*, 176–188.

Triarhou, L. C., & del Cerro, M. (2006b). Semicentennial tribute to the ingenious neurobiologist Christfried Jakob (1866–1956). 2. Publications from the second Argentina period, 1913–1949. *European Neurology, 56*, 189–198.

Triarhou, L. C., & del Cerro, M. (2007). Pioneers in Neurology: Christfried Jakob (1866–1956). *Journal of Neurology, 254*, 124–125.

Triarhou, L. C. (2010a). Final publications of Christfried Jakob: On the frontal lobe and the limbic region. In: C. E. Flynn, & B. R. Callaghan, (Eds.), *Neuroanatomy Research Advances* (pp. 165–169). Hauppauge, NY: Nova Science Publishers.

Triarhou, L. C. (2010b). Revisiting Christfried Jakob's concept of the dual onto-phylogenetic origin and ubiquitous function of the cerebral cortex: A century of progress. *Brain Structure and Function, 214*, 319–338.

Tsapkini, K., Vivas, A. B., & Triarhou, L. C. (2008). 'Does Broca's area exist?' Christofredo Jakob's 1906 response to Pierre Marie's holistic stance. *Brain and Language, 105*, 211–219.

Von Uexküll, J. (1934). A stroll through the worlds of animals and men. In: C. Schiller (Ed.), *Instinctive behavior* (pp. 5–80). New York: International Universities Press, 1957.

Von Weizsäcker, V. (1950). *Der Gestaltkreis.* Stuttgart: Thieme.

Wallace, M. T., Meredith, M. A., & Stein, B. E. (1992). Integration of multiple sensory modalities in cat cortex. *Experimental Brain Research*, 91, 484–488.

Wiener, N. (1948). *Cybernetics or control and communication in the animal and the machine.* Paris: Hermann.

York III, G. K. (2009). Localization of language function in the twentieth century. *Journal of the History of the Neurosciences, 18*, 283–290.

Cerebellum (2016) 15:395–416
DOI 10.1007/s12311-016-0790-0

The Cerebellar System and What it Signifies from a Biological Perspective, by Professor Christofredo Jakob (English Translation)

Anny Tzouma[1] · Lazaros C. Triarhou[2]

Published online: 27 May 2016
© Springer Science+Business Media New York 2016

Abstract The paper is an English translation of Christofredo Jakob's 1938 lecture on cerebellar neurobiology, rendered from the original Spanish text. Communicated at the special sessions of the Society of Neurology and Psychiatry of Buenos Aires, December 1938.

Keywords Cerebellar histophysiology · Ontogeny · Phylogeny · History of neuroscience

In the first atlas of the nervous system that I published in 1895 [1–6], I concluded my discussion on the organization of the cerebellum with the following claim: "At present, we do not know with certainty much on its organization." There is no doubt that the existence of the cerebellum and its rough configuration was already known from the ancient times of Aristotle and Galen. We obtained the first notion of its morphology (peduncles, arbor vitae, etc.) from Vincenzo Malacarne (1744–1816) and received much of the information from Thomas Willis (1621–1675). Raymond Vieussens (1741–1715)

The Introduction of Cerebellar Classic XII is available at http://dx.doi.org/10.1007/s12311-016-0789-6 and the Commentary paper related to Cerebellar Classic XII is at http://dx.doi.org/10.1007/s12311-016-0791-z.

Electronic supplementary material The online version of this article (doi:10.1007/s12311-016-0790-0) contains supplementary material, which is available to authorized users.

✉ Lazaros C. Triarhou
   triarhou@uom.gr

[1] Graduate Program in Neuroscience and Education, University of Macedonia, 156 Egnatia Ave., Thessalonica 54006, Greece

[2] Laboratory of Theoretical and Applied Neuroscience and Graduate Program in Neuroscience and Education, University of Macedonia, 156 Egnatia Ave., Thessalonica 54006, Greece

described the dentate nucleus and Félix Vicq d'Azyr (1748–1794) described the internal configuration. Pierre Tarin (1735–1761) discussed the valve (inferior medullary velum) that bears his name, Friedrich Burdach (1776–1847) described the red nucleus, and Johann Christian Reil (1759–1813) discussed the commissures and decussations. In the beginning of the 19th century, these last authors told, with great tidiness, of the macroscopic topography of the cerebellar centers and pathways.

Jan Evangelista Purkyně (1787–1869), who in 1837 discovered its efferent cells, began the study of the cerebellar histology. Subsequently, Benedict Stilling (1810–1879) examined the deep nuclei. However, it is only through the work of Albert von Kölliker (1817–1905) and Santiago Ramón y Cajal (1852–1934) that we arrived at the classical period of fine cerebellar histology. On the other hand, Paul Flechsig (1847–1929), Vladimir M. Bekhterev (1857–1927), Jules Déjerine (1849–1917), and their pupils investigated the systematization of its pathways.

Based on these known facts, toward the end of the 19th century, I summarized the biological organization of the cerebellum [1–6] as follows:

The cerebellum undoubtedly exerts an influence on static coordination and the maintenance of equilibrium of the body in the erect position and while walking…To this end, it receives, via centripetal pathways, muscle stimuli (as well as visual, tactile, and other stimuli from the periphery)…Through the restiform bodies, pathways proceeding from the posterior spinal column nuclei make their way to the cerebellum. The cerebellum further receives the lateral cerebellar tract (of unknown function) and fibers from the vestibular and trigeminal nerve (direct sensory cerebellar system).

We can only formulate speculative hypotheses to be able to interpret the cerebellar influence on such coordination…As the cerebellum has multiple connections with the anterior cerebrum, we may imagine that its regulatory influence is exerted on such motor cortical areas and not directly on muscular function, in view of the fact that direct connections with these particular centers have not been demonstrated… Pathways originating in nuclei of the contralateral side of the pons connect to the cerebellum; the fronto-, temporo- and occipito-pontine tracts terminate in those pontine nuclei…Consequently, each cerebellar hemisphere appears to be in direct relation with the frontal, temporal, and occipital lobes of the contralateral cerebral hemisphere…Moreover, each cerebellar hemisphere is connected, by means of the superior cerebellar peduncle, with the contralateral red nucleus of the tegmentum, which, in turn, is connected with the optic thalamus, etc.

We know of the existence of various such cerebrocerebellar systems; however, their function is not yet elucidated, although there is little doubt that they are, in some way, associated with coordination…It is conceivable that there is a direct relationship between the musculature and the cerebellum by means of the contralateral medullary olivary fibers, which are directed, through the restiform body, to the cerebellum. Also from this point, the central tegmental tract (mesial fillet) continues upward, whereas the olivary spinal tracts descend (tract of Hellweg) through the lateral column of the spinal cord to the cells of the anterior cervical horn (?)…In all, it appears that in the cerebellum, there exist no distinct areas like those in the forebrain…In essence, one may attribute to the cerebellum a *static*, *tonic*, and *sthenic* influence on muscle coordination. From all this, we can affirm that *we have very little definite knowledge* in this field of study.

That was the case 40 years ago. Currently, it would be very interesting to understand further the latest developments. To this end, I shall present my phylogenetic biological propositions published 10 years ago, prompted by a case of cerebellar softening [8, 9].

Only when guided by genetic facts will it be possible to elucidate, in the future, the "Gordian Knot" that rises before us today, as it did in the past: the complex problem of "cerebellar dynamics" and the analysis of its factors. Accepting Ludwig Edinger's (1855–1918) theory, with his known division into the paleocerebellum and neocerebellum, I arrived, through my own studies, at the following concepts:

In cerebellar morphofunctional phylogeny, we have to distinguish *three stages* in the successive perfection of the intensity and the extent of the cerebellar influence upon motor acts, which are initially instinctive, and subsequently become voluntary.

First Period (Static Vestibulocerebellar Dynamics) This function, which occurs from lower fish to humans (Fig. 1), characterizes the cerebellum as a suprareflex center of stabilizing correlation through stimuli derived from the lateral organs and the labyrinth (lateral and vestibular nerve). It only involves part of the inferior vermis, along with a lateral accessory lobe, a system represented, from mammals to humans, by the nodulus of the inferior vermis and the floccular lobe (flocculus and paraflocculus). So, then, the *archicerebellum* is represented by this oldest part of the *metencephalon*, and its function as a vestibular suprareflex center (the cerebellar granule cells functioning as a multiplier system in all species) links the stimuli of the ambient vibratory pressure to the position of the organism and its axial balance. No other source of stimuli is involved in this primeval static dynamics, which is effected through descending cerebellomedullary pathways that run in the posterior longitudinal fasciculus, along with neighboring zones of the reticular formation (juxtarestiform body) and, through its nuclei, toward the spinal motor nuclei of the caudal fin in fishes, in conjunction with the bulbospinal descending pathways (Müller and Mauthner fibers).

Second Period In higher fish, and further accentuated in reptiles, the labyrinthine apparatus tends to be refined, while the lateral organs disappear and the vermicular cerebellar aspect is augmented; the cerebellar vestibular root now terminates in the sub-cerebellar nuclei (the future deep cerebellar nuclei), and the cerebellar cortex is perfected and further receives muscle and tactile stimuli via the spinobulbocerebellar pathways. Hence, our organ receives, and by now associates with, various stimuli, elaborating with them its synthetic suprareflex function of dynamic correlations. Its influence is extended and becomes more intense over muscle coordination during the movements of the entire body of the organism: *dynamic cephalotruncal vestibulo-spinocerebellar function*. This constitutes our *paleocerebellar* apparatus, which, via the precursor descending pathways of the superior cerebellar peduncle, near the secondary juxtarestiform pathways, evokes coordinating regulations by means of the nuclei of the reticular formation of the hindbrain and the midbrain as well (the precursors of the red nucleus). Its dynamic influence coordinates, then, in a combined function, the position and movement of the musculature of the trunk, neck, and head (with the exception, it appears, of the extremities).

Third Period This is characterized by the appearance of the *neocerebellar* apparatus, which in mammals (Fig. 2) adds two more static-dynamic functions, elaborating on the central vermicular lobe the cortical influence over the cerebellar hemispheres via the corticopeduncular pathways, toward the pontine nuclei and through the middle cerebellar peduncle to the contralateral cerebellum, increasing, in this way, the form and function of the cerebellar hemispheres. These are added to the

 Springer

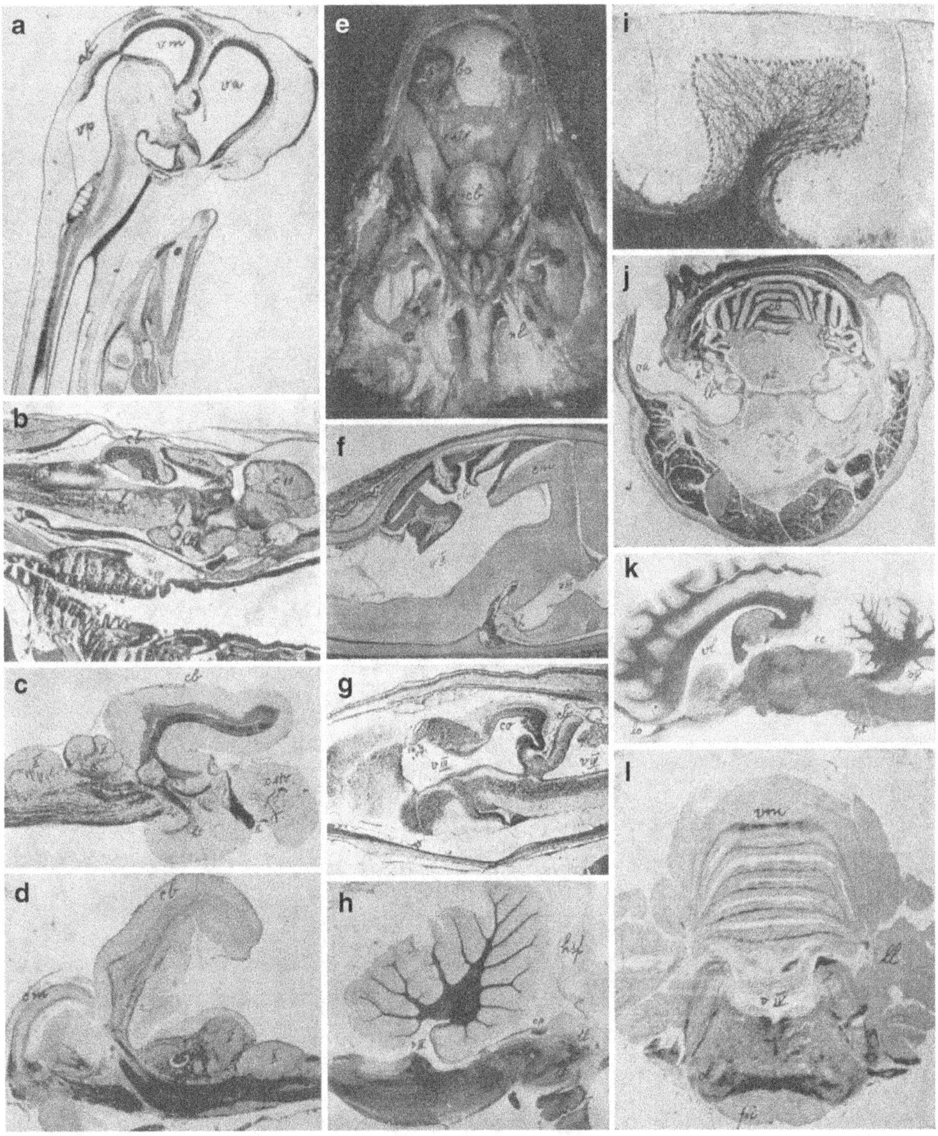

Fig. 1 **a** Shark (*Tiburón*) embryo with the three cerebral vesicles and the tectal cerebellar lamina (*cb*). **b** Cerebrum of a viviparous fish (*Jenynsia multidentata*) from the Río de La Plata basin; *cb*, cerebellum. **c** Cerebellum (*cb*) of the South American catfish (*Pseudoplatystoma tigrinum*), with its hypothalamic pathways. **d** Cerebellum (*cb*) of the ocean sunfish (*Mola mola*), with its bulbospinal pathways. **e** Shark cerebellum, dorsal aspect; *nl*, lateral nerve. **f** Shark cerebellum (*cb*); its cellular structure in midsagittal section. **g** Rudimentary cerebellum (*cb*) of an amphibian. **h** Avian cerebellar fibers in a pigeon. **i** Cerebellar lobule of a weasel (*Mustela*); row of Purkinje cells with fibers. **j** Cranial-labyrinthine (*lb*) topography of the rat cerebellum; *cb*, vermis; *ll*, lateral lobule. **k** A sagittal section of the sheep (*Oveja*) brain; fronto-cerebellar relations. **l** Cerebellum of a lion that is a few days old; myelination of the vermis, absence of myelination of the cerebellar hemispheres and the pyramidal tract

central lobe; at the same time, the superior and inferior vermis grows accordingly. This new dynamics, reinforced by new spinobulbocerebellar afferent pathways, now elaborates on the true cerebellar synergy—the neocerebellar function, which also extends to the movements of the extremities in combination with the trunk, associating its game with that of the position and support of the body during locomotion in all its phases (because any function is fulfilled in live collaboration with the *cumulative cerebellar propensity*. This is comparable to the game of the spider at any point in its intricate

🕮 Springer

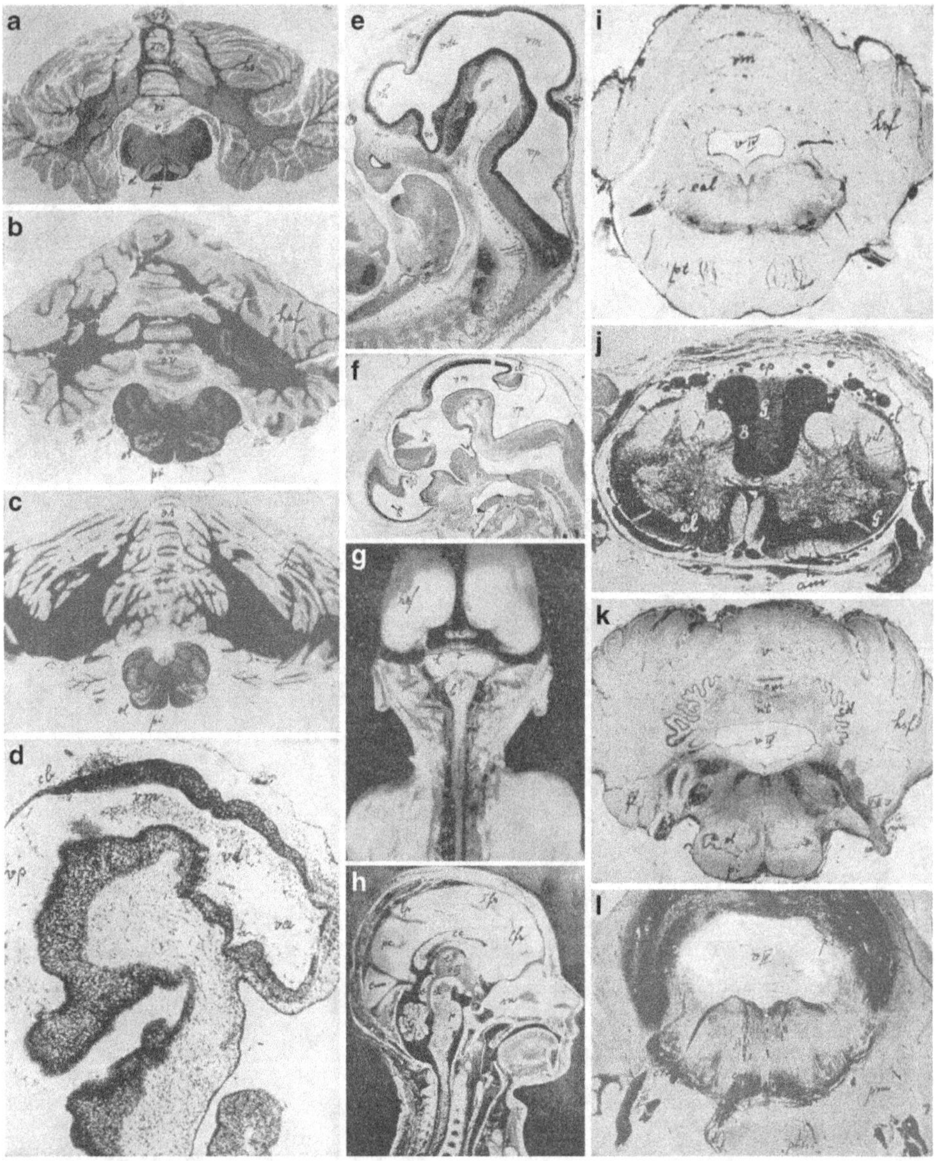

Fig. 2 a Seal cerebellum and medulla with marked hemispheric and olivary development. **b** Cerebellum of a baboon, early primate type. **c** Chimpanzee cerebellum, anthropoid type. **d** Two-week-old human embryo, cerebral vesicles with the cerebellar lamina (*cb*) in the tectum of the posterior vesicle. **e** Four-week-old human embryo with the cerebellar bud (*cb*) in the tectal lamina of the posterior cerebral vesicle (*vp*). **f** Eight-week-old human with the medial cerebellar bud (vermis, *cb*). **g** Four-month-old human fetus with the beginning of the cerebellar vermian sulci; the hemisphere is still underdeveloped and smooth in its inferior aspect. **h** Sagittal section of a 5-month-old human fetus. The vermis is fully foliated. **i** Cerebellar section of a 6.5-month-old human fetus, without cerebellar myelination; the reflex systems are already matured. **j** Medulla of a 7-month-old human fetus, with myelination of afferent cerebellar pathways (*cl*, *cv*); efferent pathways of the anterior-marginal zone (*a-m*) are not myelinated. **k** Cerebellum of an 8-month-old human fetus. Beginning of flocculonodular (*fl*) and commissural tegmental (*nt*) myelination; completed myelination of the vestibular system (*VIIIv*, to the right). **l** Fetal cerebellum at term; myelinated vermian systems (*pi*); *id*, commissural system. Absence of myelination in the middle cerebellar peduncle and the pyramidal tract

spider web). This also refines the afferent vestibulo-spinobulbar pathways, which are associated with those of the crossed corticopontine stimuli (via the middle cerebellar peduncles and the olivary pathways). Cerebellar discharge becomes intensified through the major development of the superior cerebellar peduncle and the rubral and pararubral systems (rubrospinal, pararubrospinal and rubro-hypothalamo-pallidal, the hypothalamic radiations).

Of special interest is the confirmation that the fine structure of the *cerebellar cortex* remains constant along the entire phylogenetic scale. From fish to humans, we have the three typical layers: molecular, Purkinje cell, and granule cell layers. Likewise, they do not differ in their constituent elements, but they only differ in their numbers. For example, in fish, there are only one million Purkinje cells; in humans, there are eight million, while the number of granule cells rises to seven billion (personal counts).

While the lower vertebrates only possess cerebellospinal pathways, in higher vertebrates that efferent pathway is further perfected, such that it discharges toward the diencephalon (rubral system) and eventually toward the striatum and the frontal lobe. Thus, in mammals, the rubrolenticular and thalamocortical systems appear; the influence of the cerebellar dynamics on the corpus striatum and the praxical and motor cortex, as we shall see below, is achieved in primates by means of such systems, which are especially developed, and even more perfected, in humans. In exchange, the direct cerebellospinal connections become reduced.

In human cerebellar pathophysiology (Fig. 3), as already explained, we have to take into account the partial or complete lesions of those three different dynamics, since the archicerebellar and paleocerebellar functions have been combined with the neocerebellar functions. In such a complex act, the vestibular and spinobulbocerebellar tracts mediate, as afferent pathways, the peripheral stimuli (tract of the restiform body and cerebellovestibular root). The pontocerebellar tracts mediate the cortical stimuli (frontal-opercular and temporal-parietal), as does the olivocerebellar path, while the juxtarestiform body and the superior cerebellar peduncles mediate as efferent pathways, discharging the cerebellar influence from both the vermis (trunk) and the cerebellar hemispheres (extremities). This is not accomplished directly via long pathways toward lower motor centers, but rather, as we have seen, it is accomplished with all of them terminating—from the lowest vertebrates to the mammals, except for primates—in various nuclei of the pontine (nucleus of Deiters, magnocellular nucleus, etc.) and mesencephalic (red and pararubral nuclei) reticular formation, from where the final stimuli depart to the spinal motor centers (rubrospinal and bulboreticulospinal pathways). The cerebellovestibular tract (*faisceau en crochet* of Russell) is also described as a descending cerebellopontine pathway to the tegmentum, but its function, whether afferent or efferent, remains unclear.

In primates, and especially in humans, all of these descending efferent pathways appear rather rudimentary. In exchange, the functions of the ascending pathways to the red nucleus—where a large proportion of the fibers of the crossed superior cerebellar peduncle terminate—gain in intensity. It is here where, subsequently, the fundamental rubral systems that typify the primate brain originate, reaching their maximum growth in the human brain: the *rubrothalamic tract* (to the lateral nucleus of the thalamus and thence via the anterior thalamic radiations to the *frontal* and *opercular cortex*) and underneath the *rubrolenticular tract* which, at the same time, carries another part of the cerebellar impulse via the ansa lenticularis to the lenticular and caudate nucleus.

Thus, in human cerebellar pathophysiology, it is not the direct cerebellar influences that dominate motor centers, although they do exist, but rather, it is the pronounced indirect, cortical, and striatal pathways. This is due to the fact that these higher centers begin to develop such new pathways which discharge the cerebellar contribution to the motor foci. The claudication of the cerebellar influence in humans thus manifests itself rather indirectly through the functional alteration of striatofugal and corticofugal pathways.

All this as far as phylogeny is concerned. From the study of ontogeny (Fig. 2), we know that the cerebellum originates in the lamina of the tectum in the anterior one third of the third cerebral vesicle. Initially forming a symmetrical neuroependymal bud, it is already reunited by the end of the first month, in the form of a transverse brain tissue lamina, with neuroblasts migrating from the ependyma toward the periphery (thickening vermicular plate). In the third month, the nodular bud becomes separated by the uvulonodular sulcus. The central sulcus (false "primary") appears next, between the future superior and inferior vermis. The hemispheric buds only appear in the fourth month, and in the fifth month, they acquire their first sulci (the dorsal part is acquired much earlier than is the basal). Toward the seventh month, the cerebellum is morphologically completed, but it is, nonetheless, very small. By the fifth month, the Purkinje neuroblasts have appeared, while the granule cell layer presents a unique fact in embryology: its neuroblasts, having initially migrated toward the periphery, begin, around the seventh month, to follow a reverse path toward the interior as they mature, until they settle in their final position beneath the Purkinje cell layer. And yet, in the newborn, the periphery (the marginal zone of Obersteiner) is occupied to a great extent by granule cell neuroblasts, and only after the second post-natal month are all of its elements settled in their definitive position.

This explains the fact that the cerebellum, along with the corpus striatum and the cerebral cortex, belongs to those nervous centers that become perfected right after birth (in contrast with reflex systems); here, evidently, function influences maturation, with regard to centers of "remanence," in which, in some form, functional dispositions persist, facilitating new successive analogous

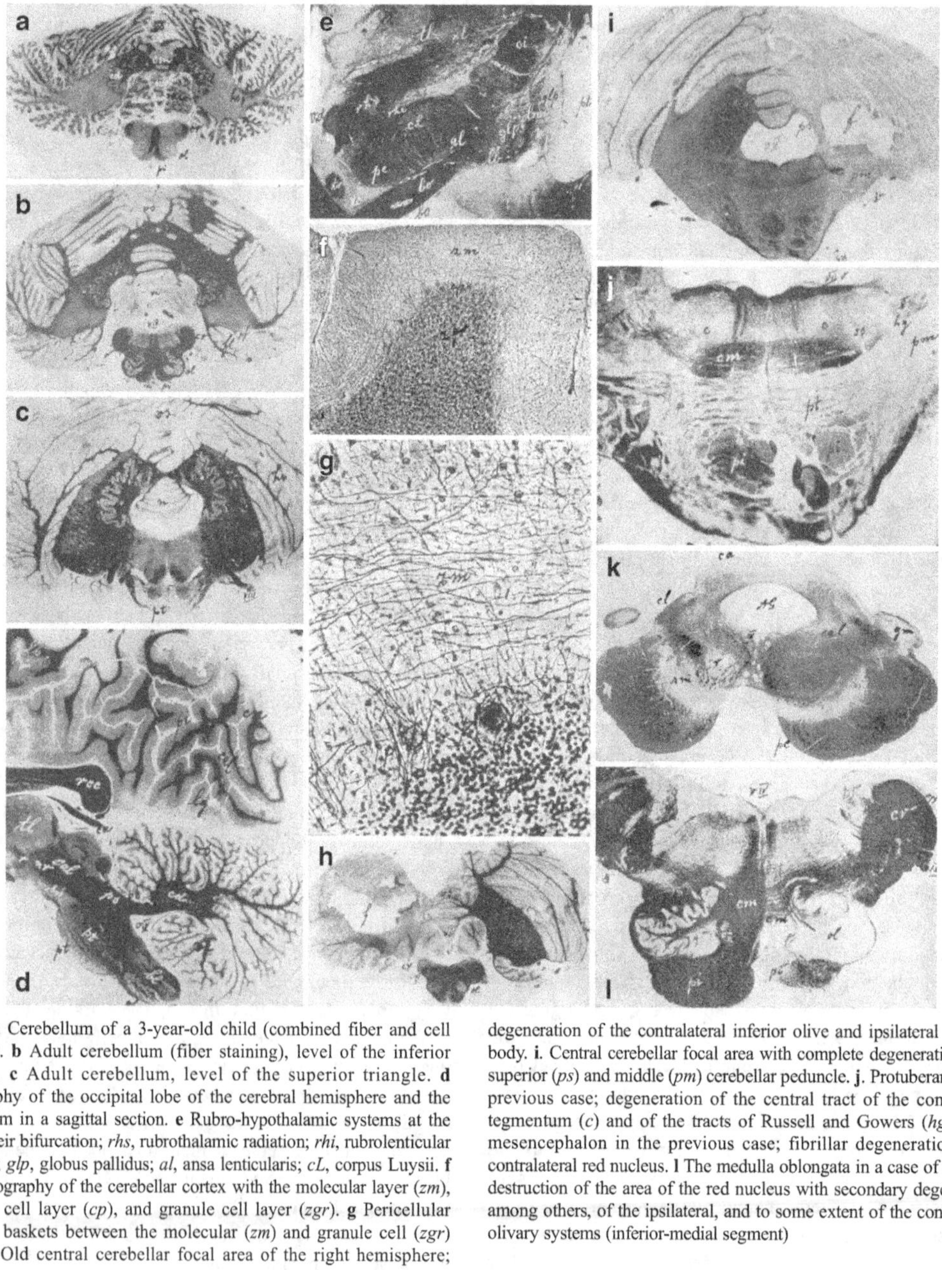

Fig. 3 a Cerebellum of a 3-year-old child (combined fiber and cell staining). **b** Adult cerebellum (fiber staining), level of the inferior triangle. **c** Adult cerebellum, level of the superior triangle. **d** Topography of the occipital lobe of the cerebral hemisphere and the cerebellum in a sagittal section. **e** Rubro-hypothalamic systems at the site of their bifurcation; *rhs*, rubrothalamic radiation; *rhi*, rubrolenticular radiation; *glp*, globus pallidus; *al*, ansa lenticularis; *cL*, corpus Luysii. **f** Histotopography of the cerebellar cortex with the molecular layer (*zm*), Purkinje cell layer (*cp*), and granule cell layer (*zgr*). **g** Pericellular Purkinje baskets between the molecular (*zm*) and granule cell (*zgr*) layer. **h** Old central cerebellar focal area of the right hemisphere; degeneration of the contralateral inferior olive and ipsilateral restiform body. **i**. Central cerebellar focal area with complete degeneration of the superior (*ps*) and middle (*pm*) cerebellar peduncle. **j**. Protuberance of the previous case; degeneration of the central tract of the contralateral tegmentum (*c*) and of the tracts of Russell and Gowers (*hg*). **k** The mesencephalon in the previous case; fibrillar degeneration of the contralateral red nucleus. **l** The medulla oblongata in a case of unilateral destruction of the area of the red nucleus with secondary degeneration, among others, of the ipsilateral, and to some extent of the contralateral, olivary systems (inferior-medial segment)

functions. The myelination of the neocerebellar system is perfected shortly after birth, along with the maturation of instinctive normokinesis. The vestibulo-nodulo-floccular system becomes myelinated first (fifth to sixth month of gestation); the spino-vermicular comes next (seven to ninth month of gestation), and the neoencephalic hemispheric

myelination (medium peduncle, etc.) follows, right after the generation, in parallel, of the pyramidal tracts.

With these general clues, I shall now discuss the human afferent, associative and efferent cerebellar systems, their origins and terminations, as well as their contacts and commissures, which would be impossible to do in a scientific manner

⬭ ⚨ Springer

without at least taking into account the essentials of their bio-dynamic functions.

A. Archicerebellar Systems

These represent the ascending and descending labyrinth-vestibulocerebellar pathways, which arrive by coursing through the fastigial nucleus and the nodulofloccular cerebellar cortex.

I. *Afferent systems*—(a) Collaterals of the vestibular nerve (still under discussion with regard to its origin in the labyrinth, ampulla, utricle, and saccule, and its cortical and subcortical termination). (b) The vestibulocerebellar tract, which is born as a second-order pathway, in the dorsal nucleus of the acoustic nerve (Deiters and triangular) to reach the nodulus, passing through the superior vestibular nucleus of Bekhterev.

II. *Efferent systems*—Cerebellomedullary tract (part of the juxtarestiform body, with its origin in Purkinje cells, nodulus and flocculus), whose termination is still debated.

III. *Associative systems*—Here, we treat the pathways from the nodulus to the flocculus and the vermis (under discussion).

   The relationship of this system to reflex and suprareflex functions, the position and movement of the head and trunk, is generally accepted, except for what amounts to details.

B. Paleocerebellar Systems

These comprise the afferent, efferent, and association pathways of the inferior and superior vermis and part of the deep cerebellar nuclei (emboliform, globose and dentate nucleus).

I. *Afferent systems*—Most of the fibers that comprise these systems arrive with the restiform body at the ipsilateral and contralateral vermis after outlining the dentate nucleus: (a) The dorsal spinocerebellar (lateral cerebellar) tract, which originates in Clarke's column from the third lumbar spinal segment to somewhere in the cervical cord, conveys stimuli (muscle, tactile?) to the superior vermis and a portion of the hemispheric region. (b) The ventral spinocerebellar tract (part of the tract of Gowers) conveys (tactile, thermal, nociceptive?) stimuli from the entire spinal cord (?), mostly crossed (?), to the vermis. (c) External arcuate fibers, contralateral and ipsilateral from the nuclei of Goll, Burdach, and von Monakow, convey muscle and joint (?) stimuli from the trunk and extremities. (d) Direct

or indirect trigeminocerebellar fibers (?) convey muscle stimuli from the neck and head. (e) Paraolivo-vermian pathways originate in the middle and dorsal paraolive and are directed, respectively, to the inferior and superior vermis (their function is unknown).

II. *Efferent system*—In addition to the reinforced archicerebellar systems (?) already mentioned, a part of the superior cerebellar peduncle is added to these, with crossed descending collaterals (tract of Cajal) that are first directed to the reticular formation and before continuing to the spinal cord (?). To these would belong the rubro-pararubral system (see below), as well as the relationships between the hypothalamic radiations and the striatum and thalamus. In short, we have the following elements:

   Firstly, a reinforced afferent system to the inferior vermis through the juxtarestiform body

   Secondly, the efferent system from the superior vermis: the dorsal portion of the superior cerebellar peduncle, the deep cerebellar nuclei, and the dentate nucleus to the red nucleus with: (a) Descending collaterals (tract of Cajal) going to the medulla and spinal cord (the "tract of Marchi" is not supported anymore). (b) The cerebellovestibular tract (*faisceau en crochet* of Russell) from the superior vermis to the tegmentum. (c) The descending rubral systems, formed by the pararubrospinal and part of the central tegmental tract. (d) The secondary rubrostriatal and rubrothalamic radiations.

III. *Associative and commissural systems*—Part of these systems is formed by the pathways of communication between the deep cerebellar nuclei and the association pathways with the archicerebellum and neocerebellum, as well as the inter-vermicular, all of which are under discussion.

Thus, completing the general review of paleocerebellar organization, the existence of uncertainties and gaps becomes evident.

Before proceeding to the neocerebellum, I shall next discuss in brief the stimuli, their topographic distribution, their contacts, and their effects.

With regard to the quality of the stimuli that the vermis receives, the admission of the muscle, and in general, the "deep" stimuli is a commonplace fact, despite the lack of solid direct evidence. The known division into primitive (proteropathic) and secondary (deuteropathic) stimuli cannot be applied to the cerebellum, because this is an organ that receives stimuli both from primary (labyrinthine) and secondary (muscular) pathways. Also, a classification is meaningful into proprioceptive stimuli (interoceptive) and exteroceptive; thus, for example, muscle pressure is exteroceptive and muscle contraction interoceptive. In addition,

 Springer

visceral stimuli are interoceptive and we undoubtedly do not accept their entry into the cerebellum. There is no denying that tactile stimuli enter, and then pressures and distensions intervene in the articulate game in both denervation and ligation experiments from the skin. All of this raises the following question: when does tactile sensitivity begin, and when does deep sensitivity begin? We could formulate a similar question concerning thermal sensitivity and nociception. It is true that it is of little importance, but I mention that the problem of the posterior collateral roots and their conduction remains unresolved.

Regarding the topographic distribution, the domination of stimuli that originate in the trunk and large joints appears certain, as well as of stimuli derived from the neck and head, binocular eye movements being part of them (for this, it is necessary to accept the existence of sensory muscular trigeminal pathways).

With respect to sensory contact, the following problem arises: where do afferent pathways terminate in the cerebellar cortex? We have two types of terminations—the mossy and the climbing fibers. The opinions of researchers are divided, with some supporting one or the other manner of termination. In addition to these types, there is also an eclectic type. My position, in this regard, has been the following for some time [7]. Most afferents—sensory in nature—terminate in the granule cell layer with the mossy fiber system, whereby this zone represents "the multiplier system of the cerebellum" in accordance with a principle of general neurodynamic organizational economy. Here, the conduction is limited and the contact is multiplied. It is only from there that the stimulus passes to the molecular layer in order to reach the Purkinje ramification, either directly or through the basket elements.

In contrast, climbing fibers are reserved for the conduction of associative stimuli, and, primarily, olivopontine. Purkinje cells represent the efferent system (to the dentate nucleus), as well as the associative system (under discussion). Afterward, these motor elements receive both multiplied and accumulated peripheral impulses (via the granule and basket cells), as well as central impulses (climbing fibers). However, there is also another cortical element, the *large stellate cell*, of which there are various types. These also receive stimuli from the longitudinal terminations of the granule cells and, in turn, concomitantly terminate among the granule cells. This cannot be explained as a system of "neuroenergetic accumulation," from which it follows that the cerebellum surely belongs to the suprareflex systems of "remanence," as does the associated striatal system. They both expand the reaction of the instinctive sphere, and especially *normokinesis*, via the autonomous regulation of the varying permanent tension that occurs in the muscular inter-play between flexors and extensors.

As far as their effects are concerned, we ignore the essence of "cerebellar neuroenergetics" because we can only discuss

its final dynamics, in correlation with those of the corpus striatum and the frontoinsular corticality, while it is certain that, in a solely rudimentary form, they themselves discharge, in humans, toward the medulla oblongata and the spinal cord.

As previously stated, the human cerebellum is starkly striato-cortico-prone as a consequence of the helio-proneness of humans due to upright walking.

Let us move on to the last chapter.

C. The Neocerebellar System

This system was recently perfected in higher mammals, reaching its culmination in primates, and above all, in humans. Under such refinement, there are the cerebellar hemispheres, which are rudimentary in most mammals (except the seal, dolphin, whale, and elephant), together with the major development of the olivary-vermicular system and the dentate nucleus. Nevertheless, attention is drawn to the fact that its cortical histological organization is exactly the same as that of more ancient systems, with the only difference being that it becomes myelinated later.

I.   *Afferent systems*—(a) Crossed (and probably ipsilateral) pontocerebellar pathways, which originate in the pontine nuclei, arrive through the middle cerebellar peduncle at the cortex of the cerebellar hemispheres (and vermis?), terminating as climbing fibers directly on Purkinje cells. As the frontal, insular, temporal, and parietal pathways terminate in the pontine nuclei, a regulatory cortico-cerebellar circuit of unknown importance is established as well. (b) Olivocerebellar pathways, which project from the olives, in inverted form, to the contralateral cerebellar hemisphere. As we ignore the class of olivary stimuli that arrive there via the central tract of the tegmentum (diencephalic and mesencephalic origin), only hypothetically could we think, among others, of trigeminal stimuli (phonetic musculature?).

II. *Efferent systems*—Only the superior cerebellar peduncle constitutes part of this system, especially reinforced in its inferior portion, which, originating in the dentate nucleus and decussating, terminates in the red nucleus (it is under discussion whether the termination in this nucleus is full or partial), and this gives origin to the hypothalamic radiations that are directed to the corpus striatum (inferior hypothalamic radiation or rubrolenticular radiation) and to those directed to the anterolateral thalamus (superior hypothalamic radiation or rubrothalamic radiation), which then continue to the thalamofrontal radiations.

III. *Associative and commissural systems*—These exist between the vermis and the cerebellar hemispheres in a limited amount. We lack information on their organizational details.

 Springer

Cerebellum (2016) 15:395–416

Whereas the coordinated regulation of the vermis fundamentally falls on the trunk axis, the hemispheres influence, in a more pronounced fashion, the extremities. I cannot be more specific in these statements because we lack precise data on localization and most likely on the existence of a cerebellar influence en bloc. The true biological situation of the cerebellum is substantiated in humans—I never get tired of insisting on this—in its intimate collaboration with the frontal lobe (a theme upon which I do not intend to elaborate here).

With regard to what the red nucleus signifies in higher primates, and especially in humans, we have to consider a fundamental fact that is still relatively unknown today: the red nucleus of other mammals is not identical to that of primates, neither in its structure nor in its function. The red nucleus in other mammals is of the "macrocellular" type and emits, in continuity with the efferent cerebellorubral system, a typical rubrospinal path, with weakened ascending rubrolenticulothalamic discharges.

Instead, the red nucleus of humans, which consists of "mesocellular" elements, does not emit any rubrospinal tract; in exchange, it gives rise, as we have seen, to the potent ascending rubrolenticular and rubrothalamic systems. From the macrocellular nucleus of mammals, only certain reduced cellular elements are conserved in humans in the reticular formation of the tegmentum (perirubral). From those also emanate other descending systems, as it happens with other magnocellular types of the reticular substance in the medulla (magnocellular nucleus of von Kölliker, its mesencephalic portion). Being in a position to talk only about the pararubral fasciculus in humans, we consequently ignore whether this bears any relationship to the superior cerebellar peduncle. Instead, in the human red nucleus, a portion of the central tegmental tract, the "rubro-olivary" system, originates at a different level, which is clearly related to the cerebellum. While it is unlikely that it reaches the spinal cord [because it is accepted that part of the central tract also terminates in the reticular substance (lateral medullary nucleus) and from there, the descending "bulbospinal" tracts could begin], it comes much closer to the cerebello-rubro-olivo-cerebellar "retrograde ring" (of unknown importance) and forms an interesting parallel with the major ring of the cerebello-rubro-thalamo-fronto-ponto-cerebellar dynamics (of no lesser importance in its ultimate function).

From all this, one can deduce that it would be risky to directly compare the results of experimental animal physiology with human mesencephalic syndromes (Fig. 3), as some neurologists have precipitously rushed to do. Nevertheless, in spite of any analogy, humans maintain their "biological reserves," the presence of which is a necessity.

Conclusion

In summary, I can affirm that the direct spinal influence on cerebellar phylogeny becomes progressively weakened, gaining, in exchange, in its hemispheric relations. Its physiological interpretation in many ways remains a cloudy terra incognita.

To finalize my opinion regarding the *role* of the cerebellum in the general peripheral-central neurodynamics, I have long upheld the following principle: our central cortical organ elaborates two categories of fundamental ontopsychisms in order to create the individual mentality of the conscious human—the *gnoses* and the *praxes*—or, in other words, the orientations and the interventions into the environs, acquired by the affective force of the person. With these, we construct our material and cognitive world.

The making of the gnoses, i.e., the "gnoseopoiesis" (in its true etymological sense, the Greek word "poiesis" means organization, production), utilizes the peripheral stuff that arrives directly at the thalamic centers (basal pathway), and from there, onto the parietal, occipital, and temporal cortex. On the other hand, the "praxiopoiesis" necessitates the "cerebellized" stuff, which, having passed through the cerebellum and red nucleus, reaches the thalamus, and from there, it continues to the frontal and opercular cortex (dorsal or indirect pathway). The terminal contact loops of both of these circuits ultimately converge into the gnoseopraxial associative cortical systems, the instigators of human ideopoietic dynamics.

Compliance and Ethical Standards

Conflict of Interest The authors declare that they have no conflict of interest.

References

1. Jakob C. Atlas des gesunden und kranken Nervensystems nebst Grundriss der Anatomie, Pathologie und Therapie desselben. München: J. F. Lehmann; 1895.
2. Jakob C. An atlas of the normal and pathological nervous systems, together with a sketch of the anatomy, pathology, and therapy of the same (Collins J, translator). New York: William Wood & Co.; 1896.
3. Jakob C. Atlas zdorovoy i bolnoy nervnoy sistemy, s kratkimi osnovami po anatomiy, fiziologiy i terapiy yeya (Traynin PD, Blyumenau MB, translators). Sankt-Peterburg: Izdanie V. S. Ettinger; 1896.
4. Jakob C. Atlas du système nerveux à l'état normal et à l'état pathologique, suivi d'un précis d'anatomie, de pathologie et de thérapeutique (Rémond A, Clavelier F, translators). Paris: A. Maloine; 1897.
5. Jakob C. Atlas-manuel du système nerveux à l'état normal et à l'état pathologique (Rémond A, Clavelier F, translators). Paris: J.-B. Baillière et Fils; 1899.

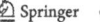

6. Jakob C. Atlante del sistema nervoso nello stato sano e nel patologico con un sunto di anatomia patologica e terapia del medesimo (Clerici A, Medea E, translators). Milano: Società Editrice Libraria; 1899.

7. Jakob C. Das Menschenhirn: Eine Studie über den Aufbau und die Bedeutung seiner grauen Kerne und Rinde. München: J. F. Lehmann; 1911.

8. Jakob C. Hemiplejía, hemiataxia y hemianestesia homolateral cerebelosa. In: Ameghino A, editor. Actas de la Primera Conferencia Latinoamericana de Neurología, Psiquiatría y Medicina Legal, vol. 1. Buenos Aires: Imprenta de la Universidad; 1929. p. 240–52.

9. Jakob C. Hemiplejía, hemiataxia y hemianestesia homolateral de origen cerebeloso. Arch Argent Neurol. 1929;4:13–31.

 Springer

Cerebellum (2016) 15:417–424
DOI 10.1007/s12311-016-0791-z

Commentary on "The Cerebellar System and What it Signifies from a Biological Perspective: A Communication by Christofredo Jakob (1866–1956) Before the Society of Neurology and Psychiatry of Buenos Aires, December 1938"

Anny Tzouma[1] · Daniel S. Margulies[2] · Lazaros C. Triarhou[3]

Published online: 27 May 2016
© Springer Science+Business Media New York 2016

Abstract This commentary highlights a "cerebellar classic" by a pioneer of neurobiology, Christfried Jakob. Jakob discussed the connectivity between the cerebellum and mesencephalic, diencephalic, and telencephalic structures in an evolutionary, developmental, and histophysiological perspective. He proposed three evolutionary morphofunctional stages, the archicerebellar, paleocerebellar, and neocerebellar; he attributed the reduced cerebellospinal connections in humans, compared to other primates, to the perfection of the rubrolenticular and thalamocortical systems and the intense ascending pathways to the red nucleus in exchange for the more elementary descending efferent pathways. Jakob hypothesized the convergence of cerebellar pathways in associative cortical regions, insisting on the intimate collaboration of the cerebellum with the frontal lobe. The extensive lines of communication between regions throughout the association cortex substantiate Jakob's intuition and begin to outline the mechanisms for substantial cerebellar involvement in functions beyond the purely motor domain. Atop a foundation of anatomical and phylogenetic mastery, Jakob conceived ideas that were noteworthy, timely, and have much relevance to our current thinking on cerebellar structure and function.

Keywords Cerebellar histophysiology · Ontogeny · Phylogeny · History of neuroscience

The academic life and the neurobiological work of Christfried Jakob (1866–1956) has resurfaced in the English scientific literature over the past decade [1, 2]. Born in Bavaria, Jakob moved to Argentina in 1899, where he spent the rest of his life, save a return to Europe in 1910–1912. In Argentina, his chosen country of residence and vocation, his forename initially became "gallicized" to Christian, and subsequently "castillianized" to Christofredo. Jakob left 50 books and 260 papers spanning from macroscopic and cellular neuroanatomy to developmental and evolutionary neuroscience and human neuropathology. Some of his landmark contributions pertain to the onto-phylogeny of the cerebral cortex [3], the emotional brain [4, 5], cortical dynamics and cognition [6–8], the neuroanatomy of language [9, 10], neurophilosophy [11], and neuroeducation [12].

Like numerous other classical neuroscientitsts, Jakob also became interested in the cerebellum. During his early years in Erlangen and Bamberg, Jakob [13, 14] had already written critiques of Frédéric Courmont's book "The cerebellum and its functions," published in Paris in 1891, and of the German edition of Ramón y Cajal's "Contribution to the study of the medulla oblongata, the cerebellum and the origin of cranial nerves," published in Leipzig in 1896.

Decades later, at the initiative of the neurologist Vicente Dimitri (1885–1955), founder and editor-in-chief of the "Neurological Review of Buenos Aires," the young Society of Neurology and Psychiatry of Buenos Aires held a special series of sessions, in December 1938, to

The Introduction article of Cerebellar Classic XII is available at http://dx.doi.org/10.1007/s12311-016-0789-6 and the original paper related to Cerebellar Classic XII is 10.1007/s12311-016-0790-0

✉ Lazaros C. Triarhou
triarhou@uom.gr

1 Graduate Program in Neuroscience and Education, University of Macedonia, 156 Egnatia Ave., Thessalonica 54006, Greece

2 Research Group for Neuroanatomy and Connectivity, Max Planck Institute for Human Cognitive and Brain Sciences, Stephanstrasse 1A, Leipzig 04103, Germany

3 Laboratory of Theoretical and Applied Neuroscience and Graduate Program in Neuroscience and Education, University of Macedonia, 156 Egnatia Ave., Thessalonica 54006, Greece

mark the successful term of Gonzalo Bosch (1885–1967) as its president for 1937–1938. The themes of the sessions were cerebellar pathology and presenile dementia. Jakob, at 72, was invited to open the series. He delivered his lecture, focusing on the cerebellar system.

In the timeline of cerebellar research, in the period between 1912 and 1962, only a couple of landmark discoveries are generally highlighted: the identification in 1937 by Olof Larsell (1886–1964) of the cerebellar lobules and fissures [15], that later became his widely recognized and unique nomenclature in birds and mammals [16, 17], and the documentation between 1938 and 1940 by Giuseppe Moruzzi (1910–1986) of the role of the cerebellum in modulating cardiovascular activity [18].

It is during that time that Jakob, in his lecture, presented the culmination of a lifetime of neurobiological research. He began with a brief overview of the known facts on the cerebellar tracts and the connections with the frontal, temporal, and occipital lobes. Prompted by his methodological emphasis on ontogeny, phylogeny, histophysiology, and neuropathology, Jakob presented both his own observations and a timely update on the much debated question of cerebellar function. The lecture transcript was expanded and published in two different journals [19, 20], after having added the illustrations (Fig. 1).

That event offered Jakob an opportunity to revisit the evolution of his own ideas on cerebellar structure and function, from his first atlas of clinical neurology [21–24] (Fig. 2), through his early lectures to medical students on the anatomy and physiology of the nervous system and the cerebellum in particular [25], to the larger atlas of the human brain [26] (Fig. 3) and the anatomic pathology of clinical cases of ataxia [27–30]. As a matter of fact, a distinct form of lower bilateral ("bibasal") cerebellar degeneration combined with dementia was given the name "Jakob type" by Aranovich [31], based on morphological criteria; it was identified in 15 of 31 cases of cerebellar atrophy in women over 50 years old. The lesions involved a loss of Purkinje cells, originating in the depth of the horizontal fissure (aka *le grand sillon circonférentiel* of Vicq d'Azyr), with a striking separation of its lips, and progressing through the destruction of neighboring lamellae and an atrophy of the subjacent white matter and the middle cerebellar peduncles.

For Jakob, structure and function in biology were one. Thus, he consistently approached neuromorphology in a functional context. This is the principle that guided him in his theorizing about the integrative function of the nervous system. Jakob considered form as "…*stabilized function*…" and function as "…*change of form*…". He insisted upon "the vital energy of an organism being a single entity, which will present

Fig. 1 The transcript of Jakob's lecture was published in the "Journal of the Argentinian Medical Association" and in the "Buenos Aires Neurological Journal" [19, 20]

Fig. 2 *Upper and middle row,* the human brainstem and rhomboid fossa ▶ seen from above. The vermiform process of the cerebellum is divided by a sagittal section, exposing the fourth ventricle. The floor of the ventricle is formed by the rhomboid fossa, which contains the ala cinerea *(a)*, calamus scriptorius *(c)*, and funiculi teretes *(f.t)*. Locus coeruleus *(lc)*; posteriorly on either side are the restiform bodies *(c.rst)* and the tuberculum acusticum *(ta)*. At the entrance of the aqueduct of Sylvius, covered by the anterior medullary velum *(v.a)*, and frenulum *(f)*, is the trochlear nerve *(IV)*. In front of the rhomboid fossa lie the corpora quadrigemina: corpus quadrigeminum anterius *(qa)* and posterius *(qp)*, and the medial geniculate body *(gm)* receiving the posterior brachium *(brp)*. In front of the corpora quadrigemina are the optic thalami *(thal.opt)*, the posterior segment of which has received the name pulvinar *(Pulv.)*, and between them the ganglion habenulae *(h)* and the taenia thalami *(t)*. The epiphysis has been removed from its pedicle. Third ventricle *(vIII)*. Behind the rhomboid fossa: columns of Goll *(f.G)* and Burdach *(f.B)*; lateral column of the spinal cord *(f.l)*; the nuclei of the columns of Goll unite above to form the clavae *(cl)*. In the vermiform process of the cerebellum: *lc*, lobulus centralis; *Cu*, culmen; *Dc*, declive; *Fc*, folia cacuminis; *Cb*, commissura brevis; *Py*, pyramis; *U*, uvula; *No*, nodulus; *lsp*, lobulus superior posterior; *lip, lim, lia*, lobulus inferior posterior, medius, anterior. From the second edition of Jakob's atlas of clinical neurology, plate 10 [23]. *Lower row,* the structure of the cerebellar cortex in two variants drawn by Jakob, based on silver impregnation preparations, for the first *(left)* and the second *(right)* edition of his atlas of clinical neurology (plate 16, Fig. 4 [21], and plate 20, Fig. 2 [23], respectively). *Lower left:* The white matter *(black)* is narrow. Close to it is the granule cell layer *(zona granulosa)* of the cerebellar cortex, on the borders of which are the large Purkinje cells. Their ramifications form, as well as those of other myelinated fibers, the molecular layer *(zona molecularis)* of the surface. Then comes the closely apposed pia covering (not shown). A portion of cells in the granule cell layer (not shown) have short branches of the Golgi type. *Lower right:* A cerebellar convolution. A few isolated fibers from the narrow white matter *(black)* can be seen passing, through the granule cell layer, to the cortical layer proper, which is also very narrow. At the foot of the cortical layer, we see a row of Purkinje cells *(P)* with their elaborate arborizations; a few mossy-like fibers *(r)* and some individual granule cells *(g)* are shown

itself as form in the latent state and as function in the kinetic state" [32]. He kept repeating that "...*form, structure and function are inseparable, if not identical. It is only scholastic science that has managed to separate them. Only a basis that is fundamentally biological, morphostructural and histophysiological at the same time, unified in an ample ontogenetic and phylogenetic context, can let us address in legitimate ways the serious questions of modern neuro- and psychobiopathology...*" [33].

The "triple-synthesis" (evolutionary, developmental, and histophysiological) has been viewed as one of Jakob's prime

Fig. 3 Schematic drawing of the histological structure of the cerebellar cortical layers: nerve cells and their afferent and efferent connections. From Jakob's large atlas of human neuroanatomy [26]

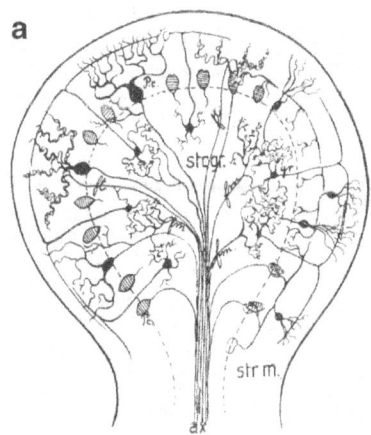

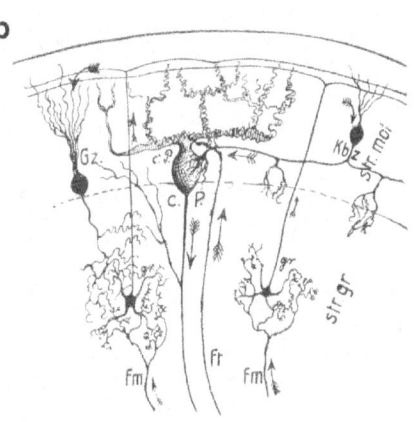

contributions to neurobiology [34–36]. Already in his early Atlas, Jakob [21, 23] had laid out his plan to study the brain; the approaches he considered meaningful to understand the nervous system were: (1) histological staining and serial section reconstruction of the adult human brain, (2) neuropathological changes and their sequelae, (3) comparative neuroanatomy and neuroembryology, (4) human brain development and myelination, and (5) experimentally induced lesions in animals. Thus, in his cerebellar lecture, Jakob remains faithful to his early methodological paradigm.

After crediting, in his introductory statement, historical personalities for their cerebellar discoveries, he details in greater depth the advances made in cerebellar neurobiology during the preceding 50 years, from Ramón y Cajal's discovery of the detailed cerebellar circuitry in 1888 [37] through Jakob's personal observations on cerebellar ataxias during the late 1930s. He divides cerebellar histophysiology into three evolutionary morphofunctional stages, according to the same scheme that he had proposed for cerebral cortical dynamics, that is, archi-, paleo-, and neo-neuronal organization. He highlights the connectivity between the cerebellum and mesencephalic, diencephalic, and telencephalic structures, and he places the cerebellum, along with the cerebral cortex and the striatum, among the nervous centers that conclude their maturation postnatally (the phenomenon we today refer to as "neoteny"). With regard to cerebellar dynamics, he underlines the "…*multiplying…*" role of the granule cells in the dynamics of the cerebellar cortex, a good two decades before the excitatory action of L-glutamate as a neurotransmitter was even discovered [38]. He similarly mentions the accumulated impulses arriving from "…*central…*" structures via the climbing fibers. He also talks about the concept of *neurodynamic economy*, a principle that Ramón y Cajal [39, 40] had introduced in his classical monographs on the nervous system by formulating the *laws of the economy of conduction time*, *of saving cellular material*, and *of the economy of space* in attempting to explain axonal and dendritic relationships.

In the phylogenetic context, Jakob analyzes neocerebellar structure and function as it appears in humans and other primates, as well as in dolphins, whales, and elephants. These latter mammals had been in the epicenter of cerebellar research by some classical neurologists [41–43]. Moreover, Jakob's comprehensive model of the gnoses and the praxes, which he had already discussed decades earlier [7], involves cortical, striatal, thalamic, and mesencephalic influences, including the red nucleus. To our day, the function of the cerebrorubral, the cerebellothalamocortical, and the corticorubral-olivary loop systems, and their convergence at the meso-diencephalic junction constitute one of the main questions of cerebellar neurobiology [44].

With a brief reference to cerebellar ontogeny and myelination, Jakob subsequently discusses in detail the afferent, efferent, and associative pathways in the archicerebellar,

paleocerebellar, and neocerebellar systems, concluding with the following principle: "…*The cerebellum is responsible for the individual mentality of the conscious human elaborating gnoseopoiesis and praxiopoiesis. These two circuits create the gnoseopraxical associative cortical systems, a framework of the dynamic workings of the human cerebral cortex…*" [2].

Based on the phylogenetic division by Ludwig Edinger (1855–1918) of the cerebellum into paleocerebellum and neocerebellum, Jakob carried on with his own studies on cerebellar morphofunctional phylogeny; he distinguished three evolutionary stages in the perfection of the cerebellar influence on motor acts and emphasized the fact that the structure of the cerebellar cortex stays the same throughout the entire phylogenetic scale. He noted the reduced cerebellospinal connections in humans, compared to other primates, as a consequence of the perfection of the rubrolenticular and thalamocortical systems and the intense ascending pathways to the red nucleus in exchange for the more elementary descending efferent pathways.

A clarification is in place. The book reviewer of the *American Journal of Insanity* (*American Journal of Psychiatry* today) criticized Jakob for attributing mental functions to the cerebellum, not purely motor coordination functions:

> …A few paragraphs are also obscure in their meaning, as, for instance, the fifth one on page 66, in which the declaration is made that …*a study of the development of foetal and childish psychic activity affords an approximate idea of the character of conscious processes in the cerebellar cortex…* Unfortunately we do not know what the function of the cerebellum is, and hence it is idle to write, even as an hypothesis, of psychic activity therein… [45].

Such confusion was brought about by an error in the translation. In the German original, Jakob mentions *Grosshirnrinde* (cerebral cortex), which was inadvertently rendered as *cerebellar cortex* in the English translation. Even the neurologist Henry H. Donaldson (1857–1938) had raised criticism about several aspects of the English translation of the atlas [46].

Since the cerebellum was first mentioned by Aristotle in the fourth century B.C. in *De Partibus Animalium* [47], its function has been on of the major concerns in neurology. Over the past 200 years, physiologists shed light on cerebellar functions. Pierre Flourens (1794–1867) made the fundamental observation that "…*all movements persist following ablation of the cerebellum……*", suggesting that the cerebellum is responsible for coordinating movement. In 1837, Jan Evangelista Purkyně (1787–1869) discovered the "Purkinje" cells. These large cerebellar projection neurons with their profusely branched dendritic trees were apparent in the entire

 Springer

extent of the folial gyrations [48]. Purkinje cells constitute the efferent system from the cerebellar cortex to the cerebellar nuclei—a discovery made in 1897 by Klimoff in Kazan, Imperial Russia [49–51].

Toward the end of the nineteenth century, Luigi Luciani (1840–1919), through improved operative techniques, mentioned three cardinal signs: asthenia, atonia, and astasia as the permanent effects of cerebellar lesions. The unitary concept that the cerebellum acted as a whole was supported by the histological studies of Ramón y Cajal [52, 53] and the view of Sir Charles Sherrington that "...*the cerebellum is the chief coordinative system of the proprioceptors, the head ganglion of the proprioceptive system...*" [54, 55]. Ramón y Cajal, besides his immense contribution to neuroscience with the neuron theory, also discovered, using the Golgi method, the mossy and climbing fibers [56]. After three years of intense work, Ramón y Cajal succeeded in deciphering the organization of the cerebellar circuitry [57, 58].

Neurobiologists have since continued to systematically delve into cerebellar circuits, contributing several new perspectives regarding its function [59–61]. While in the past the cerebellum was considered solely a regulator of movement coordination, today there is general agreement that it works in concert with regions throughout the cerebral cortex involved not only in movement, but also in emotional, language, and higher cognitive functions [62]. This broad conceptual shift emerged from converging evidence across multiple research domains, but only came to fruition with the introduction of non-invasive neuroimaging methods. Studies conducted using positron emission tomography and magnetic resonance imaging have demonstrated localized cerebellar activation, albeit unexpectedly at first, during a variety of cognitive tasks [63–67], providing a basis for expanding our understanding of its functional repertoire (for recent historical overviews, see Buckner [68] and Koziol [69]). A meta-analysis summarizing results across dozens of cognitive neuroimaging studies provides an overview of this breadth, implicating cerebellar lobules in spatial, language, working memory, emotional, and executive functions [70].

The heterogeneity of functions attributed to cerebellar areas stands in contrast with its consistent cytoarchitectonic organization, indicating that other structural factors such as connections with the cerebral cortex account for its functional diversity. Jakob hypothesized the convergence of cerebellar pathways in associative cortical regions, insisting on the "...*intimate collaboration* [of the cerebellum] *with the frontal lobe...*" [19, 20]. Mapping cortico-cerebellar connectivity, however, poses a distinct challenge due to the indirect projections between these structures. Extensive tract-tracing studies in the macaque monkey have mapped the topographic distribution of connections from cerebral association cortex to the pons, as it provides an anatomical junction for afferent connections to the cerebellum [71–75]. The circuit was further described using transsynaptic viral tract-tracing for mapping projections, which revealed projections from the dentate nucleus to prefrontal areas [76, 77] as well as a closed-loop circuit between Brodmann's middle frontal area 46 (or granular frontal area FD of Economo and Koskinas) and Crus II [78, 79]. These extensive lines of communication between regions throughout association cortex substantiate Jakob's intuition, and begin to outline the mechanisms for substantial cerebellar involvement in functions beyond the motor domain [80, 81]. Precise understanding of the extent of cerebellar connections with association cortex remains beyond the scope of these gold-standard, but challenging, techniques.

A recent methodological development from non-invasive neuroimaging—resting-state functional connectivity—has provided further insight into the distribution of connectivity between the cerebellum and association cortex. This approach is founded on the observation that regions which are co-activated during task performance also synchronize in the absence of externally driven task demands [82, 83]. Of critical importance for investigating cerebellar connectivity, this technique is largely consistent with underlying structural projections [84–86], but additionally appears capable of capturing patterns of polysynaptic connectivity with high spatial precision. As early as 2005, the potential for this methodology to be applied to mapping cerebellar-frontocortical connectivity was demonstrated [87]. Numerous studies have since confirmed the widespread functional connectivity between the cerebellum and association cortex [88–93]. One landmark discovery of this line of research is the observation of multiple topographic maps of association cortex within the cerebellum, appearing as a continuation of the somatomotor representations present in the anterior and posterior lobes [94]. The surface area of the cerebellar cortex devoted to each cerebral network also generally corresponded, providing support for the co-expansion of neocerebellar and association regions. Understanding the topographic principles of cortico-cerebellar interaction provides a foundation for future research into the complex dynamics that give rise to higher-order processing.

The cerebellum plays an actual role in driving higher-order brain function. In the rat, it modulates the prefrontal cortical activity [95]. In the clinical setting, the cerebellar circuitry seems to be involved in cognitive conditions, including autism spectrum disorders [96]. Moreover, abnormalities within the cerebellar vermis are detected in diseases such as schizophrenia and epilepsy.

Owing to these diverse lines of research, the question of whether the cerebellum contributes to cognitive function has been superseded by the investigation of how it contributes [69]. How can the role of the cerebellar circuitry in motor processing be extrapolated to cognitive domains? What is the mapping of connections between the association regions of the cerebral cortex and the cerebellum? How does this

 Springer

extensive cortico-cerebellar topography converge with dynamic mechanisms in a neurocognitive theory of cerebellar function? We thus find ourselves at a comparable vantage point, where, atop a foundation of anatomical and phylogenetic mastery, Jakob himself posed similar questions eight decades ago.

Acknowledgments The authors gratefully acknowledge the anonymous reviewers for their constructive comments, which have led to an improved manuscript; the Ibero-Amerikanisches Institut Preussischer Kulturbesitz zu Berlin; Universitätsbibliothek Kiel; Ruth Lilly Medical Library of Indiana University in Indianapolis; University of Michigan Library in Ann Arbor; the Library of Congress; and the National Library of Medicine of the United States for bibliographic sources.

Conpliance with Ethical Standards

Conflict of Interest The authors declare that they have no conflict of interest.

References

1. Triarhou LC, del Cerro M. Semicentennial tribute to the ingenious neurobiologist Christfried Jakob (1866–1956). 1. Works from Germany and the first Argentina period, 1891–1913. Eur Neurol. 2006;56:176–88.

2. Triarhou LC, del Cerro M. Semicentennial tribute to the ingenious neurobiologist Christfried Jakob (1866–1956). 2. Publications from the second Argentina period, 1913–1949. Eur Neurol. 2006;56: 189–98.

3. Triarhou LC. Revisiting Christfried Jakob's concept of the dual onto-phylogenetic origin and ubiquitous function of the cerebral cortex: a century of progress. Brain Struct Funct. 2010;214:319–38.

4. Triarhou LC. Centenary of Christfried Jakob's discovery of the visceral brain: an unheeded precedence in affective neuroscience. Neurosci Biobehav Rev. 2008;32:984–1000.

5. Triarhou LC. Tripartite concepts of mind and brain, with special emphasis on the neuroevolutionary postulates of Christfried Jakob and Paul MacLean. In: Weingarten SP, Penat HO, editors. Cognitive psychology research developments. Haupaugge, NY: Nova Science; 2009. p. 183–208.

6. Theodoridou ZD, Triarhou LC. Challenging the supremacy of the frontal lobe: early views (1906–1909) of Christfried Jakob on the human cerebral cortex. Cortex. 2012;48:15–25.

7. Theodoridou ZD, Triarhou LC. Christfried Jakob's 1921 theory of the gnoses and praxes as fundamental factors in cerebral cortical dynamics. Integr Psychol Behav Sci. 2011;45:247–62.

8. Theodoridou ZD, Triarhou LC. Christfried Jakob's late views (1930–1949) on the psychogenetic function of the cerebral cortex and its localization: culmination of the neurophilosophical thought of a keen brain observer. Brain Cogn. 2012;78:179–88.

9. Tsapkini K, Vivas AB, Triarhou LC. 'Does Broca's area exist?'— Christofredo Jakob's 1906 response to Pierre Marie's holistic stance. Brain Lang. 2008;105:211–9.

10. Vivas AB, Tsapkini K, Triarhou LC. 'Anatomo-biological considerations on the centers of language': an Argentinian contribution to the 1906 Paris debate on aphasia. Brain Dev. 2007;29:455–61.

11. Barutta J, Hodges J, Ibáñez A, Gleichgerrcht E, Manes F. Argentina's early contributions to the understanding of frontotemporal lobar degeneration. Cortex. 2011;47:621–7.

12. Theodoridou ZD, Koutsoklenis A, del Cerro M, Triarhou LC. An avant-garde professorship of neurobiology in education: Christofredo Jakob (1866–1956) and the 1920s lead of the National University of La Plata, Argentina. J Hist Neurosci. 2013;22:366–82.

13. Jakob C. Le cervelet et ses fonctions, par F. Courmont (Besprechung). Dtsch Z Nervenheilk (Leipz). 1893;3:355–6.

14. Jakob C. Beitrag zum Studium der Medulla oblongata, des Kleinhirns und des Ursprunges der Gehirnnerven, von S. Ramón y Cajal, übersetzt von J. Bresler, mit einem Vorwort von Prof. Mendel (Besprechung). Dtsch Z Nervenheilk (Leipz). 1896;9:145–6.

15. Larsell O. The cerebellum: a review and interpretation. Arch Neurol Psychiatry. 1937;38:580–607.

16. Larsell O. The development and subdivisions of the cerebellum of birds. J Comp Neurol. 1948;89:123–89.

17. Larsell O. The morphogenesis and adult pattern of the lobules and fissures of the cerebellum of the white rat. J Comp Neurol. 1952;97: 281–356.

18. Manto M, Haines D. Cerebellar research: two centuries of discoveries. Cerebellum. 2012;11:446–8.

19. Jakob C. El sistema cerebeloso y su significación biológica. Rev Asoc Méd Argent. 1939;53:198–209.

20. Jakob C. El sistema cerebeloso y su significación biológica. Rev Neurol B Aires. 1939;3:207–35.

21. Jakob C. Atlas des gesunden und kranken Nervensystems nebst Grundriss der Anatomie, Pathologie und Therapie desselben. J. F. Lehmann: München; 1895.

22. Jakob C. An atlas of the normal and pathological nervous systems, together with a sketch of the anatomy, pathology, and therapy of the same (Collins J, translator). New York: Wood; 1896.

23. Jakob C. Atlas des gesunden und kranken Nervensystems nebst Grundriss der Anatomie, Pathologie und Therapie desselben. 2nd ed. J. F. Lehmann: München; 1899.

24. Jakob C. Atlas of the nervous system, including an epitome of the anatomy, pathology, and treatment, 2nd edn. (Fisher ED, translator). Philadelphia – London: W. B. Saunders & Co. 1901.

25. Jakob C. Lecciones sobre anatomía y fisiología del sistema nervioso, en sus relaciones con la psiquiatría. Lección VI: Estructura y funciones del cerebelo. Semana Méd (B Aires) 1900;7:479–82.

26. Jakob C. Das Menschenhirn: Eine Studie über den Aufbau und die Bedeutung seiner grauen Kerne und Rinde. J. F. Lehmann: München; 1911.

27. Jakob C. Hemiplejía, hemiataxia y hemianestesia homolateral de origen cerebeloso. Arch Argent Neurol. 1929;4:13–31.

28. Jakob C. Hemiplejía, hemiataxia y hemianestesia homolateral cerebelosa. Comunicación resumida. Actas Conf Lat Amer Neurol Psiquiatr Med Leg. 1929;1:240–52.

29. Jakob C, Beretervide JJ, Caballero E. Atrofia olivo ponto cerebelosa familiar. Prensa Méd Argent (B Aires). 1934;21:1997–2017.

30. Jakob C. La sistematización del haz central de la calota como vía neoneuronal cerebelosa eferente olivobulbar. Rev Neurol B Aires. 1942;7:1–24.

31. Aranovich J. La atrofia cerebelosa marginal bibasal de Chr. Jakob Rev Asoc Méd Arg. 1937;51:115–22.

32. Jakob C. Folia neurobiológica argentina, tomos I–V. Buenos Aires: Aniceto López-López y Etchegoyen; 1941–1946.

33. Jakob C. El cerebro humano (Folia neurobiológica argentina, atlas I–III). Buenos Aires: Aniceto López; 1939–1941.

34. Moyano BA. Christfried Jakob (25/12/1866–6/5/1956). Acta Neuropsiquiátr Argent. 1957;3:109–23.

35. Orlando JC. Christofredo Jakob: su vida y obra. Buenos Aires: Editorial Mundi; 1966.

 Springer

36. Meyer L. Christofredo Jakob: a veinticinco años de su muerte. Acta Psiquiátr Psicol Amér Lat 1981;27:13–4.

37. Ramón y Cajal S. Estructura del cerebelo. Gac Méd Catal. 1888;11: 449–57.

38. Watkins JC, Jane DE. The glutamate story. Brit J Pharmacol. 2006;147:S100–8.

39. Ramón y Cajal S. Textura del sistema nervioso del hombre y de los vertebrados, tomo I. Madrid: Nicolás Moya; 1899.

40. Ramón y Cajal S. The neuron and the glial cell (de la Torre J, Gibson WC, translators). Springfield, IL: Charles C. Thomas; 1984.

41. Flatau E, Jacobsohn L. Handbuch der Anatomie und vergleichenden Anatomie des Centralnervensystems der Säugetiere. Berlin: Samuel Karger; 1899.

42. Obersteiner H. Die Kleinhirnrinde vom Elephas und Balaenoptera. Arb Neurol Inst (Wien). 1913;20:145–54.

43. Ingvar S. Zur Phylo- und Ontogenese des Kleinhirns nebst ein Versuch zu einheitlicher Erklärung der zerebellaren Funktion und Lokalisation. Folia Neuro-Biologica (Haarlem). 1918;11:205–495.

44. Voogd J. What we do not know about cerebellar systems neuroscience. Front Syst Neurosci. 2014;8:227. doi:10.3389/fnsys.2014.00227.

45. Anonymous. Atlas of the nervous system, including an epitome of the anatomy, pathology, and treatment, by Dr. Christfried Jakob, Erlangen, translated from the second German edition, and edited by Edward D. Fisher (book review). Am J Insanity. 1902;58:559–60.

46. Donaldson HH. Atlas of the nervous system, including an epitome of the anatomy, pathology and treatment, by Christfried Jacob, with a preface by Prof. Dr. Ad. v. Strümpell, authorized translation from the second revised German edition, edited by Edward D. Fisher (book review). Psychol Rev. 1901;8:622–6.

47. Clarke E, O'Malley CD. The human brain and spinal cord: a historical study illustrated by writings from antiquity to the twentieth century. San Francisco: Jeremy Norman; 1996.

48. Glickstein M, Strata P, Voogd J. Cerebellum: history. Neuroscience. 2009;162:549–59.

49. Klimoff J. On the conduction paths of the cerebellum: experimental-anatomical observations (Dissertation) [in Russian]. Kazan: Imperial University of Kazan; 1897.

50. Klimoff J. Über die Leitungsbahnen des Kleinhirns. Arch Anat Physiol Anat Abth. 1899:11–27.

51. Haines DE, Patrick GW, Satrulee P. Organization of cerebellar corticonuclear fiber systems. In: Palay SL, Chan-Palay V, editors. The cerebellum—new vistas. Berlin: Springer; 1982. p. 320–71.

52. Ramón y Cajal S. Les nouvelles idées sur la structure du système nerveux chez l'homme et chez les vertébrés (transl. by L. Azoulay). Paris: C. Reinwald. 1894.

53. Ramón y Cajal S. The Croonian Lecture—"La fine structure des centres nerveux." Proc Roy Soc London 1894;55:444–68.

54. Sherrington CS. The integrative action of the nervous system. New York: Charles Scribner's Sons; 1906. p. 347–9.

55. Sherrington CS. On the proprio ceptive system, especially in its reflex aspect. Brain. 1907;29:467–82.

56. Ramón y Cajal S. Structure et connexions des neurones (conférence Nobel faite à Stockholm le 12 décembre 1906). In: Hasselberg KB, Pettersson SO, Mörner KAH, Wirsén CD, Santesson MCG, editors. Les prix Nobel en 1906. Stockholm: Imprimerie Royale P. A. Norstedt & Söner; 1908; 1–27.

57. Sotelo C. Viewing the brain through the master hand of Ramón y Cajal. Nat Rev Neurosci. 2003;4:71–7.

58. Sotelo C. Viewing the cerebellum through the eyes of Ramón y Cajal. Cerebellum. 2008;7:517–22.

59. Leiner HC, Leiner AL, Dow RS. Does the cerebellum contribute to mental skills? Behav Neurosci. 1986;100:443–54.

60. Schmahmann JD. Rediscovery of an early concept. In: Schmahmann JD, editor. The cerebellum and cognition (International Review of Neurobiology, vol. 41). San Diego: Academic Press; 1997. pp. 3–27.

61. Manto MU, Jissendi P. Cerebellum: links between development, developmental disorders and motor learning. Front Neuroanat. 2012;6:1. doi:10.3389/fnana.2012.00001.

62. Schmahmann JD. An emerging concept: the cerebellar contribution to higher function. Arch Neurol. 1991;48:1178–87.

63. Petersen SE, Fox PT, Posner MI, Mintun M, Raichle ME. Positron emission tomographic studies of the cortical anatomy of single-word processing. Nature. 1988;331:585–9.

64. Raichle ME, Fiez JA, Videen TO, et al. Practice-related changes in human brain functional anatomy during nonmotor learning. Cereb Cortex. 1994;4:8–26.

65. Jueptner M, Rijntjes M, Weiller C, et al. Localization of a cerebellar timing process using PET. Neurology. 1995;45:1540–5.

66. Allen G, Buxton RB, Wong EC, Courchesne E. Attentional activation of the cerebellum independent of motor involvement. Science. 1997;275:1940–3.

67. Ryding E, Decety J, Sjöholm H, Stenberg G, Ingvar DH. Motor imagery activates the cerebellum regionally: a SPECT rCBF study with 99mTc-HMPAO. Cogn Brain Res. 1993;1:94–9.

68. Buckner RL. The cerebellum and cognitive function: 25 years of insight from anatomy and neuroimaging. Neuron. 2013;80:807–15.

69. Koziol LF, Budding D, Andreasen N, et al. Consensus paper: the cerebellum's role in movement and cognition. Cerebellum. 2014;13:151–77.

70. Stoodley CJ, Schmahmann JD. Functional topography in the human cerebellum: a meta-analysis of neuroimaging studies. NeuroImage. 2009;44:489–501.

71. Glickstein M, May JG IIIrd, Mercier BE. Corticopontine projection in the macaque: the distribution of labelled cortical cells after large injections of horseradish peroxidase in the pontine nuclei. J Comp Neurol 1985;235:343–59.

72. Schmahmann JD, Pandya DN. Anatomical investigation of projections to the basis pontis from posterior parietal association cortices in rhesus monkey. J Comp Neurol. 1989;289:53–73.

73. Schmahmann JD, Pandya DN. Projections to the basis pontis from the superior temporal sulcus and superior temporal region in the rhesus monkey. J Comp Neurol. 1991;308:224–48.

74. Schmahmann JD, Pandya DN. Anatomic organization of the basilar pontine projections from prefrontal cortices in rhesus monkey. J Neurosci. 1997;17:438–58.

75. Prevosto V, Graf W, Ugolini G. Cerebellar inputs to intraparietal cortex areas LIP and MIP: functional frameworks for adaptive control of eye movements, reaching, and arm/eye/head movement coordination. Cereb Cortex. 2010;20:214–28.

76. Middleton FA, Strick PL. Anatomical evidence for cerebellar and basal ganglia involvement in higher cognitive function. Science. 1994;266:458–61.

77. Middleton FA, Strick PL. Cerebellar projections to the prefrontal cortex of the primate. J Neurosci. 2001;21:700–12.

78. Kelly RM, Strick PL. Cerebellar loops with motor cortex and prefrontal cortex of a nonhuman primate. J Neurosci. 2003;23:8432–44.

79. Bostan AC, Dum RP, Strick PL. Cerebellar networks with the cerebral cortex and basal ganglia. Trends Cogn Sci. 2013;17:241–54.

80. Strick PL, Dum RP, Fiez JA. Cerebellum and nonmotor function. Annu Rev Neurosci. 2009;32:413–34.

81. Ramnani N. Frontal lobe and posterior parietal contributions to the cortico-cerebellar system. Cerebellum. 2012;11:366–83.

82. Biswal B, Yetkin FZ, Haughton VM, Hyde JS. Functional connectivity in the motor cortex of resting human brain using echo-planar MRI. Magn Reson Med. 1995;34:537–41.

83. Smith SM, Fox PT, Miller KL, et al. Correspondence of the brain's functional architecture during activation and rest. Proc Natl Acad Sci U S A. 2009;106:13040–5.

84. Honey CJ, Sporns O, Cammoun L, et al. Predicting human resting-state functional connectivity from structural connectivity. Proc Natl Acad Sci U S A. 2009;106:2035–40.

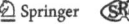
Springer

Cerebellum (2016) 15:417–424

85. van den Heuvel MP, Mandl RC, Kahn RS, Hulshoff Pol HE. Functionally linked resting-state networks reflect the underlying structural connectivity architecture of the human brain. Hum Brain Mapp. 2009;30:3127–41.

86. van den Heuvel MP, Sporns O. An anatomical substrate for integration among functional networks in human cortex. J Neurosci. 2013;33:14489–500.

87. Allen G, McColl R, Barnard H, Ringe WK, Fleckenstein J, Cullum CM. Magnetic resonance imaging of cerebellar-prefrontal and cerebellar-parietal functional connectivity. NeuroImage. 2005;28:39–48.

88. Habas C, Kamdar N, Nguyen D, et al. Distinct cerebellar contributions to intrinsic connectivity networks. J Neurosci. 2009;29:8586–94.

89. Krienen FM, Buckner RL. Segregated fronto-cerebellar circuits revealed by intrinsic functional connectivity. Cereb Cortex. 2009;19: 2485–97.

90. O'Reilly JX, Beckmann CF, Tomassini V, Ramnani N, Johansen-Berg H. Distinct and overlapping functional zones in the cerebellum defined by resting state functional connectivity. Cereb Cortex. 2010;20:953–65.

91. Lu J, Liu H, Zhang M, et al. Focal pontine lesions provide evidence that intrinsic functional connectivity reflects polysynaptic anatomical pathways. J Neurosci. 2011;31:15065–71.

92. Bernard JA, Seidler RD, Hassevoort KM, et al. Resting state cortico-cerebellar functional connectivity networks: a comparison of anatomical and self-organizing map approaches. Front Neuroanat. 2012;6:31. doi:10.3389/fnana.2012.00031.

93. Kipping JA, Grodd W, Kumar V, Taubert M, Villringer A, Margulies DS. Overlapping and parallel cerebello-cerebral networks contributing to sensorimotor control: an intrinsic functional connectivity study. NeuroImage. 2013;83:837–48.

94. Buckner RL, Krienen FM, Castellanos A, Diaz JC, Yeo BT. The organization of the human cerebellum estimated by intrinsic functional connectivity. J Neurophysiol. 2011;106:2322–45.

95. Watson TC, Becker N, Apps R, Jones MW. Back to front: cerebellar connections and interactions with the prefrontal cortex. Front Syst Neurosci. 2014;8:4. doi:10.3389/fnsys.2014.00004.

96. Reeber SL, Otis TS, Sillitoe RV. New roles for the cerebellum in health and disease. Front Syst Neurosci. 2013;7:83. doi:10.3389/fnsys.2013.00083.

 Springer

ABSTRACTS AND PRESENTATIONS
ON CHRISTOFREDO JAKOB

Conference Abstracts on Christofredo Jakob

Théodoridou Z.D., Triarhou L.C. (2010) Early views of Christfried Jakob on the cerebral cortex: Challenging the supremacy of the frontal lobe. 15th Annual Meeting of the International Society for the History of the Neurosciences, Paris. *Journal of the History of the Neurosciences 19:* 409–410.

Théodoridou Z.D., Triarhou L.C. (2010) Gnoses and praxes as fundamental factors in Christfried Jakob's theory on cerebral cortical dynamics. 2nd Hellenic Conference on Cognitive Science, Paros. *Abstracts of the Hellenic Society for Cognitive Science 2.*

Théodoridou Z.D., Triarhou L.C. (2011) Christfried Jakob's late views on cortical development, localization and neurophilosophy. 2011 Biennial Conference of the Society of Applied Neuroscience, Thessalonica. *Neuroscience Letters [Suppl] 500:* e43–e44.

Triarhou L.C. (2008) Christfried Jakob's 1911 proposition on the dual onto-phylogenetic origin and ubiquitous sensory-motor function of the cerebral cortex. Cortical Development: Stem Cells, Neurogenesis, Migration, Circuit Formation, Cortical Disorders, Mediterranean Agronomic Institute (MAICH), Chania, Crete. In: Fishell G., Kriegstein A.R., Parnavelas J.G. (Eds.) *Cortical Development Book of Abstracts 3:* 89–90.

Triarhou L.C. (2008) The books of Christofredo Jakob: Lasting treasures of evolutionary neuroscience. 38th Annual Meeting of the Society for Neuroscience, Washington, D.C. *Society for Neuroscience Abstracts 38:* 221.16.

Triarhou L.C. (2009) Alfons and Christfried (Christofredo) Jakob: Two Bavarian neuropathologists in Latin America. 13th Annual Meeting of the International Society for the History of the Neurosciences, Berlin. *Journal of the History of the Neurosciences 18:* 131–132.

Triarhou L.C. (2010) Evo-devo origin of the cerebral cortex and the triune brain. 2nd Hellenic Conference of Cognitive Sciences, Paros. *Abstracts of the Hellenic Society for Cognitive Science 2.*

Triarhou L.C. (2013) The triune brain. Keynote Lecture, 8th Hellenic Epilepsy Congress, Thessalonica. archives.erasmus.gr/el/congresses/athens/2013/epilepsycongress2013.

Triarhou L.C., del Cerro M. (2006) Semicentennial tribute to neurobiologist Christofredo Jakob (1866–1956). 20th Annual Meeting of the Hellenic Society for Neuroscience, Herakleion, Crete. *Hellenic Society for Neuroscience Abstracts 20:* 114–115.

Tsapkini K., Vivas A.B., Triarhou L.C. (2007) The 1906 Paris debate on aphasia: Two little known timely contributions from Buenos Aires. 21st Annual Meeting of the Hellenic Society for Neuroscience, Thessalonica. *Hellenic Society for Neuroscience Abstracts 21:* 210–211.

ᚸ ৡ

www.ingramcontent.com/pod-product-compliance
Lightning Source LLC
Chambersburg PA
CBHW082035190526
45165CB00021B/3326